Introduction to Polymer Rheology

Introduction to Polymer Rheology

Montgomery T. Shaw

A JOHN WILEY & SONS, INC., PUBLICATION

Published by John Wiley & Sons, Inc., Hoboken, New Jersey.
Published simultaneously in Canada.

For general information on our other products and services please contact our Customer Care Department within the United States at (800) 762-2974, outside the United States at (317) 572-3993 or fax (317) 572-4002.

Wiley also publishes its books in a variety of electronic formats. Some content that appears in print, however, may not be available in electronic formats. For more information about Wiley products, visit our web site at www.wiley.com.

Library of Congress Cataloging-in-Publication Data is available.

ISBN 9780470388440

Preface

I am keenly aware of the quote in a book written by the late Arthur Lodge saying roughly: "Who needs another rheology book"? Agreed. For teaching, I personally am a fan of *Dynamics of Polymeric Liquids*, written by R. Byron Bird, and Ole Hassager (Volume 1) and R. Byron Bird, Charles F. Curtiss, Robert C. Armstrong and Ole Hassager (Volume 2). As far as I am concerned, this book obviated forever the need for another rheology textbook.

Do I use it for teaching my graduate rheology course? No. First of all, at 600+ pages, it's expensive. Even Volume 1 retails for more than $200 on amazon.com. At this price, the average graduate student will consider either doing without, dropping the course, or buying an illegal copy. But my most serious reservation with this and many other texts is the slant. It is directed at engineers who have had basic transport phenomena, along with linear algebra, differential equations and numerical analysis. Very few of my polymer students are so privileged. Short-hand tensor notation, while convenient for the expert, is baffling to them.

Why then do polymer students, mostly with organic chemistry backgrounds, bother with a rheology course? It doesn't take them long to figure out that training in polymers carries with it the necessity for an acquaintance with their mechanical, viscoelastic and rheological properties. Sure, molecular spectroscopy, thermal analysis and microscopy are the mainstays of polymer analysis, but they hear the news from their friends in industry—learn about rheology.

The long and short of this discussion is that a textbook aimed a bit lower seemed like a valuable addition. The popular text *Introduction to Polymer Viscoelasticity* was taken as the starting point. A reproduction of the relaxed style of this text was attempted. Certainly one cannot learn about rheological behavior without dealing in some fashion with three-dimensional mechanics, but the basics are really enough. Tensors can be presented consistently in matrix form instead of with shorthand. This is tough on the author, but helps the student to sort out the important aspects of the various categories of deformation and flow.

The knowledgeable rheologists will encounter in this volume some shortcuts that they will find somewhat bothersome and certainly lacking in rigor. For example, the continuity equation is barely mentioned and rarely used. No one needs the continuity equation to figure out that the velocity in outward axisymmetric radial flow is (1) positive and (2) falls as the reciprocal of the radius. This is true with many other one- and two-dimensional flows as well. Where possible, shell force balances are used as opposed to plowing through the collection of confusing terms in the differential momentum balance.

One annoying aspect of a math-based physical science is nomenclature. In rheology, it's not just about symbols, but also there is the sign issue and a factor of two in the definition of the rate of deformation. This text sticks as closely as possible to the conventions endorsed by the Society of Rheology. Thus $\boldsymbol{\sigma} = \eta\dot{\boldsymbol{\gamma}} - p\boldsymbol{\delta}$, for example. The symbol τ is introduced for the extra (dynamic, deviatoric) stress tensor, as a convenience. The use of the "positive for tension" sign convention is noted near each equation by the abbreviation (ssc), which stands for "solids sign convention," because this convention was a product of tensile testing of solid samples. However, the student is often reminded that the "fluids sign convention" (fsc) is in wide use, and examples are given. Another annoyance for the fastidious is the use of a few symbols for two meanings. For example, the symbol τ might appear in one place to mean shear stress and another place as a time constant, even in the same equation. Confusing? Not really, as the context readily differentiates the two, especially after the student becomes familiar with the concept of a dimensionless term. Also, in most cases, τ for stress carries two subscripts, e.g., τ_{21}.

In addressing three-dimensional mechanics, the stress tensor is introduced first (Chapter 2). Stress is very physical concept that sits well with most students. The nuances of how a stress can be applied to a sample are explained in perhaps more detail than in most texts. How does one turn the tensile force in a string into a nice uniform uniaxial stress in a sample of finite size? To solve this, a construct termed "ideal clamp" is introduced (although also used in *Introduction to Polymer Viscoelasticity*, 3rd edition), which has properties that can only be approached by real mechanical devices (some of which are pictured). The complexities of "simple" shear and plane stress are worked over at some length

After a discussion of stress, one expects a discussion of strain; however, strain is left to the last possible moment (Chapter 8). There is no question that the transition from linear viscoelasticity to finite strain of fluids is difficult to learn and difficult to teach. Rather than jumping right into this treacherous topic, an entire chapter (Chapter 3) is devoted to a discussion of the rate-of-deformation tensor and the magnitude of the rate. Rate, in spite of being a derivative, is much simpler to understand and use. Certainly, most applications of nonlinear polymer rheology in industry will involve analysis using Newtonian (Chapter 4) and generalized Newtonian fluid models (Chapter 5) in steady or quasi-steady flows.

Moving to strain (Chapter 8) involves heavy use of the concept of displacement relative to the present configuration and begins with the infinitesimal strain tensor. In my experience, getting students to accept the concept of the present position as the reference condition is certainly one of the biggest hurdles of rheology instruction. This is especially true for those who have had instruction in linear viscoelasticity, but they can see that it all works out when the strain is small. However, the morphing of the Boltzmann superposition principle into finite-strain integral models still produces frustration and doubt, but this is reduced if time is spent explaining why the initial condition no longer makes sense for finite strains of fluids. However, the doubt always returns when it is explained that strains before the sample is touched ($t' < 0$) must contribute to the stress if the present position is the reference. Equally difficult is accepting the fact that the strain at $t' = t$ is zero when clearly the sample has been deformed. Several examples are provided to help with this admittedly complicated topic. Important also is pointing out, with more examples, why using the undeformed configuration as the reference leads to problems.

Most polymer science students are interested in using rheology as an analytical tool. They usually are knowledgeable about techniques based on linear viscoelasticity, but are much less familiar with steady-flow techniques and usually unaware of the attractions and difficulties of extensional and transient flows. Thus, Chapter 7 goes through how to find viscosity and normal stresses using rotational geometries, and some of the issues faced with these devices. Capillary viscometry is also addressed, along with its many problems. The section on extensional flows starts out with a brief history of this challenging measurement and moves to a description of the broad array of techniques that have been introduced.

The connection of molecular structure to rheological response is an important aspect of polymer rheology. In fact, it is so important that it is not confined to one chapter (Chapter 9), but is spread throughout so the connection with each rheological function is clear. While the basic ideas of molecular motion are discussed in Chapter 9, there is no attempt to go into the details of, for example, the exciting and rapidly growing field of molecular dynamics of chain structures.

Oddly enough, very little space is devoted to the application of rheology to polymer processing (Chapter 10). Polymer processing is an exceedingly diverse and complicated subject requiring techniques that are far removed from the interests and abilities of most polymer science students. Instead, the chapter is devoted to explanation of the lubrication approximation, and its application to the simple analysis of flows involved in common laboratory processing methods. The goal of most laboratory processing is to make a sample that can be analyzed or tested. Typical tests include infrared analysis, contact angle, x-ray diffraction, microscopy, dielectric analysis and light mechanical testing. Processing methods may be limited to solution casting, spin casting and compression molding because the sample mass may often be less than a gram. These methods are discussed and examined, with analyses often confined to Newtonian and generalized Newtonian fluids. The goal here is to provide an understanding of how rheology can help the student adjust their sample preparation methods and conditions to avoid problems with the fabricated object in subsequent characterization.

Most of the polymer students end up in an industrial position, and many call about rheology problems they are experiencing on the job. Their "rheometers" are sometimes rudimentary quality-control devices that they have never seen before they joined the company. For this reason, Chapter 11 deals with typical quality-control measurements such as melt-flow index, Mooney viscosity and Rossi-Peakes flow. The goal is not only to define and describe these measurements to the students, but to convince them that such methods have very good reasons for existing.

Typically a semester rheology course runs out of time before the last chapter which deals with the influence of polymer modification of rheological properties. Again, the idea is to describe some of the key aspects of this very broad area, which is covered thoroughly in texts such as *The Structure and Rheology of Complex Fluids* (R. G. Larson, Oxford, 1998). Short sections on fillers, crosslinking, liquid crystallinity, and physical intermolecular interactions are included. The goal is to inform the student that rheological properties are very sensitive to these structural variables.

Problems are a key feature of the book. Every chapter has at least ten and some well over twenty problems at the end of the chapter, in addition to worked sample problems with the text. As with *Introduction to Polymer Viscoelasticity*, many of the problems have solutions in the final section of the book.   These problems, and their solutions, are an intrinsic part of the presentation, and can cover aspects of rheology that are not even mentioned in the text.   Some of the problems carry tags such as "Computer," meaning a computer is essential to the solution; "Open end" meaning the answer will depend on the assumptions, simplifications or materials chosen by the student; and "Challenging."   What does "challenging" mean?   The experienced rheologist will certainly wonder why they are so labeled.   The student, however, will find the solutions may take several hours, or even more.   Thus the instructor should assign these judiciously, and perhaps with extra hints.

If the answers are available in the back, won't the students just copy the answer? Sure. Good. At least they have done something. The more serious student will work out the problem without referring to the answer, and will be able to check her result against the solution provided in the back. If the two are different, they will have a good start on finding the source of the discrepancy.

In many, many cases, students have had no experience with numerical methods of any sort. Thus, some sections contain advice and instruction on how to get an answer given readily available tools. For instance, an example is worked out on nonlinear modeling using an Excel® spreadsheet, rather than simply referring the student to, say, Mathematica®. Elementary aspects of the precision of the results and other statistical considerations are also discussed.

The trained rheologist approaches solutions in the classical fashion, but students often explore routes that lead to incorrect answers. Some of these incorrect approaches are illustrated. For example, the classical approach to the analysis of the cone and plate geometry uses spherical coordinates. So, why not use cylindrical coordinates? Well, the cylindrical coordinate system work fine for the torque, but fails badly for the normal force. The student needs to be shown why this is the case, rather than simply be told that spherical coordinates must be used.

Finally, there are many really good books that have missing or cursory indices. There is nothing more frustrating than trying to find, say, an equation of motion, but not finding that entry anywhere in the index. (This is a real example.) Thus attention has been paid to formulating a complete index that has primary entries where one expects to find these entries, and with no annoying instances of "*See...*". This is important for the student, but also important for the instructor.

While instructed in linear viscoelasticity in graduate school, I never had a formal course in rheology. Rather, I was fortunate to have several mentors along the way, including Professors Robert Bird, Morton Denn and Arthur Metzner. These academicians were consultants at Union Carbide, where I was declared a "rheologist" by the management on joining the company. Helpful rheology-oriented colleagues at the same company included Drs. Stuart Kurtz, Duane Marsh, John Miller and Lloyd Robeson. On joining the faculty at the University of Connecticut, I benefited from a long collaboration with Prof. Robert A. Weiss, who indeed did have a formal graduate course in rheology. The best way to learn any subject is to teach it, which Prof. Weiss and I did for roughly three decades. In addition, I have benefited enormously as a result of my association with and participation in the activities of The Society of Rheology.

Finally, I wish to acknowledge the constant support from the Institute of Materials Science at the University of Connecticut, which has purchased, maintained and replaced numerous rheometers and accessories. Included in this acknowledgement are the staff and students of the IMS, especially my own very special students who have become accomplished rheologists. In concluding the acknowledgments, I wish to thank my wife, Maripaz N. Shaw, for once again patiently enduring long hours of computer widowship.

Montgomery T. Shaw

Storrs, Connecticut
September 2011

Contents

1
Introduction

A. POLYMERS AND THE IMPORTANCE OF RHEOLOGY

1. General information about the structure and properties of polymers

Polymers are generally organic and share many of the physical and chemical attributes and shortcomings, including low density, low cohesion, susceptibly to oxidative degradation, and high electrical resistance and dielectric strength. As with many organic fluids, most polymers absorb only a small amount of visible light, and are therefore colorless and transparent. If the structure of the polymer chain is regular, crystallization is possible.

The unique aspect of polymers is their high molecular weight, generally achieved by linking together organic moieties into a linear chain-like structure. Other structures are possible and useful, including random linking of the starting organic molecules into continuous net-like structures that extend indefinitely in three dimensions. Much of the commercial and research effort is focused on the linear structure, as there is some hope with this of developing a universal description. This universal description would ideally capture in a simple formula all the behavior of the polymer in terms of characteristics of the chain—length, width, stiffness, and secondary interactions with neighboring chains. Rheological behavior is one aspect where progress has been made as a result of continued work on models for the chain motions and interactions, and extensive characterization of a huge number of polymer structures. The economic motivation for this effort is that the rheology ties in closely with the physical and processing characteristics of hundreds of commercially important polymers. One can say with some validity that if polymer melts and solutions were all low in viscosity, polymer rheology would receive much less attention.

With linear polymer structures, the polymer chemist strives for high molecular weight, corresponding to long chain length, because the longest chains provide the most useful mechanical properties.   Unfortunately, the longest chains also lead to the highest viscosity.   Thus the chemist strives for methods to control molecular weight: high enough for good mechanical properties, and low enough for convenient processing characteristics.

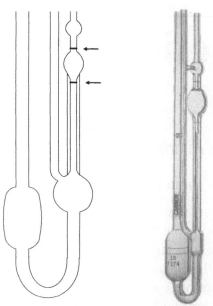

Figure 1-1. Capillary viscometer of the Ubbelohde design. The capillary is in the section just below the second bulb down on the right-most tube with the two marks (arrows), which are used to time the flow. A key design feature is the vent tube for the bulb just below the capillary. The vent tube keeps the pressure in the lower bulb constant at 1 atm. The photo on the right shows a commercial example. (Reproduced with permission of Cannon Instruments, Inc.)

Taking polyethylene as an example, at a molecular weight of 100,000 g/mol (100 kDa),[*] it is easy to process, with reasonable mechanical properties. At 1 MDa, the strength has improved, but processing becomes difficult, especially with techniques such as injection molding. At 10 MDa, the "ultrahigh" molecular weight range, the properties are extraordinary, but processing techniques are now confined to specialized methods, including machining of

---

[*] The molecular weight quantities g/mol and Dalton (Da) have slightly different origins. The former is the mass of Avogadro's number of molecules, while the latter, a dimensionless number, is derived by summing the atomic weights of all atoms in the molecule, dividing by the atomic weight of ^{12}C and multiplying by the integer 12. While the two methods may give results that are very slightly different, there is no practical difference.

shapes. Over this 100-fold molecular-weight range, the viscosity has increased by a factor of about 400,000!

2. Rheology as a method of analysis and a quality control tool

In view of the importance of molecular weight, chemists have developed many techniques for its measurement, or estimation. While many high-accuracy instrumental techniques are now available, the standby in the laboratory is solution viscosity. Let's examine this technique briefly, as it can provide a familiar example for introduction of some rheological terms.

The viscosity measurement is classically done by preparing several solutions of the polymer in a good solvent. The solutions should be of different concentration over a broad range. They must be free of particles, including "gel" particles that can result during the synthesis. The instrument is the familiar glass capillary viscometer. The design pictured in Figure 1-1 is often used, as the side tube ensures that the pressure at the bottom of the capillary is held constant at one atmosphere.

For the design pictured in Figure 1-1, the pressure at the top of the capillary is greater than one atmosphere because of the hydrostatic head developed by the fluid in the reservoir at the top. This pressure "head" or potential energy of the fluid in the upper bulb appears in two forms as the solution flows through the capillary: (1) kinetic energy of the exiting stream and (2) heat due to frictional (viscous) losses in the capillary. If the fluid were viscosity free, then the potential energy would be converted entirely to kinetic energy, as if the fluid were being dropped through the capillary without hitting the sides. If the viscosity is high, then the exit velocity is low and most of the energy is dissipated as heat. This is the situation that the operator wants, as it is the viscosity of the solution that is important to the analysis. Under these conditions, the flow time is proportional to the viscosity divided by the density of the fluid. Why the density? Because a high density means higher pressure at the bottom of the reservoir, and thus faster flow. Pressure beneath the surface of a quiescent fluid is given by

$$P = \rho g h \tag{1-1}$$

where P is the developed hydrostatic pressure at depth h, ρ is the density of the fluid, and g is the acceleration of gravity. The pressure-driven flow through the capillary is thus driven by a pressure that is proportional to the density of the fluid, and is resisted by the viscosity of the fluid. According to Poiseuille's law for flow through a capillary (which we will derive later), the flow rate Q will be given by

$$Q = \pi(\Delta P) R^4/8\eta L \qquad (1\text{-}2)$$

where ΔP is the pressure drop through the capillary of length L and radius R and η is the viscosity of the fluid.[†] The polymer chemist measures the time t_f it takes the fluid to leave the upper reservoir. This time will be lengthened by decreasing flow rate as the fluid height drops in the reservoir. However, the flow time will be proportional to the viscosity and inversely proportional to the density, i.e.,

$$t_f \propto \frac{\eta}{\rho} \qquad \text{or} \qquad \eta \propto \rho\, t_f \qquad (1\text{-}3)$$

The ratio η/ρ, called the kinematic viscosity, has dimensions of $[L]^2/[t]$ where $[L]$ and $[t]$ signify length and time, respectively. In SI units, this amounts to m^2/s, a type of diffusivity. Mass and thermal diffusivities have the same units.

The usual way of handling the capillary experiment is to eliminate the density by dividing by the flow time of the solvent. Elimination of concentration effects is done by extrapolating to zero concentration, or by interpolating to some fixed concentration such as 0.1%. This procedure is most easily seen by examining the expected effect of polymer concentration on solution viscosity η via the familiar Huggins equation:

$$\eta = \eta_s \{1 + [\eta]c + k_H[\eta]^2c^2 + \cdots\} \qquad (1\text{-}4)$$

where η_s is the solvent viscosity, c is the concentration (often g/dL) and $[\eta]$ is the intrinsic viscosity. Naturally enough, the constant k_H is called the Huggins' constant. High $[\eta]$ means that the polymer will have a strong effect on the solution viscosity; and, indeed, theory indicates that $[\eta]$ scales as molecular weight to a power of about 0.8 for good solvents, but less for poor solvents. As Poiseuille's equation has convinced us that the viscosity is proportional to ρt_f, then

$$t_f = t_{f,s} \{1 + [\eta]c + k_H[\eta]^2c^2\} \qquad (1\text{-}5)$$

where $t_{f,s}$ is the flow time for pure solvent, and where the ρ's on each side have been cancelled out. There is an assumption here that the density of all the solutions is the same, which is reasonable for dilute mixtures. The classical approach to finding $[\eta]$ is to divide both sides by $t_{f,s}$, subtract 1 from both sides, and finally divide by c. A plot of the modified left-hand side against c would thus give an intercept of $[\eta]$ according to the relationship

---

[†] This solution was also described earlier by Gotthilf Hagen.

$$\eta_{sp}/c = (t_f/t_{f,s} - 1)/c = [\eta] + k_H[\eta]^2 c \qquad (1\text{-}6)$$

Another approach, which has some statistical advantages, is to fit the observed flow times vs. concentration with the quadratic form of equation (1-5), i.e., $y = a_0 + a_1 x + a_2 x^2$, where $y = t_f$ and $x = c$. Once a_0, a_1 and a_2 are found, the intrinsic viscosity is just a_1/a_0. Note that with this method the observation of flow time for the solvent is not necessary. With equation (1-6), a mistake in $t_{f,s}$ will impact directly the value of $[\eta]$; with equation (1-5), $t_{f,s}$ is simply another data point and counts no more than any of the others.

It should be mentioned that many routine quality-control protocols call for a single measurement of solution viscosity at a specified concentration, say, 1%. This is generally reported as simply η_{sp} or η_{inh}. The latter is determined as $\ln(t_f/t_{f,s})/c$. Clearly, this method will work fine for distinguishing changes in molecular weight for a given polymer/solvent systems as long as the concentration is exactly right.

Solution methods for polymer analysis have moved in the direction of chromatography, especially size exclusion chromatography (SEC). This method, in its simplest form, uses columns of swollen gel particles with different crosslink densities. When a sample is injected into the column, the gel particles offer extra volume to the small molecules, while the largest molecules are restricted to the void volume between the particles. The size of the molecules is determined by their size in the solvent used in the instrument, and this size is directly related to the product $M[\eta]$. The relationship that leads in this direction is

$$[\eta] = \Phi \alpha^3 \frac{\left\langle r_{g,0}^2 \right\rangle^{3/2}}{M} \qquad (1\text{-}7)$$

where M is molecular weight, α is the linear chain-expansion factor, and $\left\langle r_{g,0}^2 \right\rangle$ is the mean square end-to-end distance of the chains in their unperturbed states. As Φ is considered a universal constant, the product $M[\eta]$ is proportional to the chain volume, and consequently should be largely independent of the polymer's structure.

As one might expect, rheology plays a role in analysis of polymer melts as well as solutions. While melt rheology may not have the precision of solvent-based analysis, it is without doubt much quicker and more convenient for process control and quality control (QC) purposes. Classical QC methods have included the widely used "melt index" measurement, which is the melt equivalent of the single-point inherent viscosity measurement for solutions. More will be said about melt index in Chapter 11. Most progress, however, has been made correlating the linear viscoelastic properties of the polymer melt with the molecular weight distribution (MWD). The analyses, both empirical and semi-theoretical, have proven capable of detecting small changes in the MWD, and indeed can be more sensitive than GPC to the high-molecular-weight part of distribution.

3. Rheology as a predictor of processing performance

The other side of rheology, aimed in the direction of polymer application, is as a predictor of polymer processing performance. The standard questions that the engineer and equipment designers attempt to answer are the flow patterns and pressure values throughout a process. The goals are to design processes that control molecular orientation at desired levels, and reduce or eliminate flaws in the final parts. As always in commercial processing, the economics favor faster process speeds, but with acceptable product quality. Process rheology attempts to find ways to increase productivity, while minimizing problems.

Processing flows can be divided into four categories:

- Transient flows where the melt is largely confined by surfaces. A prime example is injection molding.
- Transient flows where the melt is exposed to the air during processing. Bottle blowing is a good example.
- Continuous flows where the melt is largely confined by solid surfaces. Certain pipe extrusion processes fit this category.
- Continuous flows in which the melt is exposed to the air during a critical part of the process. Most extrusion processes, notably blown film extrusion, fall under this heading.

The challenge in process rheology is to use both polymer structure and the characteristics of the process to get the desired product and economics. Many times problems arise because of changes in resin supplier or grade of resin.[‡] Often the rheologist can detect these changes, and suggest process changes to accommodate the change. However, there are important exceptions, for a logical reason, which is as follows: For economic reasons, the process is often finely tuned to push the polymer as hard as possible, and even the minutest change in machine settings or polymer can upset the balance. In fact, there are times when this balance is so delicate that the machine becomes the only device sensitive enough to detect changes in the polymer. This is certainly a humbling

---

[‡] The term "resin" is widely used in industry instead of "polymer." Why? Aside from being slightly easier to say (two syllables vs. three) the word "resin" has some history. Since prehistoric times, technologists formulated coatings (e.g., varnish) using resins from plants. Synthetic substitutes were simply called by the same term.

and undesirable situation, as it automatically means that acceptance tests on a resin shipment will fail to detect problems before they occur—a continuing challenge for polymer scientists.

4. Complex flows

The quantities that are needed in process flows include the pressure, the values of the stress, the magnitude and direction of the velocity, and the temperature at every location throughout the process. This is a tall order. With flows that have velocity variations in only one direction, analytical solutions are relatively straightforward. Flow through a straight cylindrical tube is an example. By "analytical solution" we mean an answer in the form of an equation, as with Poiseuille's equation for flow in a tube; it predicts the pressure drop through the tube, a useful process variable. From this we can predict the power required to push the polymer through the tube at a given rate.

If the geometry is more complicated, which is nearly always the case, analytical solutions are either very complex, or impossible. The approach, then, is to use numerical methods. Numerical approaches generally follow either finite-difference or finite-element methods. The finite-difference method starts with the defining differential equations for the flow and substitutes numerical finite-difference approximations for the differentials. For example, if we have evenly spaced values of the variable x, the derivative dy/dx evaluated at $x = x_0$ is given by the equation

$$\left.\frac{dy}{dx}\right|_{x=x_0} \cong \frac{1}{2}\left(\frac{y_1 - y_0}{x_1 - x_0} + \frac{y_0 - y_{-1}}{x_0 - x_{-1}}\right) = \frac{y_1 - y_{-1}}{2\Delta x} \qquad (1\text{-}8)$$

where y_1 is the value of y one step ahead of x_0, i.e., at $x = x_1$, whereas y_{-1} is the value one step behind. With uniform steps, $\Delta x = x_1 - x_0 = x_0 - x_1$. It can be seen that the derivative approximation is just the average of the two-point slopes in front of and behind the point at which the derivative is needed. It can also be seen that the points of evaluation are most conveniently equally spaced, a limitation of the method. The net result of transferring all the derivatives to their finite-difference approximations is that the defining differential equations applicable to each point in the process become linear algebraic equations. The solution to the problem then is matter of solving a huge set of simultaneous linear equations. While this may sound easy, it can in fact be very difficult.

Because of the limitations of the finite-difference method, the favorite approach is currently the finite-element method. This method does not require uniform spacing of the points at which solutions are sought, nor does it require that the elements be rectangular. The gain is significant, in that points can be spaced closely where the flow is rapidly changing, and widely spaced where not much is happening. The penalty is that the derivative approximations become complex and difficult.

Although numerical methods have been successfully applied to many steady and transient flows, the majority of problems have involved fairly simple fluid models. For transient flows involving highly elastic polymer melts, the field is still open for plenty of innovation.

5. Polymers as complex fluids

The typical commercial polymer used in fabrication processing has the complication of high molecular weight with the attendant high viscosity and elasticity. In addition, the response is often further complicated by a broad or bimodal molecular weight distribution. Aside from these complications, commercial polymers often are mixtures containing second polymers, solid particles or fibers, and flow-modifying additives called lubricants.

Aside from the complications of molecular-weight distribution, the flow of complex polymeric fluids often involve additional factors such as:

a. *Suspended hard or soft particles or needles* which increase viscosity and can introduce paste- or gel-like behavior.

b. Equilibrium or stress-induced formation of organized domains can occur in melts. These phases involve highly aligned molecules. Although such has been demonstrated, little is known about the influence of stress-induced nematic phases on the rheology of melts.

c *Strong interactions with dissolved or micellized additives* can either promote or reduce intermolecular interactions. Generally molecular motions of both ingredients (polymer and additive) are not eliminated in the mixture, but the rate of these motions can be changed profoundly. The resulting effects are termed *plasticization* or *anti-plasticization*.

d. *Influence of blockiness* in vinyl polymers is exploited commercially for making soft vinyl products. Without the interaction between blocks of similar structure, e.g., tactic sequences, these vinyl products would flow at room temperature. Blocks that don't crystallize can separate into microphases due to unfavorable interaction with the other parts of the molecule. Extreme examples include alternating blocks of silicone or fluorocarbon with a more polar polymer. In the case of fluorocarbon, even very short sequences are enough to influence the flow behavior.

e. *Strong secondary interactions* between the polymer chains include ion-ion, ion-dipole and hydrogen bonding. These interactions, because of their additive influence on a chain, can increase viscosity markedly.

B. RHEOLOGY IN ITS SIMPLEST FORM

1. What needs to be measured?

Measurement of rheological properties is of understandably great importance to anyone dealing with the production or scientific study of materials with unusual flow characteristics. The process of measurement, often called "rheometry," should meet several requirements:

1. The result of the measurement should be a fundamental property of the material, untainted by the peculiarities of the measurement method. Thus, at the very least, a different instrument or procedure should produce the same result for the same material. At best, the measurement process should be able to reproduce the certified results for standard materials. Some such materials, called standard reference materials (SRM),, are available from the National Institutes of Standards and Technology (www.nist.gov). The most thoroughly characterized rheological standards are a silicone melt (2491) and a polyisobutylene solution (2490). Also available are a few other polymers that have received less extensive rheological characterization, including several polyethylene resins that have certified melt indices.

2. All rheometry requires the measurement of three fundamental dimensions: length, mass, and time. The fundamental dimension mass appears as force ($1 \text{ N} = 1 \text{ kg m/s}^2$), which means that somewhere along the way a value of gravity is assumed, either directly, because a weight is used to produce the force, or indirectly during calibration of a load cell.[§] Recall that gravity does vary from place to place on the earth, with a range of about 0.4%. Small, but not insignificant. Do you know the value of gravity in your lab?

3. One other fundamental dimension—temperature—is of huge importance for interpretation of rheological measurements. All those working with polymer melts or solutions appreciate the enormous effect temperature has on rheological properties. Unfortunately for rheologists, measurement and control of temperature has been one of the most difficult aspects of instrumentation. For example, the temperature range over the surface of plates in an expensive rheometer has been observed to vary by over 10 °C, and is invariably significantly lower than the set temperature of the oven

---

[§] The term "load cell" is used frequently in this and other rheology books. At the simplest level, it's a device that measures force or torque and produces a recordable signal.

(Figure 1-2). Furthermore, deformation of viscous polymers leads to heat generation. This is unavoidable, and can lead to serious underestimation of the viscosity or other rheological properties under some conditions.

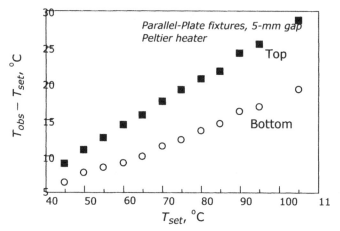

Figure 1-2. Variation of temperature in a parallel-plate fixture in a commercial rheometer. T_{set} and T_{obs} are the setpoint of the rheometer and the observed temperature, respectively. The gap is 5 mm and the sample is a filled thermoplastic. [Data from D. A. Barker and D. I. Wilson, "Temperature profiles in a controlled-stress parallel plate rheometer," *Rheol. Acta*, **46**, 23–31 (2006). Adapted with permission of Springer-Verlag, © 2006.].

2. Observations from particular flow tests

A basic material property is important to the polymer scientist in that connections can, in principle, be established between these properties and molecular structure. On the other hand, rheological characterization to obtain material functions can be tedious and often requires trained operators and expensive instrumentation. Thus, it is not surprising that industry has developed a number of standardized tests that are used extensively for quality control. Many of these have been codified by the American Society for Testing Materials (ASTM), and full descriptions can be purchased from that organization (see www.astm.org). The premier example is "melt index," which measures the ease of flow of a melt through a short capillary. Such tests are specific to a standardized instrument and procedure. To measure melt index, for example, a company will need to purchase the exact instrument, which is not surprisingly called a melt indexer. While this instrument looks superficially like a capillary rheometer, the use of the latter to measure melt index would not produce an acceptable melt-index result.

Some of the tests developed over the years are aimed directly at processibility, the ability of a polymer to be passed through certain equipment and end up as an acceptable product. These tests attempt to reproduce to the extent possible the flow conditions involved in the processing equipment, but often fall short. For some processes, it is clear that no single instrument, and probably no combination of existing tests, can provide a measure that is as sensitive as the process itself to slight changes in the polymer. Thus resin suppliers often find themselves obliged to purchase at huge expense the exact equipment used by their customer. It is the long-range hope of the rheologists interested in numerical methods to eliminate this situation by modeling precisely every step in the process.

3. Overview of the scope of rheology as it applies to polymer science and engineering

As mentioned earlier, polymer scientists and engineers are interested in rheology mainly as a characterization tool that will provide material property data over a wide range of conditions. This information can then be used to test molecular and empirical formulations for the dynamics of polymer melts and solutions. Much can be said from a relatively simple rheological test that would be difficult to deduce from any other measurement. An example is physical gelation of polymer solutions. Rheologically, the difference between a fluid solution and a gelled solution is profound, whereas it can be very difficult to see any differences using, say, spectroscopy. Another example is the phenomenon of delayed elastic recovery of deformed melts, which is plainly obvious from rheological tests yet difficult to correlate with any other analysis. The source of elastic recovery is a small amount of very high molecular weight polymer. Sometimes this component can be seen using size exclusion chromatography (SEC), but often the resolution of SEC does not extend to a high enough molecular weight.

In this regard, a popular activity among rheologists is the quantitative prediction of the entire molecular weight distribution from rheological properties alone. The connection between viscosity and molecular weight goes back to the earliest theories of polymer dynamics. For example, the Rouse theory predicts that molecular weight of a single component of the distribution will be given by

$$M = \frac{\rho R T \pi^2}{6\eta} \tau_1 \qquad (1\text{-}9)$$

where ρ and η are the polymer's density and viscosity, respectively, and τ_1 is the maximum relaxation time of the polymer. For polymers featuring a broad distribution of molecular weights, the relaxation behavior will be similarly broadly distributed—the two can be connected, in principle, by a suitable transform.

4. Historical complications that make life difficult

Mechanics was developed very early on by those interested in properties of solids, specifically elastic solids. For purely elastic solids, time was not a variable. Tensile properties were of paramount importance because structural members such as beams were likely to "fail" to fulfill function due to excessive tensile stress. While logically the pulling aspect of tension should be associated with a negative pressure and thus receive a negative sign, the historical importance of tension in the technology of material strength led to its familiar positive sign. This "accident" of history has also created a factor-of-two discrepancy in definitions of deformation, which will be explained in more detail in the next sections.

Geometrical mechanics was hardly necessary for describing simple fluids obeying Newton's viscosity law:

$$\sigma = \eta\dot{\gamma} \qquad\qquad \text{(ssc)   (1-10)}$$

where σ is the shear stress, η is the viscosity and $\dot{\gamma}$ is the shear rate. Flow in simple geometries was handled using the Stokes equations, which relates the stresses to measurable pressure changes. Discussion of the Stokes equations and the more general equations of motion will be provided in Chapter 2. These equations, in principle, allow one to relate pressure changes to flow patterns in arbitrarily complex geometries for Newtonian fluids.

PROBLEMS

1-1. Conventional wisdom holds that a suction cup should be only able to lift a weight equal to atmospheric pressure times the area of the suction cup. Thus, a suction cup with an area of 5 cm^2 should be not be able to pull with more force than about 5 cm^2 × 10 N/cm^2 = 50 N (= 11 lb$_f$). Under what circumstances might this estimate be too low?

1-2. A solution of an SBS block copolymer exhibits the following flow times in a glass capillary viscometer:

Concentration, g/dL	Flow time, s
2.5	669
1.2	255
0.625	141
0.3125	94
0.156	73

Calculate the intrinsic viscosity in dL/g for this polymer.

1-3. The data in Problem 1-2 were gathered using a viscometer with a calibration constant of 0.010 (mm^2/s)/s. From this information and the data from Problem 1-2, calculate the kinematic viscosity η/ρ, where η is the viscosity and ρ is the density.

Given that the temperature was about 30 °C, which solvent was most likely to have been used for this experiment? Confine your search to common polymer solvents, e.g., those listed in the *Polymer Handbook*, Section III.

1-4. Visit the NIST website and download the description of SRM (standard reference material) 2491, which is a poly(dimethyl siloxane) resin. Using the data in this article, calculate the molecular weight using the Rouse theory [equation (1-9)] and assuming the maximum relaxation time τ_1 is given by the reciprocal of the frequency at which G' and G'' cross. Compare this with the stated molecular weight. What value of τ_1 would be required to give exact agreement?

1-5. Electronic balances have been used for various rheological experiments because they are, in effect, a readily available load cell. Balances are often checked using a secondary standard mass traceable to the standard kilogram stored in a vault located in France. Does the local gravitational constant need to be known for this standardization? If the balance is carefully moved from Red River in Canada ($g = 9.824$ m/s^2) to Liberia ($g = 9.782$ m/s^2), will the standardization need to be repeated?

1-6. Repeat Problem 1-5 assuming the balance is used for measuring forces developed by stretching polymer solutions between to disks.

SUGGESTED REFERENCES, WITH COMMENTARY

R. B. Bird, R. C. Armstrong and O. Hassager, *Dynamics of Polymeric Liquids*, Vol. 1, 2nd edition, Wiley-Interscience, New York (1987). Content varies from introductory to highly advanced. For polymer scientists, Chapters 2, 3 and 10 are particularly appropriate. Uses "fluids" sign convention (see Chapter 2 for explanation).

J. M. Dealy and K. F. Wissbrun, *Melt Rheology and Its Role in Plastics Processing*, Van Nostrand Reinhold, New York (1990). In spite of its title, is very much "polymer rheology" oriented, with only 1/5 of the volume devoted to processing. Nomenclature follows strictly the conventions suggested by The Society of Rheology.

C. W. Macosko, *Rheology Principles, Measurements, and Applications*, VCH Publishers, New York (1994). Several chapters are written by other authors, including one entitled "Rheology of Polymeric Liquids" by M. Tirrell. The "tensile" nomenclature convention is used for deformation (see Chapter 3 for explanation).

P. J. Carreau, D. C. R. De Kee and R. P. Chhabra, *Rheology of Polymeric Systems*, Hanser/Gardner Publications, Cincinnati, OH (1997). Uses the same nomenclature and sign system as Bird et al. (1987), except that σ is substituted for τ for the deviatoric stress (see Chapter 2 for explanation). More of a general rheology text—there is relatively little direct treatment of polymers.

F. A. Morrison, *Understanding Rheology*, Oxford University Press, New York (2001). Written with the same nomenclature and sign system and at about the same mathematical level as Bird et al. (1987). The chapter on "Experimental Data" is largely polymer-based.

J. M. Dealy and R. G. Larson, *Structure and Rheology of Molten Polymers*, Hanser Gardner Publications, Cincinnati, OH (2006). The beginning chapters read very much like a polymer textbook, covering basic concepts of structure, polymerization methods, molecular weight, etc. The subsequent chapters deal with a large collection of practical and theoretical topics on rheology, including extensive treatment of tube models for polymer flow.

2

Stress

Stress is the result of the application of force to a material, be it fluid or solid. Conceptually, this is very straightforward; we do this routinely when stretching a rubber band, or spreading paint. The force can be applied by pulling or pushing on the surfaces of the sample,* either with a solid **clamp** or an immiscible fluid (gas or liquid). The former can, in principle, be used to apply the force in any direction with respect to the interface: push, pull, twist, slide, etc. The immiscible fluid usually works only to push against the surface.

Force can also be applied by exploiting the mass of the body and acceleration, often gravity. Pouring is a common example. Spinning pizza dough to thin it out is a good example where gravitational acceleration is enhanced to increase the stress over that achievable by simply hanging the dough and letting gravity do the work. In principle, forces can also be applied using magnetic or electrical fields, but these are often weak effects with pure polymers or homogeneous polymer solutions. Strong magnetic effects can be achieved by introducing ferromagnetic particles into the fluid, and are exploited in seals for computer hard drives. In fact, these fluids, referred to as ferrofluids, have been considered for seals of pressure vessels.

---

* The word "sample" will be used throughout the text to designate a unit of material prepared for testing. Synonyms include "specimen" and, occasionally, "test unit." However, in sections discussing the scatter of observations, the word "sample" will assume its statistical meaning.

A. STRESS AND PRESSURE

Stress and pressure both express the application of force per unit area. The implication is that the "area" refers to a surface of the sample, or to an interface between the sample and another material such as a steel piston, a gas, or an immiscible fluid. However, we can be more general and assume that stress refers to force applied uniformly over any area (as small as we like) within or on the sample. The only difference between stress and pressure is that the former is associated with a force applied in a particular direction, whereas pressure is assumed to act in all directions equally. If a surface is subjected to pressure, the resulting stress on the surface will always be directed normal to the surface. On the other hand, the force creating the stress can be in any direction. As the force is a vector, it can be resolved into two components: one parallel to the surface and the other perpendicular to the surface. The component perpendicular to the surface plane (parallel to the surface normal) results in a **normal stress**, whereas the component in the surface plane produces a **shear stress**.

1. Dimensions, units and nomenclature

Stress and pressure have the same dimensions and common units. However, the nomenclature of stress, because of its association with a direction, is a bit more complicated.

The dimensions of stress or pressure are $[M]/[L][t]^2$, where $[M]$, $[L]$ and $[t]$ signify mass, length and time, respectively. The fundamental SI units are $\text{kg}\,\text{m}^{-1}\,\text{s}^{-2}$, which can be expressed in the derived units N/m^2, where N is the abbreviation for Newtons, the SI unit of force. A 100-g weight exerts roughly 1 N of force. The next level of derived unit for stress is the Pascal, abbreviated Pa, where $1\ Pa = 1\ N/m^2$. Other common units for stress include $lb_f/in.^2$, or psi, and dyn/cm^2, where dyn is the abbreviation for dyne. The dyn/cm^2 was very widely used until the 1990's. Two conversions to remember are:

- $1\ Pa = 10\ dyn/cm^2$
- $1\ psi = 6.89\ kPa$ (roughly 7 kPa)

Thus, 1 psi is bigger than 1 Pa, which is bigger than 1 dyn/cm^2. Lesser used, but still seen, are units such as kg_f/cm^2, which is a gravity-based system similar to pounds mass vs. pounds force. Thus $1\ kg_f = (g/g_c)\ (1\ kg_m)$, where g is the local gravitational constant, and g_c is the standard acceleration of gravity.[†]

[†] The value of $g_c = 9.80665\ m/s^2$ has been chosen to represent the values of gravitational acceleration at sea level at intermediate latitudes. In fact, this value matches g in only two major cities to four significant figures—Atlanta and Zurich.

Interestingly, the dimensions of stress are the same as energy concentration, that is, J/m^3 in SI units.

While the quantification of stress and pressure are fairly straightforward, the question of signs is not, at least for stress. Nearly everyone agrees that a compressive pressure should have a positive sign. How do we know a pressure is positive or compressive? Such a pressure will tend to decrease the volume of the sample, if the sample is at all compressible. All real materials are, to some extent.

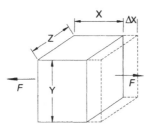

Figure 2-1. Sample with forces applied to opposite surfaces in the outward direction.

With stress, however, the agreement on sign is far from universal. Examining Figure 2-1, we see forces applied perpendicular to the opposite faces of a sample. It is important to understand that the arrow denotes uniform application of the force F over the entire area shown, which could be very small. Conceptually, this might be achievable if the forces were applied as shown to the faces of the sample, but the faces were intimately covered with a substance that resisted bending or bowing outward, yet did not resist at all motion along the surface. We shall refer to such a substance as an *ideal* or *virtual clamp*. The type of force shown, referred to as a tensile force, was given a positive sign by those involved with tensile testing of solid materials. This is very understandable. However, note that the tensile force will tend to pull the sample as if there were a negative pressure applied to the faces. Thus, we have, with this convention, a disagreement between the senses of a positive stress and a positive pressure. This inconvenience can't be wished away—it's here to stay. We will refer to this convention as the *solids sign convention*, to denote its origin in the testing of solid samples. The opposite will be called the *fluids sign convention* for reasons that will be explored in Chapter IX. Of course, either sign convention may be used with either or both states of matter. When we use the solids sign convention in an equation, the equation will carry the notation (ssc). Similarly, (fsc) will denote the fluids sign convention.

Before pursuing the "sign problem" in depth, let's take a look at the rules for directing the force relative to the area to which it is applied. As shown in Figure 2-1, the tensile force must be applied perpendicular to the area; otherwise, it would have a component in the plane, which would complicate the situation by producing a shear stress. This direction of the normal force is the same as the outward-pointing normal to the surface, i.e., the force and outward-pointing normal are parallel. Thus, the tensile force shown is also called a "normal" force, and the resulting tensile stress is referred to as a *normal stress*. The stress is simply the force per unit area. If we want to stretch the sample in the x direction, we apply the pair of normal stresses in opposite directions, as shown. To identify this stress completely, we specify the coordinate direction of *both* the surface normal and the force, i.e., $\sigma_{xx} = F_x/A_x$, where σ_{xx} is the stress, F_x is the force in the x direction, and A_x is the area pointed, via its normal, in the x direction. By convention, the first subscript in σ_{xx} refers to the direction of the area normal, while the second refers to force direction.

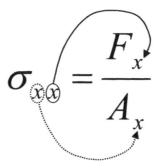

Figure 2-2. Association of the subscripts with the outward-pointing normal and the force direction, both pointed in the positive x direction.

While this sounds a bit redundant, it is important for cases where the force points in, say, the y direction. In that case, according to our rule, the stress would be σ_{xy}, and would be called a shear stress. More about shear later.

It is evident at this point that the forces applied to the sample act over an area, which can be conveniently small. The cube of material on which the forces act must be in static equilibrium; that is, all the forces and moments applied to cube must balance to zero. Otherwise, the sample will accelerate and body forces will enter the picture, as in spinning pizza dough.

Note that the sign of the stresses is always what we apply to the sample, not what the sample applies to us. Thus if we squeeze a sample between two plates, the stress is acting inward on the sample, but the sample fights back with a force acting outward on the plates. The sign is determined by the former. With the solids sign convention (ssc), the stress in this example is negative; with the fluids sign convention (fsc) the sign is positive.[‡]

[‡] The fluids sign convention (fsc) has an added important advantage in that it gives the right sign for the transfer of momentum from high velocity regions to low velocity regions.

2. Measurement of pressure and stress

The measurement of pressure and stress are often done in a similar fashion. We are so familiar with pressure-measuring devices such as manometers and Bourdon gauges that we don't stop to think much about what they are actually doing. In quiescent fluids, we know that the pressure we measure doesn't depend on the direction in which it is measured, and the possibility that situations might arise where this was not true never entered our consciousness. Nevertheless, we are now armed with enough knowledge to state that pressure-measuring devices are really measuring normal stress, and the direction they point can make a difference.[§]

Figure 2-3. Photo of a diaphragm "pressure" transducer. The diaphragm covers the right-hand circular surface. This device measures the normal stress perpendicular to the diaphragm surface, and is insensitive to shear stresses along its surface. (Photo courtesy of Omega Engineering, Inc.)

To understand this more clearly, let us consider a diaphragm pressure transducer commonly found on polymer processing equipment. While there are many design differences, most comprise a flexible diaphragm, the deflection of which is measured in some fashion, e.g., with strain gauges either directly on the diaphragm or attached to a beam that is bent by the diaphragm via a rod.

Figure 2-3 shows a typical transducer. The transducer shown is intended to be mounted flush with the surface containing the fluid, thus avoiding disturbance of the flow. While the outside surface of the diaphragm is exposed to the fluid, the inside surface may be exposed to either the atmosphere, a vacuum, or some other reference pressure. In practice, this makes little difference, as the output signal can be zeroed at any pressure, applied to the outside, of your choice. Most frequently, the output value is zeroed when the transducer is exposed to the atmosphere. With a vacuum behind the diaphragm, the signal will drift with atmospheric pressure changes, whereas

[§] Prof. John M. Dealy has provided a definitive essay on this topic; the reference is: "Misuse of the Term Pressure," *Rheology Bulletin*, January 2008, pp. 10–13 and p. 26. A copy of the article can be accessed at www.rheology.org/sor/publications/rheology_b/issues.htm.

nothing will happen if the back is exposed to the atmosphere. With the vacuum reference, the *absolute pressure* is measured.

If pressure gauges are really stress gauges, then how does one determine the pressure? This question is far more complicated than it sounds, but there is an easy answer: measure the stress in three orthogonal directions, and average the results. Thus, isotropic pressure *p* can be defined as

$$p = -\frac{\sigma_{xx} + \sigma_{yy} + \sigma_{zz}}{3} \qquad \text{(ssc) (2-1)}$$

In practice, this could be very cumbersome to do, and virtually impossible without interfering with the flow. By using very simple flow geometries, pressure differences can be deduced easily by measuring stress differences at two points. For example, for flow through a tube, the measurement of radial stress at two separate points along the tube[**] will give the pressure drop as long as the flow is steady (not varying with time) and fully developed (not varying in any fashion along the length of tube under consideration).

It could be argued that pressure will result in thermodynamic changes that might be measured if we were clever enough. The most direct is a change in density of the melt or solution. However, such "density transducers" are not readily available. Less direct would be the change in some optical or spectral feature of the material that responds only to pressure and not stress. The problem is that these effects tend to be so small that they lack the sensitivity needed for practical rheology problems, and are most useful only at super high pressures. Thermodynamic pressure effects are important in gas-polymer mixtures, where the formation and growth of bubbles responds to the isotropic pressure. This is important in the formation of foam products. Pressure effects may also be important in miscible polymer blend systems or solutions.[1]

3. Applying stress—The concept of the ideal clamp

A *clamp* is a term we will use as a general tool for applying force to a sample surface. Practical clamping devices sold with instruments are usually referred to as "fixtures," meaning that they hold a sample as a light fixture holds a light bulb and attach the sample to the outside world. An *ideal clamp* is in imaginary clamp that has the following properties:
- Has no influence on the properties of the material in the sample.

[**] Transducers of the type shown in Figure 2-3 are available only with flat diaphragms, and thus are not able to be mounted in a tube without disturbing the flow. However, if both upstream and downstream transducers are installed in exactly the same fashion, the errors due to the disturbances should cancel out when the two readings are subtracted.

- Transmits force to the sample uniformly over the entire contact surface.
- Holds its normal direction steady and resists bending.
- Freely shrinks or expands in area as needed to assure uniformity of the stress applied to the sample over the entire area of the clamp.

The last requirement is, of course, the one that it most difficult to achieve. For example, if we wish to stretch a thin, square piece of elastomer uniformly, we can grab the edges all around and pull. The problem with this is that the edges get longer as the sample stretches, so a rigid clamp won't work. However, the ideal clamp can do this, as it can expand with the width and shrink in the thickness direction as the sample stretches and thins. The device pictured in Figure 2-4 is designed to do this in an approximate fashion.

Figure 2-4. Device used for biaxial stretching of a rubber sheet with an attempt to make ideal clamps. The individual clamps along the edges of the sample are designed to move freely along the rails to eliminate forces applied parallel to the edge. The entire frame can be tilted at various angles to stress the sample in different modes. [Reprint of Fig. 2 from G. W. Becker and O. Kruger (1973)[2] © 1973 Plenum Press, with kind permission from Springer Science+Business Media B.V.])

Obviously, ideal clamps do not exist, but they are a useful construct and sometimes can be approached by real fixtures.

4. Types of stress that can be applied to a sample, conceptual vs. practical

Application of stress to a material is a critical process that needs to be understood thoroughly. As discussed in the Introduction, stress can be applied using a clamp of some sort. Familiar clamps featuring solid surfaces in contact with the polymer melt or solution always create problems that may be quite significant. Imagine, for a moment, attempting to stretch a polymer melt or solution using a pair of tweezers. Two problems arise:

- The tweezers squeeze the sample too much, which causes the sample to pinch off and break.
- If a lighter pinch is used, the material between the tweezers feeds out during the stretching process, and thus the effective length of the sample is changed.

Figure 2-5. Photo of counter-rotating roller device designed to stretch polymer melts. The melt adheres to the roll surfaces at the point of tangency, leading to a sample of constant length. The force applied to the sample is obtained by measuring the torque required to turn the rolls. (Photo courtesy of Dr. M. Sentmanat, Xpansion Instruments, LLC.)

Suffice it to say, that these problems can only be reduced but not eliminated. The most successful designs are those that gobble up the sample by the ends with gear sets,[3] or by rolling it up on a capstan. Figure 2-5 shows one design using dual capstan rollers.[4]

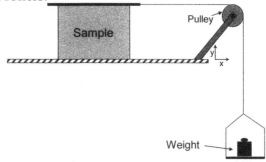

Figure 2-6. Simple method of applying shear stress to a sample. To maintain simple shear, the top and bottom plates must remain parallel and with a constant spacing. This may require application of a normal force to the top plate, which means a pair of normal stresses is applied to the sample. We are assuming at this point that constraint in and out of the page is not needed.

While application of normal stresses to a polymer melt requires considerable ingenuity, shear stresses are relatively easy to apply, although we will see that there are complications also. The definitional way of applying shear stress is illustrated in Figure 2-6, wherein a weight is shown applying force to a plate on top of the sample. While this may work for some samples, for others the plates must be constrained by the application of a matched pair of normal stresses to the sample to prevent the top plate from moving up or down (in practice the latter is rare). If the force applied to the top plate is in the positive x direction, and the direction of the top plate (outward-pointing normal) is in the y direction (these choices are conventional but arbitrary), then we can follow the convention shown in Figure 2-2 to name the shear stress as

$$\sigma_{yx} = \frac{F_x}{A_y} \qquad (2\text{-}2)$$

In the case shown in the figure, the force F_x will be the mass on the pan (plus the small contribution of the pan and the string) times the local acceleration of gravity. An advantage of this method of applying the force is obvious: gravity never quits.

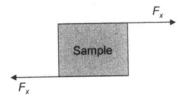

Figure 2-7. Free-body diagram of sample in Figure 2-6. Note the twisting tendency in the clockwise direction because the forces producing the shear are not aligned.

In Figure 2-6, there is a moment applied to the sample by the top and bottom plates. This moment, produced by a *shear couple*, tends to twist the sample in the clockwise direction, which we will consider to be positive. To balance this torque, the plates must apply a moment to the sample by virtue of their constraint by, say, the frame of the apparatus. Let's cut the sample and its plates free and examine the result, which is illustrated in Figure 2-7. Now the unbalanced torque is clear, and it is evident that the entire "sandwich" would turn, which would ruin the state of simple shear. We must apply a balancing torque, which most simply could be done by fastening virtual ideal clamps to the exposed faces (x and $-x$ directions) and pulling on them with a shear couple equal and opposite to that applied by the bottom and top plates. This is illustrated in Figure 2-8. However, as simple shear does not allow motion in vertical direction, the bottom and top plates are mechanically not allowed to rotate or move vertically; the balancing moment is there, but the motion is restricted. At this point, one can appreciate that "simple shear" is not that simple!

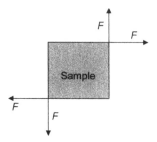

Figure 2-8. Applying an opposite moment with an added virtual clamp pair to which is applied a balancing pair of forces.

An important but somewhat different situation occurs if we have ideal clamps on all four faces, and apply the two moments as shown in Figure 2-8 but allow the clamps to move freely. This state is shown in Figure 2-9. Note that the sample and the applied moments are symmetrical about the plane depicted by the dotted diagonal line. Also note that the forces applied can be resolved into a pair of tensile forces along this diagonal line and a compressive pair at right angles to this diagonal line. (Again we are assuming that no forces are applied or needed in an out of the plane of the page.) This illustrates an important point: a shear couple applied in this fashion results in normal stresses at 45° to the directions of surfaces carrying the shear stresses.

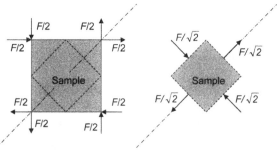

Figure 2-9. Illustration of opposing shear stress couples applied to a sample. The forces are then divided in half and applied to each corner. The two vectors in each corner are then added and applied (uniformly) to the surfaces of the smaller sample shown by the dotted lines. The pressure in the sample is zero [equation (2-1)]. This is a state of *pure shear*.

The presence of the tensile stress is dramatically illustrated by applying shear to a brittle material. Brittle materials are very much inclined to fail in tension, not in shear or compression, whereas tough materials tend to fail in shear. Figure 2-10 shows this using a piece of chalk that has been twisted to apply shear (imagine a stack of coins that is twisted; the coins rotate over each other in a sliding fashion, which is characteristic of simple shear.) If the chalk is twisted carefully, without applying any bending, the failure plane is invariably inclined at 45° to the shear planes.

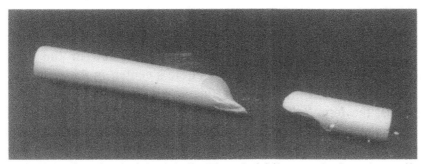

Figure 2-10. Fate of a twisted stick of chalk. The spiral failure results from the tensile stress at 45° to the shear couple.

Returning to Figure 2-9, the forces applied to the corners of the original block have been resolved along the new axes and spread evenly over a new and smaller block by exploiting the triangular pieces of material at each corner as clamps. The forces are $1/\sqrt{2}$ times the original forces used to produce the shear couples. If the original block had a size of 1.0, simple geometry gives a size of $1/\sqrt{2}$ for the new block. Thus the ratio of the force to area turns out to give normal stresses that are the same magnitude as the applied shear stresses. In effect, we have re-derived *Mohr's circle* using this simple geometrical argument. More on Mohr's circle below.

5. Relationships between pressure and stress—stress state as a combination of pressure and "extra" stress

At this point, it should be relatively clear that by proper choice of the direction of the x and y axis, any state of two-dimensional stress can be reduced to two normal-stress pairs. The shear stress is thereby "eliminated." In three dimensions, the situation is the same, albeit much more difficult to visualize. So, when the coordinate system is rotated to exactly the correct angle, the two balanced pairs of normal stresses are called the *principal stresses*. The average of these two will be the pressure in the sample, with a negative sign for the "solids" sign convention. If the set of principal stresses comprises a tension and a compression of equal magnitude, then the stress state is one of pure shear. This sounds very odd, but examination of Figure 2-9 shows clearly that by merely rotating the axes (not the sample) by 45°, the stress state comprises two equal and opposite shear couples, and no normal stresses at all. Pure shear can also be achieved by application of a tensile-stress pair in one direction and an equal compressive-stress pair at right angles.

The relationships for two-dimensional or *plane stress* have historically been summarized in a construct called Mohr's circle, shown in Figure 2-11. This construct can be helpful in visualizing the state of stress at arbitrary angles to the principal-stress directions, or to the coordinates for any known state of stress. Thus in Figure 2-11, if the stress state at 20° counterclockwise from the known state is needed, a reading located 40° counterclockwise (Point B in Figure 2-11) will be the correct result. The pressure in the sample is conveniently given by the location of the circle's center on the principal-stress axis, again assuming that the forces producing all stress lie in the plane of the drawing; thus, the term "plane stress." In Figure 2-11, the pressure is zero, which is convenient; otherwise, we might have to do something about the sample bulging out of the plane.[††]

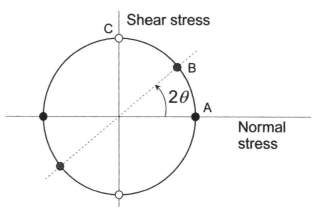

Figure 2-11. Depiction of pure shear using Mohr's circle. The axes are not spatial coordinates; rather, they show type and magnitude of stress. To find the stress combination at any angle θ away from the spatial principal-stress axis, move along Mohr's circle by 2θ, e.g., from A to B. The coordinates of point B give the magnitudes of the shear and normal stress.

Figure 2-11 depicts Mohr's circles for plane stress. "Plane stress" means that there is no component of stress in the third coordinate direction. Thus, all forces producing the stress components are located in a plane. The reason pure shear is called pure shear is that one location on Mohr's circle has the maximum shear stress located at zero normal stress. Thus this state is free of any normal stresses. As described in Figure 2-11, this is at 45° to the principal stress direction, thus 90° around the circle. This stress state corresponds to Point C on the Mohr's circle drawn in Figure 2-11.

An important outcome of this discussion is that while it is easy to find an angle that eliminates shear stresses, the elimination of normal stresses occurs only in the case of pure shear. That's why it's called "pure shear." Pure shear

---

[††] If the sample is very large in an out of the plane, we might be inclined to think that the center part of the sample will at least not be allowed to deform out of the plane. If, on the other hand, the sample is very thin, we can reason that the stress in an out of the x-y plane is zero. Thus the film being stretched in Figure 2-4 has no stress components out of the plane of the film and the negative pressure merely causes the film to get thinner.

stress can be established in solids by subjecting a solid to compression in one direction and pulling on it at right angles with equal stress. This is commonly done to study the shear yielding behavior of solid polymers to avoid a possible influence of pressure on the results.[‡‡]

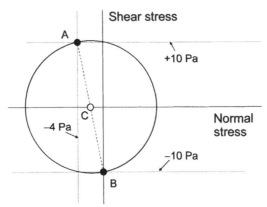

Figure 2-12. Mohr's circle construction for example given in text. A and B are known stress states at 90° to each other. Connecting these gives the center of Mohr's circle. The center point C represents the pressure, and its location at −2 Pa means the pressure is <u>positive</u> (ssc).

An example showing the construction of an arbitrary Mohr's circle is shown in Figure 2-12. Let's say we have a state of shear with a known shear stress $\sigma_{xy} = 10$ Pa produced by a positive (clockwise) shear couple and a known compressive normal stress $\sigma_{yy} = -4$ Pa. (We remind ourselves that we are using the "solids" sign convention because Mohr's circle was developed for solids.) This is Point A in Figure 2-12. To balance torques, there is a counteracting shear-stress couple of -10 Pa, but with a normal stress of 0 Pa applied to the x- and $-x$-directed surfaces (Figure 2-12). This is the usual case for a free surface, such as that depicted in Figure 2-6. Although these known stress states are on surfaces that are at 90° to each other, the prescription says that they will be diametrically opposite each other, that is, 180° apart on Mohr's circle. Knowing this, a diameter can be drawn (it is shown at the dash-dot line) and the circle is defined! Drawing in the circle completes the exercise, and stress states at any angle can be found on the perimeter.

As describe above, the location of the center of Mohr's circle on the normal-stress axis is the pressure in the sample. Its value is invariant with orientation of the reference coordinate system. If its location is in the positive direction,

---

[‡‡] For convenience, we are ignoring the contribution of atmospheric pressure. Generally, atmospheric pressure has relatively little influence of the properties of melts and solutions, although higher gas pressures can have significant effects. As the atmosphere presses in all directions on the sample, its influence on the measured stresses is removed.

the pressure is negative, and *vice versa*. This is, of course, consistent with our working definition of pressure given in equation (2-1).

Note once again that the stresses in the **neutral direction** are assumed to be zero. Referring to Figure 2-6, the neutral direction is in and out of the page, or the *z* direction of the coordinate system set up in the figure. Our result of a positive pressure, then, is bothersome, as this pressure will tend to make the sample bulge out in the *z* direction, violating the conditions of simple shear. We can attempt to constrain that direction with walls, but this is tantamount to applying an out-of-plane stress. Thus, in general, we cannot have both motions and stresses completely confined to a single plane for real materials, except at vanishingly small strains. As a result we must choose to have a state of plane stress or plane strain.

While we can recognize stress states of various types and attempt to apply such states to a sample, the sample may deform in some unexpected fashion, as suggested above. For this reason, the process of simple shear is best defined by insisting that the motions of the material points are all in *x-y* planes, and take care of the stresses as they arise.

At this point, we need to return to the problem of dealing with the resolution of applied stresses into the portion attributable to the deformation of the material and an isotropic component. The former is of most interest to the polymer scientist, of course. We can think of the total stress applied to the sample as a combination of (1) the pressure[§§] and (2) an extra amount due to the deformation of the material. The material reacts to deformation by creating this "extra" stress which adds to the pressure to give the total stress. For example, if we stretch a sample, the measured tensile stress consists of a contribution due to the negative pressure in the sample (imagine pulling a piston out of a sealed cylinder to produce a vacuum) and a contribution due to the deformation of the material. The latter is what tells us about the material, so it makes sense to give it a name. Common choices are ***extra stress*** and ***deviatoric stress***. On many occasions we will refer to it as the "material-generated stress" because of its dependence on the structure of the material. A common but not universal symbol for this stress is τ. There are a couple of good things about this choice, and some bad ones, too. The latter include the

---

[§§] As we have suggested, there are some problems with defining the isotropic term. While we call it the pressure and define it as the mean applied stress, we must recognize that it is not necessarily the true thermodynamic pressure, nor any component of the total stress less the real material-generated stress. These complications can usually be safely ignored. The important rule to remember is that during flow, a pressure transducer does not measure the pressure; it measures the total stress in the direction the transducer is pointed.

common (among the solids folks) custom of using τ for shear stresses only. As a shear stress is not sensitive to pressure changes, which simply move Mohr's circle along the normal-stress axis, the use of τ for the extra stress is at least consistent with the fact that shear stresses have no pressure contribution.[***] The other not-so-good aspect is that the symbol τ does not appear in the official nomenclature published by the Society of Rheology. The authors of these standards were perhaps pointing out that a new term and symbol are not needed, as pressure, or more properly, the isotropic term, can always be subtracted out by subtracting any two normal stresses. This will be pointed out more precisely next. The final "bad" aspect of τ is that it is widely used to denote a relaxation time, especially in the literature on linear viscoelasticity. In fact, we will depend upon the reader to recognize from the context when τ is used to denote a relaxation time rather than the extra stress. Usually (but not always) the latter will come equipped with two subscripts, as in equation (2-3) below.

With the use of τ, we can write relationships for the observed normal stresses in the following fashion:

$$\sigma_{xx} = \tau_{xx} - p \qquad \text{(ssc) (2-3a)}$$

for "solids" sign convention, or

$$\sigma_{xx} = \tau_{xx} + p \qquad \text{(fsc) (2-3b)}$$

when using the "fluids" sign convention. Again, this can be applied in all directions, i.e., the subscript x could be y or z as well.

Let's attempt to get a physical feeling for the sign difference. With the "solids" convention, a tensile stress applied to a sample is positive. The material-generated stress is also positive. If the material is stiffer, τ_{xx} will be higher at a given strain value and, consequently, so will σ_{xx}. However, if the pressure is positive due to squeezing from the sides, the sample will be stretched more easily and σ_{xx} will decrease. An extreme example of this is the "stretching" of a sample by wrapping a long, very taut rubber band around it many times. The pressure builds with each layer, but always $0 = \tau_{xx} - p$ from equation (2-3a) and $\tau_{xx} = p$. The pressure p is positive due to the squeezing action, and so is τ_{xx}. τ_{xx} represents the resistance of the material to being stretched by the squeezing action. In effect, the material is being stretched by

---

[***] For simplicity, we are ignoring the slight influence of pressure on the constitution of the polymer. Pressure dependence of material properties is important, and will need to be taken into an account in high-pressure processes such as injection molding.

the pressure inside the sample.[†††] With the fluids sign convention, a tensile stress σ_{xx} applied to the sample is considered to be negative, as will be τ_{xx}. A positive pressure will then help to reduce the magnitude of this negative total stress, as can be seen from equation (2-3b). For example, if a tensile stress of $\sigma_{xx} = -6$ kPa (fsc) is applied to the sample but the deforming material is known to produce a resistance of $\tau_{xx} = -10$ kPa, then there must be a positive pressure component inside the sample of $p = \sigma_{xx} - \tau_{xx} = -6 - (-10) = 4$ kPa (fsc).

6. Homogeneity of stress—a simplifying but unrealistic concept

In real life, ideal clamps don't exist. Solids, melts and solutions are routinely subjected to stresses that result in stress states that vary with position. Fortunately, in most cases we can still think in terms of a small cube with forces acting on its faces, but the cube must be very small. The clamps, then, become the surrounding material, which pushes and pulls on the small volume. In effect, the surrounding material serves admirably as an ideal clamp, pointing out the importance of measuring stresses and deformations far away from the actual clamps. A simple example of this concept in practice is the use of gauge marks on tensile samples. If the gauge marks are far enough from the clamps, the material between the gauge marks is subjected to uniform stress, assuming the section and the material are uniform.

This concept is carried to an extreme by attempting to map out, using material markers of some sort, the entire velocity field in the sample. The general term for this process is "flow visualization," although the "visualization" aspect can be very indirect. There are many flow-visualization methods reported in the literature, some of which are described in Chapter 7.

B. ORGANIZATION OF THE STRESS COMPONENTS

As can already be sensed, dealing with stresses in three directions can be a messy job. For this reason, those dealing with the mechanics of materials have long recognized that matrix notation can simplify the expression of important concepts. However, this can be also carried to an extreme, whereby the notation becomes so compact that connection with the real world of material flow comes only with years of experience. Let's explore some of the methods that will help us, and ignore the more elegant, but difficult-to-follow aspects.

[†††] Such an experiment was run by Griswold et al. using a polyethylene sample and wrapping it on a lathe with tensioned Lycra® fiber. [P. D. Griswold, R. S. Porter, R. J. Farris and C. R. Desper, *Polym. Eng. Sci.*, **18**, 537–545 (1978)]

C. COPING WITH SUBSCRIPTS

Understanding and using stress in real-world, three-dimensional problems requires the use of matrix notation and, consequently, subscripted variables. Fortunately, many materials of interest to polymer scientists are describable with stress states that are relatively simple, even in three dimensions. Common exceptions are long-fiber composites, and some liquid-crystal polymers. To introduce matrix notation, we need to display an elementary matrix from high school algebra:

$$\mathbf{X} = \begin{pmatrix} x_{11} & x_{12} \\ x_{21} & x_{22} \end{pmatrix} \tag{2-4}$$

Here $\mathbf{X}$ (bold type) stands for the entire matrix, and the x_{ij} are the elements of this 2×2 matrix. In the case of three-dimensional stress the relevant matrix will be

$$\boldsymbol{\sigma} = \begin{pmatrix} \sigma_{11} & \sigma_{12} & \sigma_{13} \\ \sigma_{21} & \sigma_{22} & \sigma_{23} \\ \sigma_{31} & \sigma_{32} & \sigma_{33} \end{pmatrix} \tag{2-5}$$

where $\boldsymbol{\sigma}$ is called the *total stress tensor*.

Why the word "tensor" instead of "matrix"? The answer to this question is a bit more complicated than "convention," as tensors enjoy some properties that not all matrices share. These will be introduced below, but all stem from the fact that tensors describe motions in real, three-dimensional space where the 1, 2 and 3 are associated with orthogonal directions. By orthogonal, we mean that it is possible to move along, say, the 1 direction without influencing your position in the 2 or 3 directions. For example, if you travel due north or south along the earth's surface, your position east and west does not change.

Thus, the 1, 2 and 3 in equation (2-5) are often replaced by the common initials for the coordinates. However, beware that while the meanings of say r, z and θ in cylindrical coordinates are well accepted, their assignments to positions in equation (2-5) are not. As an example, suppose we have a cylindrical sample to which a constant tensile stress σ_T is applied along the z direction, a classical creep experiment. One school of thought is that because the z direction is the primary flow direction, it should be given the prominent 1 direction. Then the mapping results in the array:

$$\boldsymbol{\sigma} = \begin{pmatrix} \sigma_{zz} & \sigma_{zr} & \sigma_{z\theta} \\ \sigma_{rz} & \sigma_{rr} & \sigma_{r\theta} \\ \sigma_{\theta z} & \sigma_{\theta r} & \sigma_{\theta\theta} \end{pmatrix} = \begin{pmatrix} \sigma_T & 0 & 0 \\ 0 & 0 & 0 \\ 0 & 0 & 0 \end{pmatrix} \qquad \text{(ssc)} \qquad \text{(2-6)}$$

where $\sigma_{11} = \sigma_{zz} = \sigma_T$. The assignment of r to the 2 position is because material points move in the r direction to preserve the volume of the sample, but the motion is less, so the r direction is less important than the z. As there is no motion at all in the θ direction, it drops to the 3 position. The problem with this approach is that the assignment might change if the sample is deformed in a different manner, for example, twisted instead of stretched.

The other way of mapping is to assign in the usual order, i.e., x, y, z for 1, 2, 3. Table 2-1 shows one such mapping convention. But many others are used, depending upon the importance assigned to each direction.

Table 2-1. Examples of mapping of number subscripts to coordinate indices

Index → Coordinate System ↓	1	2	3
Rectangular	x	y	z
Cylindrical	r	θ	z
Spherical [a]	r	ϕ	θ

[a] There is a bit of confusion about the assignment of the directions in spherical coordinates. Math books may use θ for the azimuthal direction (from west to east) in accord with cylindrical coordinates, but the ISO 31-11 standard calls for ϕ to be thus used, with θ assigned to the inclination direction (north to south).

D. TYPICAL STRESS TENSORS

In the beginning of this chapter we introduced and discussed the two sign conventions for stress, and referred to these as the solids sign convention (ssc) and fluid sign convention (fsc). The (ssc) and (fsc) are associated, respectively, with positive and negative signs for a tensile normal stress applied to a sample. The same convention is applied to the stress tensor, as suggested by equation (2-6).

1. Stress tensor for extensional deformation

Following the example above where a tensile stress σ_T is applied to a cylindrical sample in the axial direction (Figure 2-13), the total stress tensor can be expressed as a sum of the isotropic and extra stress values as follows:

$$
\sigma = \begin{pmatrix} \sigma_T & 0 & 0 \\ 0 & 0 & 0 \\ 0 & 0 & 0 \end{pmatrix} = \begin{pmatrix} \tau_{zz} - p & 0 & 0 \\ 0 & \tau_{rr} - p & 0 \\ 0 & 0 & \tau_{\theta\theta} - p \end{pmatrix} \qquad \text{(ssc)} \quad \text{(2-7a)}
$$

$$
\sigma = \begin{pmatrix} \sigma_T & 0 & 0 \\ 0 & 0 & 0 \\ 0 & 0 & 0 \end{pmatrix} = \begin{pmatrix} \tau_{zz} + p & 0 & 0 \\ 0 & \tau_{rr} + p & 0 \\ 0 & 0 & \tau_{\theta\theta} + p \end{pmatrix} \qquad \text{(fsc)} \quad \text{(2-7b)}
$$

Note that in this case all the shear-stress components (τ_{rz}, τ_{zr}, $\tau_{r\theta}$, $\tau_{\theta r}$, $\tau_{z\theta}$, etc.) are zero. These are referred to as the *off-diagonal components*.

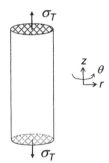

Figure 2-13. Cylindrical sample, equipped with ideal clamps at the ends, subjected to axial tensile stress. The curved surface is free of stress.

The result in equation (2-7) can be split into two tensors as follows:

$$
\sigma = \begin{pmatrix} \tau_{zz} & 0 & 0 \\ 0 & \tau_{rr} & 0 \\ 0 & 0 & \tau_{\theta\theta} \end{pmatrix} - p \begin{pmatrix} 1 & 0 & 0 \\ 0 & 1 & 0 \\ 0 & 0 & 1 \end{pmatrix} \qquad \text{(ssc)} \quad \text{(2-8a)}
$$

$$
\sigma = \begin{pmatrix} \tau_{zz} & 0 & 0 \\ 0 & \tau_{rr} & 0 \\ 0 & 0 & \tau_{\theta\theta} \end{pmatrix} + p \begin{pmatrix} 1 & 0 & 0 \\ 0 & 1 & 0 \\ 0 & 0 & 1 \end{pmatrix} \qquad \text{(fsc)} \quad \text{(2-8b)}
$$

The first tensor is called the *extra-stress tensor*, and the values of its terms are determined, as we have seen previously, by the properties of the material and the nature of deformation. The second matrix has diagonal values equal to 1, and is known as the *identity* or *unit matrix*. It is multiplied by the scalar pressure, p. In the literature, the identity matrix is given various symbols I, $\mathbf{I}$, $\mathbf{I}_3$, $\mathbf{1}$ etc.

The reason for splitting the total stress in this fashion is to emphasize the relationships between stress and pressure.   In particular, if there is no deformation of the sample, then there is no extra stress.[‡‡‡]   Equation (2-8) also emphasizes the difference between applied or total stress σ, and material-generated or extra stress τ.   The difference is the scalar pressure which modifies the diagonal elements where the normal stresses are recorded.   However, equation (2-8) does not reveal how much pressure should be subtracted (fsc) from or added (ssc) to the total stress to yield the extra stresses.

All elements of the total stress tensor σ can be determined experimentally, at least in principle.   But for characterizing the material, we need to know the elements of the extra stress tensor τ.   For shear stress, this is not a problem because the measured shear stress, say σ_{21}, is identical to the τ_{21}, as the pressure influences only the normal stresses.   Pressure always acts normal to a surface. In many cases, the best we can do with the normal stresses is to subtract out the pressure by subtracting two of the observed normal stresses.   With our example of the stretching of a cylinder, we might subtract σ_{rr} from σ_{zz}.   As σ_{rr} is zero (neglecting atmospheric pressure), the observed pulling stress σ_{zz} is equal to $\tau_{zz} - \tau_{rr}$.   However, in spite of the seeming simplicity of this geometry, we cannot isolate a value for either τ_{zz} or τ_{rr} directly.   Let's try a couple of ways to work around this problem.

First, let's assume that the pressure is the mean normal stress, i.e., as given by equation (2-1), but for cylindrical coordinates:

$$p = -\frac{\sigma_{zz} + \sigma_{rr} + \sigma_{\theta\theta}}{3} \qquad \text{(ssc) (2-9)}$$

Now both σ_{rr} and $\sigma_{\theta\theta}$ are zero.   While the former is fairly evident from the lack of forces applied to the outside surface of the sample, the latter is perhaps not as obvious.   One way of investigating this empirically is to take a piece of latex rubber hose and put a short longitudinal slit in one side.   Stretching the tube does not open the slit and does not force it closed.   So, with this in mind, we have the relationship

$$p = -\sigma_{zz}/3 \qquad \text{(ssc) (2-10)}$$

Applying this result to our problem gives

[‡‡‡] This assumes that the material properties have finite values.   Clearly the amount of deformation with very stiff materials (diamond, glass, metal, etc.) will be very small with the magnitudes of stresses common to this book.

$$\sigma_T = \sigma_{zz} = \tau_{zz} - p = \tau_{zz} - (-\sigma_{zz}/3) \qquad \text{(ssc)} \quad \text{(2-11)}$$

Thus, apparently, we can solve for τ_{zz} to give

$$\tau_{zz} = \frac{2}{3}\sigma_{zz} \qquad \text{(ssc)} \quad \text{(2-12)}$$

Pursuing another direction, we can examine σ_{rr}. Starting with the relationship

$$\sigma_{rr} = \tau_{rr} - p \qquad \text{(ssc)} \quad \text{(2-13)}$$

gives

$$p = \tau_{rr} \qquad \text{(ssc)} \quad \text{(2-14)}$$

if the outside of the cylinder is free of applied stress. The first path to the pressure p gives the result that the *trace* of the matrix τ, i.e., $\tau_{zz} + \tau_{rr} + \tau_{\theta\theta}$, is zero, something that we cannot readily check experimentally, but does obtain for many model fluids. The second path gives another route to the pressure, again something we cannot measure. However, if the cylinder maintains its cylindrical shape (it just gets longer and thinner), then we expect τ_{rr} to equal $\tau_{\theta\theta}$, implying that $\tau_{zz} = -2p$ and $\sigma_{zz} = \tau_{zz} - p = -3p$. The latter states, as does equation (2-10), that the pressure magnitude inside the sample is 1/3 of the tensile stress used to stretch the sample. The sign, though, is negative. Thus, for this simple geometry, with the assumptions of no volume change and properties that are independent of direction, the two paths give consistent results. In fact, with these assumptions, the two paths are completely equivalent. But for real materials, we cannot depend on these simple relationships.

2. Stress tensor for shearing deformations

The state of shear has been introduced quite thoroughly above in connection with the discussion of Mohr's circle for plane stress. In general, however, we will have the state of stress given by equation (2-15) when a uniform sample is subjected to shear, with motion in the 1 direction and a velocity gradient in the 2 direction:

$$\boldsymbol{\sigma} = \begin{pmatrix} \sigma_{11} & \sigma_{12} & 0 \\ \sigma_{21} & \sigma_{22} & 0 \\ 0 & 0 & \sigma_{33} \end{pmatrix} = \begin{pmatrix} \tau_{11} - p & \tau_{12} & 0 \\ \tau_{21} & \tau_{22} - p & 0 \\ 0 & 0 & \tau_{33} - p \end{pmatrix} \qquad \text{(ssc)} \quad \text{(2-15)}$$

The subscripts here refer to the orthogonal directions. It is easiest to imagine that the sample is a cube, as shown in Figure 2-7, with the shearing forces deforming the sample in the x (or 1) direction. As they are pulling on the planes defined by outward-pointing normals in the y (or 2) direction, the shear stresses are given the subscript xy, or 12, as in equation (2-15). This convention is defined in Figure 2-2. If the sample is not subject to torques (moments) that vary spatially,[§§§] then the values of σ_{12} and σ_{21} will be the same. Again, the *diagonal elements* of the tensor are the stresses acting normal to the faces, i.e., pushing or pulling perpendicular to the surfaces. If these can be reduced to zero by suitable rotation of the coordinate system, then the stress state is referred to as *pure shear*, as described earlier in connection with Mohr's circle. This state, shown in equation (2-16) for plane stress, can also obtain for some simple fluids subjected to simple-shear deformation, but generally not for polymer melts or solutions.

$$\boldsymbol{\sigma} = \begin{pmatrix} 0 & \sigma_{12} & 0 \\ \sigma_{21} & 0 & 0 \\ 0 & 0 & 0 \end{pmatrix} = \begin{pmatrix} 0 & \tau_{12} & 0 \\ \tau_{21} & 0 & 0 \\ 0 & 0 & 0 \end{pmatrix} \qquad \text{(ssc) (2-16)}$$

For three-dimensional stress, all off-diagonal can be occupied and still the stress state will have no normal stresses. For such states of stress, the pressure in the sample is zero. Needless to say, this does not happen often in practice.

3. Principal stresses for plane stress

The geometry illustrated by Mohr's circle in Figure 2-11 leads to a simple relationship between the principal stresses (point A, and the diametrically opposite point in Figure 2-11) and any combination of shear and normal stress. The equations are

$$\sigma_1 = \frac{\sigma_{11} + \sigma_{22}}{2} + \sqrt{\left(\frac{\sigma_{11} - \sigma_{22}}{2}\right)^2 + \sigma_{21}^2} \qquad (2\text{-}17a)$$

$$\sigma_2 = \frac{\sigma_{11} + \sigma_{22}}{2} - \sqrt{\left(\frac{\sigma_{11} - \sigma_{22}}{2}\right)^2 + \sigma_{21}^2} \qquad (2\text{-}17b)$$

[§§§] The most common example of this occurs while bending a bar-like sample by applying opposing forces at the ends balance by an opposing for at middle. This geometry is referred to as 3-point bending.

These two equations differ only by the sign in front of the second term. If the stress state is that depicted in Figure 2-12 with the dash-dot line, the principal stresses are dominated by shear stresses. The calculation proceeds as follows:

- Data: $\sigma_{21} = 10$ Pa; $\sigma_{22} = -4$ Pa; $\sigma_{11} = 0$.
- $\sigma_1 = (0-4)/2 + [(4/2)^2 + 10^2]^{1/2} = 8.2$ Pa
- $\sigma_2 = (0-4)/2 - [(4/2)^2 + 10^2]^{1/2} = -12.2$ Pa

These are the intercepts on the horizontal axis, with σ_1 being the one on the right, and σ_2 being the one on the left. In three dimensions, there will be three principal stresses. Only one subscript is needed to designate each of the three principal stresses.

The importance of principal stresses is that polymer chains tend to align in the direction of the maximum principal stress difference. Often, but not always, this direction will also be the direction of maximum optical retardance difference (birefringence). During polymer processing this direction will have consequences for shrinkage, strength, and coefficient of thermal expansion, in addition to the optical properties of a molded part. The magnitude of the stress difference will determine the magnitude of these anisotropic properties.

4. The equations of motion

The so-called equations of motion comprise two sets of differential equations. The easiest to understand is the equation of continuity, which simply describes the preservation of mass for a differential element of material, i.e., what goes into the element must come out if the element's mass is to stay constant. But of key interest for examining the relationships between stress and pressure are the differential equations describing the balance of momentum. In a sense, this balance is a force balance on the element. The value of these equations is that they connect properly the measurable quantities (pressure, total stress) with material-generated stresses due to deformation, forces due to material entering the volume element, and body forces such as gravitational pull. They are differential equations, and thus must be integrated to get the results.

APPENDIX 2-1: COMPILATION OF EQUATIONS OF MOTION (ssc)[****]

Rectangular (Cartesian) Coordinates

x component:

$$\rho\left(\frac{\partial v_x}{\partial t}+v_x\frac{\partial v_x}{\partial x}+v_y\frac{\partial v_x}{\partial y}+v_z\frac{\partial v_x}{\partial z}\right)=-\frac{\partial p}{\partial x}+\left(\frac{\partial \tau_{xx}}{\partial x}+\frac{\partial \tau_{yx}}{\partial y}+\frac{\partial \tau_{zx}}{\partial z}\right)+\rho g_x$$

y component:

$$\rho\left(\frac{\partial v_y}{\partial t}+v_x\frac{\partial v_y}{\partial x}+v_y\frac{\partial v_y}{\partial y}+v_z\frac{\partial v_y}{\partial z}\right)=-\frac{\partial p}{\partial y}+\left(\frac{\partial \tau_{xy}}{\partial x}+\frac{\partial \tau_{yy}}{\partial y}+\frac{\partial \tau_{zy}}{\partial z}\right)+\rho g_y$$

z component:

$$\rho\left(\frac{\partial v_z}{\partial t}+v_x\frac{\partial v_z}{\partial x}+v_y\frac{\partial v_z}{\partial y}+v_z\frac{\partial v_z}{\partial z}\right)=-\frac{\partial p}{\partial z}+\left(\frac{\partial \tau_{xz}}{\partial x}+\frac{\partial \tau_{yz}}{\partial y}+\frac{\partial \tau_{zz}}{\partial z}\right)+\rho g_z$$

Cylindrical Coordinates

r component:

$$\rho\left(\frac{\partial v_r}{\partial t}+v_r\frac{\partial v_r}{\partial r}+\frac{v_\theta}{r}\frac{\partial v_r}{\partial \theta}-\frac{v_\theta^2}{r}+v_z\frac{\partial v_r}{\partial z}\right)$$

$$=-\frac{\partial p}{\partial r}+\left(\frac{1}{r}\frac{\partial}{\partial r}(r\tau_{rr})-\frac{\tau_{\theta\theta}}{r}+\frac{1}{r}\frac{\partial \tau_{r\theta}}{\partial \theta}+\frac{\partial \tau_{rz}}{\partial z}\right)+\rho g_r$$

θ component:

$$\rho\left(\frac{\partial v_\theta}{\partial t}+v_r\frac{\partial v_\theta}{\partial r}+\frac{v_\theta}{r}\frac{\partial v_\theta}{\partial \theta}+\frac{v_r v_\theta}{r}+v_z\frac{\partial v_\theta}{\partial z}\right)$$

$$=-\frac{1}{r}\frac{\partial p}{\partial \theta}+\left(\frac{1}{r^2}\frac{\partial}{\partial r}(r^2\tau_{r\theta})+\frac{1}{r}\frac{\partial \tau_{\theta\theta}}{\partial \theta}+\frac{\partial \tau_{\theta z}}{\partial z}\right)+\rho g_\theta$$

[****] These equations assume that τ is symmetrical.

z component:

$$\rho\left(\frac{\partial v_z}{\partial t}+v_r\frac{\partial v_z}{\partial r}+\frac{v_\theta}{r}\frac{\partial v_z}{\partial \theta}+v_z\frac{\partial v_z}{\partial z}\right)$$

$$=-\frac{\partial p}{\partial z}+\left(\frac{1}{r}\frac{\partial}{\partial r}(r\tau_{rz})+\frac{1}{r}\frac{\partial \tau_{\theta z}}{\partial \theta}+\frac{\partial \tau_{zz}}{\partial z}\right)+\rho g_z$$

Spherical Coordinates[††††]

r component:

$$\rho\left(\frac{\partial v_r}{\partial t}+v_r\frac{\partial v_r}{\partial r}+\frac{v_\theta}{r}\frac{\partial v_r}{\partial \theta}+\frac{v_\phi}{r\sin\theta}\frac{\partial v_r}{\partial \phi}-\frac{v_\theta^2+v_\phi^2}{r}\right)$$

$$=-\frac{\partial p}{\partial r}+\left(\frac{1}{r^2}\frac{\partial}{\partial r}(r^2\tau_{rr})-\frac{\tau_{\theta\theta}+\tau_{\phi\phi}}{r}+\frac{1}{r\sin\theta}\frac{\partial}{\partial \theta}(\tau_{r\theta}\sin\theta)+\frac{1}{r\sin\theta}\frac{\partial \tau_{r\phi}}{\partial \phi}\right)+\rho g_r$$

θ component:

$$\rho\left(\frac{\partial v_\theta}{\partial t}+v_r\frac{\partial v_\theta}{\partial r}+\frac{v_\theta}{r}\frac{\partial v_\theta}{\partial \theta}+\frac{v_\phi}{r\sin\theta}\frac{\partial v_\theta}{\partial \phi}+\frac{v_r v_\theta}{r}-\frac{v_\phi^2\cot\theta}{r}\right)$$

$$=-\frac{1}{r}\frac{\partial p}{\partial \theta}+\left(\frac{1}{r^3}\frac{\partial}{\partial r}(r^3\tau_{r\theta})+\frac{1}{r\sin\theta}\frac{\partial}{\partial \theta}(\tau_{\theta\theta}\sin\theta)+\frac{1}{r\sin\theta}\frac{\partial \tau_{\phi\theta}}{\partial \phi}-\frac{\tau_{\phi\phi}\cot\theta}{r}\right)+\rho g_\theta$$

ϕ component:

$$\rho\left(\frac{\partial v_\phi}{\partial t}+v_r\frac{\partial v_\phi}{\partial r}+\frac{v_\theta}{r}\frac{\partial v_\phi}{\partial \theta}+\frac{v_\phi}{r\sin\theta}\frac{\partial v_\phi}{\partial \phi}+\frac{v_r v_\phi}{r}+\frac{v_\theta v_\phi\cot\theta}{r}\right)$$

$$=-\frac{1}{r\sin\theta}\frac{\partial p}{\partial \phi}+\left(\frac{1}{r^3}\frac{\partial}{\partial r}(r^3\tau_{r\phi})+\frac{1}{r}\frac{\partial \tau_{\theta\phi}}{\partial \theta}+\frac{1}{r\sin\theta}\frac{\partial \tau_{\phi\phi}}{\partial \phi}+\frac{2\tau_{\theta\phi}\cot\theta}{r}\right)+\rho g_\phi$$

[††††] Be sure to review the footnote on Table 2-1 concerning the nomenclature assignment for directions in spherical coordinates. Here ϕ is the azimuthal angle, whereas θ is now the inclination angle, from the positive z axis toward the equator.

APPENDIX 2-2: EQUATIONS OF MOTION—CURVILINEAR QUICK LIST (ssc)

This set of equations is for special use where the following assumptions are met:

1. Body and inertial forces are negligible.
2. Symmetrical extra-stress tensor.
3. For cylindrical coordinates, no changes in θ direction, i.e., axial symmetry.
4. For spherical coordinates, no changes in the ϕ direction.

These assumptions cover 100% of the one- and two-dimensional steady-flow problems that we will attempt.

Cylindrical (r, q, z):

$$0 = \frac{\partial \sigma_{rr}}{\partial r} + \frac{\tau_{rr} - \tau_{\theta\theta}}{r} + \frac{\partial \tau_{rz}}{\partial z} \tag{r}$$

$$0 = \frac{\partial \tau_{r\theta}}{\partial r} + \frac{2\tau_{r\theta}}{r} + \frac{\partial \tau_{\theta z}}{\partial z} \tag{θ}$$

$$0 = \frac{\partial \sigma_{zz}}{\partial z} + \frac{\partial \tau_{rz}}{\partial r} + \frac{\tau_{rz}}{r} \tag{z}$$

Spherical (r, θ, ϕ):

$$0 = \frac{\partial \sigma_{rr}}{\partial r} + \frac{\tau_{rr} - \tau_{\theta\theta}}{r} + \frac{\tau_{rr} - \tau_{\phi\phi}}{r} + \frac{1}{r}\left(\tau_{\theta r}\cot\theta + \frac{\partial \tau_{\theta r}}{\partial \theta}\right) \tag{r}$$

$$0 = \frac{1}{r}\frac{\partial \sigma_{\theta\theta}}{\partial \theta} + \frac{3\tau_{r\theta}}{r} + \frac{\partial \tau_{r\theta}}{\partial r} + \frac{\cot\theta}{r}\left(\tau_{\theta\theta} - \tau_{\phi\phi}\right) \tag{θ}$$

$$0 = \frac{3\tau_{r\phi}}{r} + \frac{\partial \tau_{r\phi}}{\partial r} + \frac{2\tau_{\theta\phi}\cot\theta}{r} + \frac{1}{r}\frac{\partial \tau_{\theta\phi}}{\partial \theta} \tag{ϕ}$$

PROBLEMS

2-1. Calculate the stress, in Pascals, on a cylindrical sample 1 cm in diameter suspended by an ideal clamp on the top end and a 100-g weight attached to a second ideal clamp at the bottom. What is the stress after the sample has stretched to twice its length?

2-2. One way of subjecting samples to very low tensile stresses is to use the weight of the sample itself to apply the force. Repeat Problem 2-1 but with no weight attached. Assume the sample is 10 cm long and has a mass density of 1 g/cm³. Note that the stress will not be uniform, so a complete answer will describe the stress level at each point along the sample with an equation. Solve only for the initial shape, i.e., cylindrical. Sketch what you think the sample will look like after some time.

2-3. (Challenging) Describe how to make a sample with a roughly conical shape that will have a reasonably uniform stress along its entire length. Assume as in Problem 2-2 that the mass density is 1 g/cm³, and the top is fastened with an ideal clamp.

2-4. Repeat Problem 2-3, but assume the sample is cut from a thin sheet.

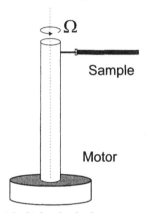

Figure 2-14. Using body forces to apply stress

2-5. To increase the stress level using body forces, the value of gravity can be increased by spinning the sample by grabbing one end with an ideal clamp, fastened to the spinning axle, as shown in Figure 2-14. Describe the stress level in a uniform, cylindrical sample of initial length L, clamped at the inner end with an ideal clamp fastened to a rapidly spinning axle at a distance R from the center.

2-6. Inflation of a thin sheet of polymer melt stretched over a hole has been used to apply a known and controllable stress to the sample. At small degrees of inflation, the resulting dome shape is roughly spherical. Describe the stress level in the sheet as a function of the height of the dome H, the radius of the hole R, the original thickness of the sheet δ, and gas pressure P.

2-7. Using plane geometry, derive the relationships for the principal stresses shown in equations (2-17a) and (2-17b).

2-8. Dalhquist et al.[5] described a cone-shaped "bob" (Figure 2-15) that when hung on a sample and immersed in a fluid would apply a constant stress to the sample. The shape of the bob, if its density was exactly that of the fluid, was given as

$$r(z) = \frac{1}{L_0 + z} \sqrt{\frac{W_0 L_0}{\pi \rho}} \qquad (2-18)$$

where r is the radius of the bob at height z above its base, L_0 is the initial length of the sample, W_0 is the full weight of the bob, and ρ is the density of bob and fluid.

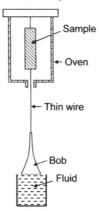

Figure 2-15. Apparatus for applying a constant tensile stress to a sample.

 (a) Design a bob for a 3-cm-long sample that will apply a stress of 10 kPa to the sample assuming the sample has a cross-sectional area of 0.1 cm^2 and water ($\rho = 1$ g/cm^3) is used as the fluid. Assume the experiment starts with the bottom of the bob exactly touching the surface of the water. Also assume the level of the fluid does not change with immersion of the bob. (Dahlquist et al. accomplished this by having the fluid spill over the rim of its container.)

(b)   Verify that the stress will still be 10 kPa when the sample is 5 cm long.

(c)   Machining hyperbolic axisymmetric bobs may be difficult for most shops. Redesign the bob such that it can be cut from a flat sheet of material, a much simpler operation. Again, assume the fluid and the bob have equal densities.

2-9. (Open end) Search the internet for equations describing the stress state in uniformly loaded beams with simple constraints: pivot at one end, roller at the other. Use these to estimate the maximum principal-stress difference in the stone bench shown in Figure 2-16, but before the central brick column was installed.

Figure 2-16. Stone bench in Christ Church Burial Ground in Philadelphia. The central brick column is modern—designed to stop the creep of the stone slab. While the land for the burial ground was purchased in 1719, much of the activity was more toward the middle to late 1700's. The stone is reportedly "Philadelphia Marble."

2-10. Rheologists interested in finding the yield stress in ER (electrorheological) fluids have used the simple setup shown. The ER fluid sample will hold itself and the weight of the center plate against the force of gravity until the voltage is decreased. If the sample density is 2.6 g/cm^3 and the spacing between the plate and the walls is 2 mm, calculate the weight of the plate needed per cm^2 of area to cause flow if the yield stress is 1 kPa. Where will the sample start deforming?

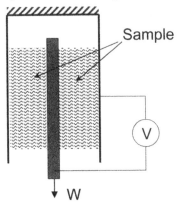

Figure 2-17. Schematic of setup for measuring yield stress in ER fluids.

2-11. For the ϕ component of the equation of motion in spherical coordinates, the terms involving the shear stress $\tau_{r\phi}$ are reported by one author to be

$$\frac{1}{r^3}\frac{\partial}{\partial r}\left(r^3\tau_{r\phi}\right)$$ (a)

while another author writes the form

$$\frac{1}{r^2}\frac{\partial}{\partial r}\left(r^2\tau_{r\phi}\right)+\frac{\tau_{r\phi}}{r}$$

(b)

Is there a mistake? Show your reasoning.

2-12. Find the dimensions of the terms in the equation motion (Appendix 2-1), and express these in SI units as well as in terms of the fundamental dimensions $[t]$, $[L]$ and $[M]$.

2-13. (Open end) In dealing with polymers, it is important to develop your senses such that your fingers can be used to estimate stress levels and, consequently, mechanical and rheological properties. Stretch as rapidly as possible a rubber band and estimate the relative magnitude of the key terms of the equation of motion.

REFERENCES

1. M. Beiner, G. Fytas, G. Meier, and S. K. Kumar, "Pressure-Induced Compatibility in a Model Polymer Blend," *Phys. Rev. Lett.*, **81**, 594–597 (1998).

2. G. W. Becker and O. Krüger, "On the nonlinear biaxial stress-strain behavior of rubberlike polymers," in *Deformation and Fracture of High Polymers* (H. H. Kausch, J. A. Hassell, and R. I. Jaffee, eds.), Plenum Press, New York, 1973.

3. H. M. Laun and H. Münstedt, "Comparison of the elongational behaviour of a polyethylene melt at constant stress and constant strain rate," *Rheol. Acta*, **15**, 517–524 (1976).

4. M. Sentmanat, B. N. Wang, and G. H. McKinley, "Measuring the transient extensional rheology of polyethylene melts using the SER universal testing platform," *J. Rheol.*, **49**, 585–606 (2005).

5. C. A. Dahlquist, J. O. Hendricks and N. W. Taylor, "Elasticity of Soft Polymers—Constant-stress elongation tests," *Ind. Eng. Chem.*, **43**, 1404–1410 (1951).

3

Velocity, Velocity Gradient, and Rate of Deformation

A. WHY VELOCITY IS SIMPLER THAN LOCATION— SPEEDOMETERS VS. GPS

It seems strange that *rate of deformation* should be discussed before deformation itself. To understand this, think of the difference between measuring the speed of your car and finding its location. For speed, a simple speedometer is all that's required. For location, if your past location is known, finding your present location would require a careful record of speed and direction at all times. This process was referred to as "dead reckoning" by the early navigators. Now, of course, location on the earth's surface is easily established with GPS, an immensely sophisticated, complicated and expensive system that does not translate well to polymer rheology, at least on the laboratory scale. In fact, rheology measurements are most often done with the equivalent of driving along a very straight road at a strictly constant speed.

In the limited world of simple shear, the rate of deformation is merely the *shear rate*, which is equal to the velocity gradient dV_x/dy. It is given the symbol $\dot{\gamma}$, where the dot on top of the gamma symbolizes a time derivative. This symbol is difficult to type, especially when making graphs, which may help to explain the wide use of other symbols, including D, Γ, Δ, to mention a few. *Strain rate* is an alternative description for "rate of deformation," but it arose primarily for extensional deformation of solids. For the latter, a different symbol, $\dot{\varepsilon}$, is used. This has origins in solid mechanics, where the symbol ε for strain is very common.

B. VELOCITY GRADIENTS

1. Velocity of what? The concept of a material point

In the introductory paragraphs above we used the terms rate of deformation, shear rate and velocity gradient as if they were all the same. In the case of simple shearing deformation, they are. This is convenient and allows those using shear rheometers to describe correctly their results. Unfortunately, things get more complicated with, say, extensional flows.

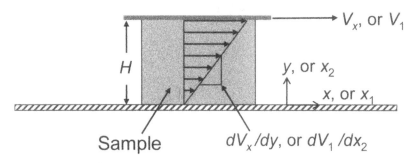

Figure 3-1. Velocity and velocity gradient in simple shear. By definition, there are no other velocity components and, consequently, no other velocity gradients.

Before delving into complications, let us consider what is moving to create a velocity gradient. For simple shear we have the situation shown in Figure 3-1. The sample is securely fastened to the immobile base, and is dragged to the right at a velocity V_x by a plate stuck to the top. This plate must be held strictly parallel to the base, and with a constant spacing, shown as H in the figure. No flow in the third direction (in or out of the paper) is allowed. (We will investigate how these criteria can be met in Chapter 7.) Given this setup, we expect the sample to flow in a laminar fashion with a velocity V_x at the surface of the top plate. Velocities within the sample logically should decrease uniformly to zero at the bottom of the plate, giving a uniform velocity gradient, or shear rate given by

$$\dot{\gamma} = \frac{V_x}{H} = \frac{dV_x}{dy} \qquad \text{(ssc) and (fsc)}^* \quad (3\text{-}1)$$

How do we know that the material is moving in this fashion? The argument is mainly based on geometry and symmetry. For example, because the stress is uniform throughout, certainly we expect a homogeneous fluid to respond in a uniform fashion.[†] There are many ways of confirming this with experiments, the simplest being the observation of tiny particles in the fluid. But for more complex fluids, this simple picture has often been observed to fail. For example, slip planes or vortices may develop in some fluids, even at low stresses. As long at the particles are small and sparse, it is reasonable that they will follow the material exactly, and represent in a visible fashion a material point, i.e., a point attached to the material. In talking about material points, it is important to understand that these points are not subject to Brownian motion and thus diffusion through the material. For this reason, a tiny particle may be a better representation of the continuum than, say, a macromolecule with a molecular tag.

Note the choices of nomenclature illustrated in Figure 3-1. As with stress, we can stick with the conventional abbreviations for the coordinate system—in this case, rectangular—or use the numbers as shown in Table 2-1. For the latter, although the x direction becomes x_1, and so forth, the velocity is not V_{x_1} by analogy, but simply V_1.

As with stress, where we combine force and area directions according to Figure 2-2, we can combine velocity and velocity-gradient directions to create a matrix in a systematic fashion to cover three-dimensional cases with velocities in many directions. Such a matrix is shown below:

$$\nabla \mathbf{V} = \begin{pmatrix} dV_1/dx_1 & dV_1/dx_2 & dV_1/dx_3 \\ dV_2/dx_1 & dV_2/dx_2 & dV_2/dx_3 \\ dV_3/dx_1 & dV_3/dx_2 & dV_3/dx_3 \end{pmatrix} \qquad (3\text{-}2)$$

Note carefully the sequence of the subscripts. While this matrix, called the velocity-gradient tensor, appears to provide a reasonable description of shear

---

[*] See Chapter 2 for definitions of (ssc) and (fsc).

[†] If the top plate is yanked (accelerated) rapidly, fluid inertia will change the stress applied to the fluid, and the gradient will no longer be linear. If the sample is vertical, gravity will also need to be considered. The equation of motion helps to keep track of these effects. We also know that the fluid will heat up, so the fluid in the center will become a bit hotter than that near the thermostated plates.

rate in simple shear ($\dot{\gamma} = dV_1/dx_2$) and extension rate ($\dot{\varepsilon} = dV_1/dx_1$), it has some very real problems as a measure of rate of deformation of the material, as will be shown next. For this reason, it has relatively little use in rheology.

C. RATE OF DEFORMATION

In spite of the dire warnings expressed in connection with equation (3-2), let's apply it and see what happens. We will consider two examples.

1. Whole-body rotation

Consider a rotating disk made from a glassy polymer such as polycarbonate. While rotating such a disk is easy, deforming it is difficult. But let's set up this case with a rectangular coordinate system, as shown in Figure 3-2.

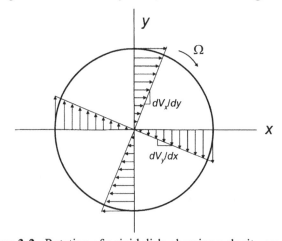

Figure 3-2. Rotation of a rigid disk, showing velocity gradients.

Note the presence of a velocity gradient, which looks superficially like that in Figure 3-1. Assuming no motion in the z direction, we have velocity gradients filling the 1,2 and 2,1 positions in equation (3-2), whereas we know that there is no deformation whatsoever. The magnitude of these gradients is just Ω, the rotation rate.

2. Extension of a square bar

For extensional flows, rheologists have settled on the definition of extensional strain rate, $\dot{\varepsilon}$, as

$$\dot{\varepsilon} \equiv \frac{\partial V_1}{\partial x_1}$$

(3-3)

where the subscript 1 denotes the direction of stretching. Thus the extensional strain rate is the velocity gradient along the stretch direction.

If we consider the terms in equation (3-2) for stretching a square bar, the result would be

$$\nabla \mathbf{V} = \begin{pmatrix} \dot{\varepsilon} & & \\ & -\dot{\varepsilon}/2 & \\ & & -\dot{\varepsilon}/2 \end{pmatrix} \tag{3-4}$$

As a reminder, the negative sign signifies an inward velocity on the sides of the bars as it thins. For a Newtonian fluid with viscosity η, we might then expect the observed stress, σ_T, to be

$$\sigma_T = \sigma_1 - \sigma_2 = \tau_1 - \tau_2 = \eta[\dot{\varepsilon} - (-\dot{\varepsilon}/2)] = 3\eta\dot{\varepsilon}/2 \qquad \text{(ssc)} \quad (3\text{-}5)$$

whereas in fact the measured stress is *twice* this result.

The fix for both these problems is evident from Figure 3-2; we need to add the two velocity gradients shown to get rid of the confusion of rotation vs. deformation. Thus, the form of the rate-of-deformation tensor, in rectangular coordinates, should be

$$\dot{\gamma} = \begin{pmatrix} \dfrac{\partial V_1}{\partial x_1} + \dfrac{\partial V_1}{\partial x_1} & \dfrac{\partial V_2}{\partial x_1} + \dfrac{\partial V_1}{\partial x_2} & \dfrac{\partial V_3}{\partial x_1} + \dfrac{\partial V_1}{\partial x_3} \\ \dfrac{\partial V_1}{\partial x_2} + \dfrac{\partial V_2}{\partial x_1} & \dfrac{\partial V_2}{\partial x_2} + \dfrac{\partial V_2}{\partial x_2} & \dfrac{\partial V_3}{\partial x_2} + \dfrac{\partial V_2}{\partial x_3} \\ \dfrac{\partial V_1}{\partial x_3} + \dfrac{\partial V_3}{\partial x_1} & \dfrac{\partial V_2}{\partial x_3} + \dfrac{\partial V_3}{\partial x_2} & \dfrac{\partial V_3}{\partial x_3} + \dfrac{\partial V_3}{\partial x_3} \end{pmatrix} \tag{3-6}$$

where $\dot{\gamma}$ stands for the rate-of-deformation or rate-of-strain tensor.

Equation (3-6) looks hopelessly complicated! However, examine it carefully; it is of regular and easy-to-remember form. The key is the following:

$$\dot{\gamma}_{ij} = \frac{\partial V_i}{\partial x_j} + \frac{\partial V_j}{\partial x_i} \tag{3-7}$$

In this equation, i and j can take any of the three values (1, 2 or 3). As can be seen by comparing equations (3-7) and (3-6), it works for both $I \neq j$ or $I = j$. For example, with the disk rotation, the terms that would show up the velocity-gradient matrix, equation (3-2), are $\partial V_1/\partial x_2 = \Omega$ and $\partial V_1/\partial x_2 = -\Omega$ in the 1,2 and 2,1 positions, respectively. As can be seen in equation (3-7), the two would cancel out and the rate of deformation would be zero, which we knew all along. For the extension of the square bar, the terms according to equation (3-7) would be given by the following equations:

$$\dot{\gamma}_{11} = 2\frac{\partial V_1}{\partial x_1} = 2\dot{\varepsilon} \tag{3-8}$$

$$\dot{\gamma}_{22} = 2\frac{\partial V_2}{\partial x_2} = 2\left(\frac{-\partial V_1 / \partial x_1}{2}\right) = -\dot{\varepsilon} \tag{3-9}$$

The result for the tensile stress would then be the expected value of $3\eta\dot{\varepsilon}$.

The rate-of-deformation tensor for simple shear is certainly worth considering. It becomes

$$\dot{\gamma} = \begin{pmatrix} 0 & \partial V_1 / \partial x_2 & 0 \\ \partial V_2 / \partial x_1 & 0 & 0 \\ 0 & 0 & 0 \end{pmatrix} = \begin{pmatrix} 0 & \dot{\gamma} & 0 \\ \dot{\gamma} & 0 & 0 \\ 0 & 0 & 0 \end{pmatrix} \tag{3-10}$$

Note that the bold symbol $\dot{\gamma}$ stands for the entire rate-of-deformation tensor, while the symbol $\dot{\gamma}$ with no subscripts symbolizes the shear-rate magnitude. In simple shear, $\dot{\gamma} = \dot{\gamma}_{12} = \dot{\gamma}_{21}$. On occasion we also use the symbol $\dot{\gamma}_{ij}$ to represent the entire rate-of-deformation tensor, as well as any component ij of that tensor. All this tends to be a bit confusing, but one can usually tell by the context. The impetus for shortened nomenclature is to save time and paper, both worthy causes, but never at the expense of clarity.

3. Curvilinear coordinates

Things get a bit more complicated for cylindrical or spherical coordinate systems, and one resorts to tabulations. Such a tabulation is provided in Appendix 3-1. Accept these equations as fact. Don't bother to derive them and certainly do no accept any substitutes for the curvilinear coordinates.

An example of the application of these tables can be seen by considering the flow of water from a huge reservoir to a single point, called the sink, in the very center of this body. We assume the boundaries of the reservoir are sufficiently far enough away from the sink such that they do not interfere with the water's flow pattern. We also must assume—and this is a bit more difficult—that the drain of the sink (e.g., a pipe to the surface) also does not interfere with the flow pattern. With these assumptions, we can immediately conclude that spherical coordinates are appropriate, and write down the velocity field for the problem as:

$$\mathbf{V} = \left(v_r, v_\theta, v_\phi\right) = \left(v_r(r), 0, 0\right) \tag{3-11}$$

where $\mathbf{V}$ is the velocity vector, v_r is the component in the r direction,[‡] etc. Furthermore it is obvious that the radial dependence of v_r is

$$v_r(r) = -\frac{Q}{4\pi r^2} \tag{3-12}$$

where Q represents the magnitude of the volumetric flow rate. Note that the velocity magnitude increases rapidly as the fluid approaches the drain but its sign is negative because the fluid is flowing inward. Oddly enough, the velocity gradient is positive, that is, the velocity gets less negative as r increases. One can also see this from the algebra involved in taking the derivative, i.e.,

$$\frac{\partial v_r}{\partial r} = \frac{dv_r}{dr} = \frac{\partial}{\partial r}\left(-\frac{Q}{4\pi r^2}\right) = \frac{Q}{2\pi r^3} \tag{3-13}$$

The total derivative in the above equation is written in recognition that the velocity changes only in the radial direction.

The intuitive approach taken so far follows a pattern that comprises these three easy steps:

(1) look for symmetry according to the geometry of the problem
(2) choose the coordinate system according to the symmetry
(3) write down the velocity field by inspection assuming constant density

If these steps can't be followed, we are probably in trouble; we will need to apply some additional assumptions to be described later.

Continuing on this problem, we use the handy tables in Appendix 3-1 to find the rate-of-deformation tensor. Inspect carefully every equation in the section marked Spherical Coordinates. You will note that many of the terms are zero because either (1) the velocity component involved is zero, and/or (2) the direction of the derivative does not have any changes. For example, the component $\dot{\gamma}_{\phi\phi}$ is listed in the table as

$$\dot{\gamma}_{\phi\phi} = 2\left(\frac{1}{r\sin\theta}\frac{\partial v_\phi}{\partial\phi} + \frac{v_r}{r} + \frac{v_\theta\cot\theta}{r}\right) \tag{3-14}$$

[‡] Both upper and lower case V are used for velocity. There is no difference in meaning. Any rationale? The lower case v uses less space, so more equation can be put on one line. However, it looks a bit like the Greek letter nu (ν), so its use is restricted to equations dealing only with velocity. Nu is used in this text for kinematic viscosity and Poisson ratio.

Note that it is not necessary to figure out which direction is which other than the r direction because of the high symmetry of the problem. Examining this equation more carefully shows that the only term that remains on the right-hand side is the second one: v_r/r. However, it is good practice to cross out variables that have values of zero and explain why they are zero, as shown in Figure 3-3:

0, no flow in ϕ direction

0, no flow in θ direction

$$\dot{\gamma}_{\phi\phi} = 2\left(\frac{1}{r\sin\theta}\frac{\partial v_\phi}{\partial\phi} + \frac{v_r}{r} + \frac{v_\theta\cot\theta}{r}\right)$$

Figure 3-3. Process of simplifying a component of the rate-of-deformation tensor for inward radial flow to a point sink. The first term could also be cancelled because there are no changes in the ϕ direction.

Applying this process to the other terms gives the rate-of-deformation tensor

$$\dot{\gamma} = \begin{pmatrix} \dfrac{Q}{\pi r^3} & 0 & 0 \\ 0 & -\dfrac{Q}{2\pi r^3} & 0 \\ 0 & 0 & -\dfrac{Q}{2\pi r^3} \end{pmatrix} = \frac{Q}{2\pi r^3}\begin{pmatrix} 2 & 0 & 0 \\ 0 & -1 & 0 \\ 0 & 0 & -1 \end{pmatrix} \qquad (3\text{-}15)$$

The form here we recognize by now as uniaxial extension. Thus the material is stretched in the radial direction at a sharply increasing rate as it moves toward the sink. This has some very important implications in the processing of polymers in converging dies.

APPENDIX 3-1: COMPONENTS OF THE RATE-OF-DEFORMATION TENSOR

Rectangular coordinates: x, y, z

$$\dot{\gamma}_{xx} = 2\frac{\partial v_x}{\partial x} \qquad\qquad \dot{\gamma}_{xy} = \dot{\gamma}_{yx} = \frac{\partial v_x}{\partial y} + \frac{\partial v_y}{\partial x}$$

$$\dot{\gamma}_{yy} = 2\frac{\partial v_y}{\partial y} \qquad\qquad \dot{\gamma}_{xz} = \dot{\gamma}_{zx} = \frac{\partial v_x}{\partial z} + \frac{\partial v_z}{\partial x}$$

$$\dot{\gamma}_{zz} = 2\frac{\partial v_z}{\partial z} \qquad\qquad \dot{\gamma}_{yz} = \dot{\gamma}_{zy} = \frac{\partial v_y}{\partial z} + \frac{\partial v_z}{\partial y}$$

Cylindrical coordinates: r, θ, z

$$\dot{\gamma}_{rr} = 2\frac{\partial v_r}{\partial r}$$

$$\dot{\gamma}_{r\theta} = \dot{\gamma}_{\theta r} = r\frac{\partial}{\partial r}\left(\frac{v_\theta}{r}\right) + \frac{1}{r}\frac{\partial v_r}{\partial \theta}$$

$$\dot{\gamma}_{\theta\theta} = 2\left(\frac{1}{r}\frac{\partial v_\theta}{\partial \theta} + \frac{v_r}{r}\right)$$

$$\dot{\gamma}_{\theta z} = \dot{\gamma}_{z\theta} = \frac{\partial v_\theta}{\partial z} + \frac{1}{r}\frac{\partial v_z}{\partial \theta}$$

$$\dot{\gamma}_{zz} = 2\frac{\partial v_z}{\partial z}$$

$$\dot{\gamma}_{zr} = \dot{\gamma}_{rz} = \frac{\partial v_z}{\partial r} + \frac{\partial v_r}{\partial z}$$

Spherical coordinates: r, θ, ϕ

$$\dot{\gamma}_{rr} = 2\frac{\partial v_r}{\partial r}$$

$$\dot{\gamma}_{r\theta} = \dot{\gamma}_{\theta r} = r\frac{\partial}{\partial r}\left(\frac{v_\theta}{r}\right) + \frac{1}{r}\frac{\partial v_r}{\partial \theta}$$

$$\dot{\gamma}_{\theta\theta} = 2\left(\frac{1}{r}\frac{\partial v_\theta}{\partial \theta} + \frac{v_r}{r}\right)$$

$$\dot{\gamma}_{\theta\phi} = \dot{\gamma}_{\phi\theta} = \frac{\sin\theta}{r}\frac{\partial}{\partial \theta}\left(\frac{v_\phi}{\sin\theta}\right) + \frac{1}{r\sin\theta}\frac{\partial v_\theta}{\partial \phi}$$

$$\dot{\gamma}_{\phi\phi} = 2\left(\frac{1}{r\sin\theta}\frac{\partial v_\phi}{\partial \phi} + \frac{v_r}{r} + \frac{v_\theta\cot\theta}{r}\right)$$

$$\dot{\gamma}_{r\phi} = \dot{\gamma}_{\phi r} = \frac{1}{r\sin\theta}\frac{\partial v_r}{\partial \phi} + r\frac{\partial}{\partial r}\left(\frac{v_\phi}{r}\right)$$

APPENDIX 3-2: COMPONENTS OF THE CONTINUITY EQUATION

Coordinate system	Constant volume[a]
Rectangular (Cartesian)	$\dfrac{\partial V_x}{\partial x} + \dfrac{\partial V_y}{\partial y} + \dfrac{\partial V_z}{\partial z} = 0$
Cylindrical	$\dfrac{\partial V_r}{\partial r} + \dfrac{V_r}{r} + \dfrac{1}{r}\dfrac{\partial V_\theta}{\partial \theta} + \dfrac{\partial V_z}{\partial z} = 0$
Spherical	$\dfrac{\partial V_r}{\partial r} + \dfrac{2V_r}{r} + \dfrac{1}{r}\dfrac{\partial V_\theta}{\partial \theta} + \dfrac{V_\theta\cot\theta}{r} + \dfrac{1}{r\sin\theta}\dfrac{\partial V_\phi}{\partial \phi} = 0$

[a] If the density is not constant, replace the "zero" with $-\partial\ln\rho/\partial t$

APPENDIX 3-3: NOMENCLATURE AND SIGN CONVENTIONS USED IN POPULAR RHEOLOGY TEXTS

Text	Total stress symbol	Extra stress symbol[a]	Stress sign[b]	Front factor[c] on $\dot{\gamma}$
(this book)	σ	τ	(ssc)	1
Society of Rheology[d]	σ	[e]	(ssc)	1
Macosko, 1994	T	τ	(ssc)	1/2
Morrison, 2001	Π	τ	(fsc)	1

Text	Total stress symbol	Extra stress symbol [a]	Stress sign [b]	Front factor [c] on $\dot\gamma$
Dealy & Wissbrun, 1990	σ	τ	(ssc)	1
Larson, 1999	T	σ	(ssc)	1/2
Carreau et al., 1997	π	σ	(fsc)	1/2
Bird et al., 1987	π	τ	(fsc)	1

[a] Also called deviatoric stress.

[b] (ssc) stands for solids sign convention, i.e., tension is positive. (fsc) is fluids sign convention, compression is positive. In both cases, pressure p is given a positive sign for compressive.

[c] A factor of 1 means $S_{ij} = (dV_i/dx_j + dV_j/dx_i)$ in rectangular coordinates, where S is the symbol used. A factor of 1/2 means $S_{ij} = \frac{1}{2}(dV_i/dx_j + dV_j/dx_i)$, but "rate of deformation" is called $2S$.

[d] "Official" nomenclature, as specified in Dealy et al.[1]

[e] Not specified.

PROBLEMS

3-1. Following the procedure shown in Figure 3-3, simplify the $r\phi$ term of the rate of deformation for radial inward fluid to a point sink.

3-2. While similar to a point sink in concept, a line sink creates a radial axisymmetric flow into what might be imagined as a long porous pipe pumping water out of a huge reservoir. See Figure 3-15 below. Derive all components of the rate-of-deformation tensor for this flow in terms of Q/L, the flow rate per unit length of porous pipe.

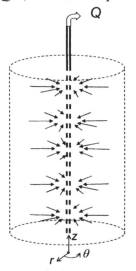

Figure 3-4. Radial flow into a line sink. The flow rate to the sink is uniform along the sink.

3-3. Imagine that a cylindrical tube of polymer is stretched in the axial direction z with a constant extension rate dV_z/dz. Now, normally if this were done, the tube diameter would tend to decrease. However, the pressure inside can be carefully and continuously controlled to keep the diameter constant.[§2] In this case, and assuming the volume of the sample itself does not change, calculate all components of the rate of deformation. Also assume the tube is thin compared to its diameter.

3-4. Twisting of cylinders is referred to as *torsional flow*. This versatile deformation can be done easily on melts by confining a short cylinder between two discs, and on solids by twisting a rod gripped by clamps at each end. The effect is to subject the material to simple shear if the length of the sample is held constant. Assuming the material is uniform and maintains its volume constant, derive all components of the rate-of-deformation tensor for this flow.

3-5. Compare the rate-of-deformation tensors for simple shear vs. pure shear, often called planar extension. (See Figure 2-9 for a drawing of a pure-shear stress state; simply replace the force vectors with velocity.)

3-6. Many investigators have attempted to produce a state of steady extension by stretching a rod or strip of polymer. By steady extension, we mean here that the rate of deformation is invariant with time, at least after the experiment is started. This has even been done with polymer melts with some success.

(a) If a cylindrical rod, initially of length L_0, is stretched such that the velocity gradient dV_z/dz is held constant at a value $\dot{\varepsilon}_0$, what must be done to the length $L(t)$ to maintain this constant rate?

(b) Assuming the volume of the material in the rod is constant, derive the behavior of the rod radius with time if the initial radius is R_0.

(c) Show that $\partial V_r/\partial r$ does not depend on radius.

(d) Using the appropriate equation of continuity, derive the θ component of the rate of deformation.

(e) The rate of stretching in the θ direction might be defined as $d\ln C/dt$, where C is the circumference at any position r. Relate this value to the θ component of the rate of deformation. Does it vary with radius?

3-7. The Brabender is a small batch mixer used by many laboratories. A sectional drawing of one of its mixing heads is shown in Figure 3-5. Goodrich and Porter,[3] and very slightly later Blyler and Daane,[4] claimed that the shear rate in this device is strictly proportional to rotor speed N, i.e., $\dot{\gamma} = KN$, where K is a constant independent of rotor speed. If the roller and cavity diameters are 3.81 and 3.97 cm, respectively, derive a dimensional equation[**] relating shear rate to N in RPM (revolutions per

---

[§] Equipment to do this experiment has been designed and made by Chung and Stevenson.[2] Needless to say, this is not an easy experiment, but it does allow one to vary independently the axial vs. the radial rates of deformation.

[**] A "dimensional equation" is one in which the units of the variables must be specified. Coefficients in the equation will, in general, have units, i.e., they are "dimensional." The equation will give the wrong answer if the correct units are not used for the variables.

minute). Compare your value of K with that found by Goodrich and Porter of 1.14 s^{-1}/rpm.

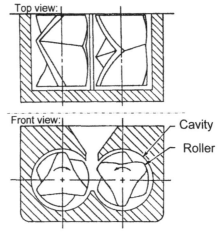

Figure 3-5. Cross-sectional depictions of the Brabender mixing head with "roller" blades. (Adapted with permission of John Wiley & Sons, Inc. from Blyler and Daane[4])

3-8. For Newtonian fluids, the velocity profile for flow through a tube is parabolic, with a maximum value V_0 located on the center line of the tube. The velocity then falls with radius to a value of 0 at the wall.

(a) From this information alone, express the velocity profile $V_z(r)$ with an equation.

(b) From this information derive the shear rate profile across the tube.

(c) Integrate the velocity profile to obtain the flow rate Q is terms of V_0.

(d) Show that the shear rate at the wall $\dot{\gamma}_w$ is given by the expression $\dot{\gamma}_w = 4Q/\pi R^3$.

3-9. Many processes involve the flow of the polymer through a slit-like channel, shown schematically in Figure 3-6. As with tube flow, the velocity profile in the slit is parabolic if the fluid is exhibiting Newtonian behavior. Following the sequence suggested in Problem 3-8, derive the expression for the shear rate at the wall in terms of the slit gap H and the flow rate per unit width Q/W.

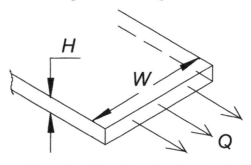

Figure 3-6. Schematic of flow through a slit channel. $W/H \gg 1$.

3-10. Stretching a glassy polymer at a temperature slightly above the T_g can impart significant orientation to the material. To this end, a bar with a square cross section is stretched in its long direction in a tensile-testing machine with a constant crosshead velocity V_{XH} of 1 cm/s. (See Figure 3-7.) The gauge length L_0, or initial distance between the grips, is 10 cm, while the bar width and thickness are both initially W_0 and D_0. Derive expressions for all components of the rate-of-deformation tensor as functions of time, assuming the material is incompressible. Is the extension rate $\dot{\varepsilon}$ constant? (Be sure to check on the definition of $\dot{\varepsilon}$.)

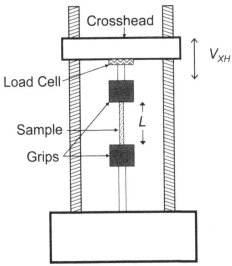

Figure 3-7. Schematic of tensile-testing machine equipped with screw drives to move the crosshead up and down at a constant velocity. In this design, the load cell, which measures the force developed by the stretching of the sample, is carried on the crosshead.

3-11. Craftsmen who install tiles will spread the mastic (roughly speaking, a thick, rubbery latex paint) on the wall with a notched trowel. The trowel leaves ridges of mastic with a height comparable to the width. Pressing the tile onto the mastic ridges forces the mastic to spread. For an isolated square ridge, describe with sketches the expected velocity profile, keeping in mind conservation of mass and the information in Problem 3-9.

REFERENCES

1. J. M. Dealy, "Official nomenclature for material functions describing the response of a viscoelastic fluid to various shearing and extensional deformations," *J. Rheol.*, **39**, 253–265 (1995).

2. S. C.-K. Chung and J. F. Stevenson, "A general elongational flow experiment: inflation and extension of a viscoelastic tube," *Rheol. Acta*, **14**, 832–841 (1975).

3. J. E. Goodrich and R. S. Porter, "A rheological interpretation of torque rheometer data," *Polym. Eng. Sci.*, **7**, 45–51 (1967).

4. L. L. Blyler and J. H. Daane, "An analysis of Brabender torque rheometer data," *Polym. Eng. Sci.*, **7**, 178–181 (1967).

4

Relationship Between Stress and Rate of Deformation: The Newtonian Fluid

A. MATERIAL IDEALIZATIONS IN RHEOLOGY

In studying viscoelasticity and rheology, it is extremely useful to invent materials with simplified or *ideal* behavior to aid in the understanding and cataloging of real-material behavior. Of course, as with all fiction, this practice can be carried to ridiculous extremes, so the inventions must obey certain constraints to be useful. For example, we cannot in good conscience define materials which depend upon the nature or orientation of the coordinate system. An example of a useless idealized fluid would be one that had a finite viscosity if stretched in the x direction, but no viscosity if stretched in the y direction, for arbitrary placement of this rectangular system. On the other hand, a material could have viscosity differences in particular material directions—a liquid crystal is an example.

The usefulness of ideal materials derives not only from their application in education. They are also important in that they often represent limiting behavior, and set boundaries or extremes that cannot be surpassed. They may also represent limits of analytical behavior, e.g., a complicated equation describing material behavior may simplify algebraically to ideal behavior when certain parameters are set equal to 0 or 1.0. Examples will be encountered below and subsequent chapters.

1. Analogies to ideal gas and ideal rubber elasticity

Throughout science and, for that matter, life in general, ideal behavior is sought for the primary attributes of simplicity and limiting behavior, and for facilitating comparisons. In government, we have the limiting behavior of a benevolent dictator as the ideal behavior against which other dictators can be compared. In fluid dynamics, the "ideal fluid" is defined as having density, but no viscosity. In thermodynamics, we refer to the idea gas as the limiting behavior of a material at very low densities. Closer to home is the concept of ideal elastic behavior, as described at very low strains by the Hookean solid and at higher strains by the ideal elastomer. In fact, the mechanical response of the ideal elastomer is often referred to as neo-Hookean behavior. Ideal behavior even exists for more complicated materials such as paste, which can be described most simply as a Bingham plastic. And, of course, the quintessential viscous material is the Newtonian fluid, which has the useful feature of being a reasonable description of some very real materials. We will learn below how to describe Newtonian behavior in three dimensions, and use this behavior as a baseline against which more complicated materials can be compared.

2. Ideal material behavior vs. limiting behavior

Ideal materials are just that—so perfect that real materials can only approach their simplified response to stress and strain. As such, the ideal material can often be regarded as a boundary beyond which no material can pass. However, the limiting behavior of a real material may fail to approach asymptotically the ideal behavior because of some unusual nonlinear behavior. For example, we often assume that at very low strain or strain rate, any material will approach linear behavior such that stresses are linearly proportional to the strains or strain rates. With the latter, the limiting proportionality constant is called the *zero-shear-rate viscosity*[*] and is given the symbol η_0.

---

[*] Also called "zero shear viscosity," which might be confused with the viscosity at zero shear strain, regardless of the imposed shear rate. (See "stressing viscosity" for more information about this concept.)

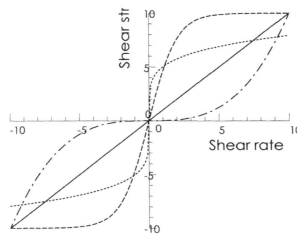

Figure 4-1. Possible responses for materials when subjected to steady shearing at various shear rates. The solid line represents Newtonian response, while the dashed line represents pseudoplastic behavior. The dash-dot and dotted line are power-law materials. The initial slope, η_0, of the former is zero, while the initial slope of the latter is infinite. A slope of zero is not expected for real materials.

However, consider the case of slow simple shear. Whether one shears to the right or to the left should not make any difference except for the sign of the resulting stress. Thus the shear stress must be an odd function of the shear rate. It seems only logical that the stress vs. rate graph should travel through the origin in a straight line, as shown in Figure 4-1. And indeed it does, but in some cases the curve is curved all the way to the origin where finally the second derivative drops to zero. Thus the ideal behavior called for by the Newtonian model would have limited usefulness. One can also imagine materials that exhibit power-law properties such that the viscosity at the origin is infinite (fine dotted line in Figure 4-1) or zero (dash-dot line). The infinite slope would signify a solid material but only at zero shear rate and stress.

3. Applying stress vs. applying deformations to a material

An aspect of rheology that is bothersome to many is the question of what is the applied to the material—stress or strain—and what is measured. To some extent, this is a chicken-and-egg question. However, for the design of rheometers, the applied variable is a very important consideration, as some actuators are "naturals" when it comes to applying deformations, while others are best at applying a force or stress. For example, a dead weight can be used to apply a very steady force to a sample,[†] whereas a stepping motor[‡] is good at deforming the material in a precise fashion. From the point of view of discussing material behavior, we can assume that a stress or strain, as well as any time variation of these, can be applied to the material. While there may be experimental advantages of controlling stress rather than strain, or *vice versa*, these will be largely ignored.

4. Materials do what they want!

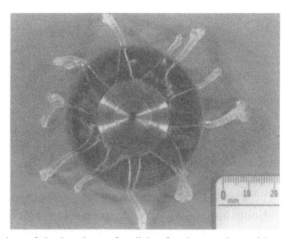

Figure 4-2. Illustration of the breakup of a disk of polymer sheared between two disks. The upper disk is transparent except for the steel hub in the center. (Picture from Remediakis et al.[1] Reproduced with permission from *Rubber Chemistry and Technology*. Copyright © 1997, Rubber Division, American Chemical Society, Inc.)

We can only do so much to control the response of a material to our applied stress or strain program. Ultimately, the material may react in a fashion that is inconsistent with what is expected. For example, if a fixed torque is applied to a disk of a polymer melt, the disk could shear in an orderly manner (the expected response) or it could slip at the surfaces of the disk. It might even break up (Figure 4-2). However, for the purposes of making some progress, we will find it convenient to assume the materials obey a few simple rules. These are discussed below:

[†] True only if accelerations are low. If the weight is very massive and overwhelms the resistance due to the sample, then the experiment switches to a constant acceleration of the movable clamp attached to the weight.

[‡] A stepping motor moves in precise increments in response to pulses generated by a controller unit. Linear and rotating stepping motors are available. A drawback of stepping motors is that they indeed move in small steps, which at low speeds can result in unwanted vibration.

- *Incompressible.* We will require the fluids to be incompressible. This means that the volume of a sample stays the same and that volumetric flow rates (m^3/s) of steady-flow processes are independent of position.
- *No slippage or fracture.* Our fluids will not be allowed to break or slip at the confining walls.
- *Isothermal.* We know that heat is generated when a viscous material is deformed. However, we will assume that heat is transferred away such that the temperature, and therefore the properties of the material do not vary with time or position.
- *Pressure* will vary throughout the system, but will have no influence on the material. (This is consistent with the incompressibility rule.) Note that pressure is distinct from stress—our materials will be allowed to change their properties with changes in stress, but only with the *invariants* of the stress tensor. Please refer to Chapter 5 for a discussion of the invariants.

There may be other rules needed in special cases, but these are the main ones. The impact of these will be examined in the problems at the end of this chapter.

B. THE NEWTONIAN FLUID

The Newtonian fluid is an ideal material that is approached very closely by most simple fluids and by many polymer solutions and melts at very low deformations and deformation rates. This ideal material is thus a very useful model that actually does predict flow rates and pressure drops very closely. The reason for the upper-case N in "Newtonian" is that this model was named after Isaac Newton (1643-1727) in recognition of his early investigations of the flow of fluids.

Table 4-1. Names and symbols used to describe viscosity.

Variable name	Symbol	Dimensions [a]	SI Units	Other Units
Dynamic viscosity [b]	η	$[M][L]^{-1}[t]^{-1}$	Pa s [=] kg/m s	poise centipoise
Newtonian viscosity	μ	"	"	"
Kinematic viscosity	ν	$[L]^2[t]^{-1}$	m^2/s	centistokes

[a] $[M]$ = mass, $[L]$ = length, $[t]$ = time.
[b] Not to be confused with the oscillatory measurement of linear viscoelasticity.

The Newtonian fluid is characterized completely by a single material property—viscosity. The symbol for viscosity is η, the Greek letter *eta*. Many texts also use the symbol μ (Greek letter *mu*) for the viscosity of a Newtonian fluid. Quite often we see the symbol ν (Greek letter *nu*) used for the kinematic viscosity, $\nu = \eta/\rho$, where ρ (Greek letter *rho*) is the density of the fluid. The dimensions and units of these variables are summarized in Table 4-1. As can be seen, the kinematic viscosity has the same dimensions as diffusivity, and characterizes the diffusion of momentum from high- to low-velocity regions.

The definition of viscosity for any fluid is the ratio of stress to strain rate. For simple shear, then

$$\eta = \frac{\sigma}{\dot{\gamma}} \qquad \text{(ssc)}^{\S} \quad (4\text{-}1)$$

where we use the symbol σ without subscripts to mean the shear stress in simple shear, and the symbol $\dot{\gamma}$ without subscripts to mean the shear rate in simple shear. The reason for the emphasis and the special symbols is that simple shear is so widely used that many probably think viscosity *must* be measured in simple shear. As we know, this is a gross simplification.

Referring back to Chapters 2 and 3, we can see that a reasonable definitional equation for viscosity and a Newtonian fluid is the model

$$\tau_{ij} = \eta \dot{\gamma}_{ij} \qquad \text{(ssc)} \quad (4\text{-}2a)$$

$$\tau_{ij} = -\eta \dot{\gamma}_{ij} \qquad \text{(fsc)} \quad (4\text{-}2b)$$

where the symbol τ_{ij} represents the ij'th component of the extra stress tensor, and $\dot{\gamma}_{ij}$ represents the corresponding component of the rate-of-deformation tensor. Thus equation (4-2) becomes our three-dimensional model of the Newtonian fluid. Note that the viscosity is a scalar and multiplies separately each component of the rate-of-deformation tensor to give the corresponding extra stresses. There are no interactions between the components; for example, the stress required to stretch a Newtonian rod at a given rate will not be influenced by the application of twist to the rod.

The viscosity of any fluid, including Newtonian fluids, will vary with the thermodynamic variables, principally concentration and temperature. Often the dependence on temperature is Arrhenius in nature over a short range of temperatures, i.e.,

---

§ See Appendix 3-3 for definitions of (ssc) and (fsc).

$$\eta(T) = \eta_R e^{-\frac{E^*}{R}\left(\frac{1}{T_R} - \frac{1}{T}\right)} \tag{4-3}$$

where T_R and η_R are the temperature and viscosity at some reference condition, while E^* is the flow activation energy. For polymer melts and concentrated solutions, this equation may not work very well, especially near the glass transition temperature.[2] The dependence of viscosity on concentration of solutions is quite complex, and will be discussed separately in Chapter 9.

1. Can polymer solutions and melts be Newtonian?

The question of the Newtonian nature of polymer melts and solutions can be addressed easily by separating Newtonian behavior from the Newtonian fluid. The Newtonian fluid is defined by equation (4-2) where η is a constant; any behavior other than that predicted by equation (4-2) means the fluid is not a Newtonian fluid. On the other hand, polymer solutions and melts can exhibit *Newtonian behavior* under some conditions, as shown in Figure 4-3. This figure illustrates the typical decrease of viscosity of a melt or solution with shear rate. The fact that the viscosity is not constant means the solution shown is not Newtonian; however, this solution does exhibit Newtonian behavior at low shear rates where the viscosity levels off and becomes constant.

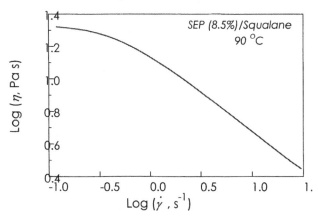

Figure 4-3. Viscosity response with shear rate measured using a rheometer for a solution of a block copolymer poly(styrene-b-ethylene-co-propylene) in squalane.[3] Note the Newtonian behavior at low shear rates, even though the solution is not Newtonian. (Reproduced with permission. © 2006, Applied Rheology.)

Things can get even more peculiar than this. By mixing a small amount of high-molecular-weight polymer into a fluid with a relatively high viscosity, one can sometimes create a fluid that shows an extended range of Newtonian behavior, but simultaneously exhibits distinct normal stresses, suggesting elastic behavior. Other combinations of fluid and polymer also appear to work.

Such fluids are called Boger fluids. The Newtonian model equation (4-2) predicts that there will be no normal stresses in shear for a Newtonian fluid (Example 4-1). Thus, the Boger fluid cannot be classified as a Newtonian fluid, although one would not know this if only the shear viscosity were examined.

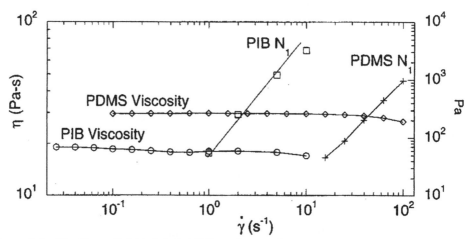

Figure 4-4. The Boger fluid is labeled PIB, and comprises a small amount of high-molecular-weight PIB dissolved in PIB of much lower molecular weight. When compared with the "normal" PDMS melt, we see that although the viscosity of the PIB is ½ that of the PDMS, its first normal-stress difference N_1 is nearly 100 times higher! (From K. B. Migler.[4] Copyright © 2000 by The Society of Rheology, Inc. All rights reserved.)

Example 4-1: *Show that a Newtonian fluid cannot exhibit normal stresses when subjected to any simple-shear deformation.*

Examining equation (4-2) we see that we will need to express the components of $\dot{\gamma}_{ii}$ using rectangular coordinates. Conveniently, these components are listed in Appendix 3-1. We will assign the flow direction to be in the x direction (or x_1 direction) and the gradient of the velocity to be in the y direction (or x_2 direction). Our velocity vector is thus: $v = (v_x(y), 0, 0)$. Writing out the tensor components in matrix form gives the rather intimidating result below:

$$\begin{pmatrix} \tau_{xx} & \tau_{xy} & \tau_{xz} \\ \tau_{yx} & \tau_{yy} & \tau_{yz} \\ \tau_{zx} & \tau_{zy} & \tau_{zz} \end{pmatrix} = \eta \begin{pmatrix} \dfrac{\partial V_x}{\partial x}+\dfrac{\partial V_x}{\partial x} & \dfrac{\partial V_y}{\partial x}+\dfrac{\partial V_x}{\partial y} & \dfrac{\partial V_z}{\partial x}+\dfrac{\partial V_x}{\partial z} \\ \dfrac{\partial V_x}{\partial y}+\dfrac{\partial V_y}{\partial x} & \dfrac{\partial V_y}{\partial y}+\dfrac{\partial V_y}{\partial y} & \dfrac{\partial V_z}{\partial y}+\dfrac{\partial V_y}{\partial z} \\ \dfrac{\partial V_x}{\partial z}+\dfrac{\partial V_z}{\partial x} & \dfrac{\partial V_y}{\partial z}+\dfrac{\partial V_z}{\partial y} & \dfrac{\partial V_z}{\partial z}+\dfrac{\partial V_z}{\partial z} \end{pmatrix} \quad \text{(ssc) (4-4)}$$

Note the viscosity η multiplying the entire rate-of-deformation matrix on the right. Referring to our velocity vector, we can see that all terms containing V_y or V_z must be zero, as there is no velocity in those two directions. Additionally all derivatives of anything in the x and z directions must be zero, as there are no changes in those directions. Applying these simplifications gives

$$\begin{pmatrix} \tau_{xx} & \tau_{xy} & \tau_{xz} \\ \tau_{yx} & \tau_{yy} & \tau_{yz} \\ \tau_{zx} & \tau_{zy} & \tau_{zz} \end{pmatrix} = \eta \begin{pmatrix} 0 & \dfrac{\partial V_x}{\partial y} & 0 \\ \dfrac{\partial V_x}{\partial y} & 0 & 0 \\ 0 & 0 & 0 \end{pmatrix} \qquad \text{(ssc) (4-5)}$$

which produces the simple result

$$\tau_{xy} = \tau_{yx} = \eta \frac{\partial V_x}{\partial y} = \eta \dot{\gamma} \qquad \text{(ssc) (4-6)}$$

Clearly all normal-stress differences are zero, as the corresponding components of the rate-of-deformation tensor are all zero. Remember, though, the total normal stresses need not be zero, but they will all be equal to the $-P$ for the isotropic pressure (ssc) in this example. For (fsc), they will be equal to P.

The conclusion from the example above is that if a fluid produces normal stresses in simple shear, it cannot be a Newtonian fluid. However, we can still assert that as the shear rate goes toward zero, the polymer solution or melt will eventually behave in a Newtonian fashion in all respects.[**]

A Newtonian fluid does not exhibit normal stresses in simple shear and shows no change in viscosity with deformation rate in any mode of deformation (shear, extension, etc.). An outcome of the latter is that Newtonian fluids cannot exhibit any time-dependent effects. For example, if a sample with a viscosity of 1 kPa s is at rest, all stresses must be zero; there can be no residual stress. If the sample is then subjected suddenly to a constant shear rate of 1 s^{-1}, the shear stress will jump instantly to a value of $\eta \dot{\gamma} = 1$ kPa. In terms of viscoelastic models, the relaxation time of the material is zero.

[**] This is not to say all fluid-like polymer solutions must be Newtonian at vanishing shear rate. The once-popular soft drink Orbitz® is an example; it shows very slight elastic effects at very low stresses.

PROBLEMS

4-1. A square plate of a Newtonian fluid is subjected instantly to forces (Figure 4-5) applied to ideal clamps on each edge. The resulting normal stresses on the edges have magnitude $\sigma = 10$ kPa, with the directions as shown. The faces of the plate are not subjected to constraints or forces

(a) Describe the stress state in the plate using the rectangular coordinates shown in Figure 4-5.

(b) Will the z-direction faces bulge out, sink in, or do nothing as the sample is deformed.

(c) If the viscosity of the material is $\eta = 2$ kPa s, what will be the extra or material-generated stresses τ_{ij} in all directions?

(d) Find all components of the rate-of-deformation tensor.

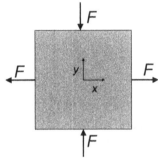

Figure 4-5. Square plate subjected to balanced pairs of forces on each edge.

4-2. Plot the flow curve (stress vs. shear rate) on linear-linear and log-log scales for a Newtonian fluid with a viscosity of 3 kPa s. Compare this with similar graphs for Bingham plastic following the relationship $\dot{\gamma}_{21} = 0$ for stresses less than $\sigma_Y = 1$ kPa, and $\tau_{21} = \sigma_Y + \eta\dot{\gamma}_{21}$ for stresses greater than 1 kPa. Comment on the shape, slopes and intercepts of these curves.

4-3. A tube made from a Newtonian fluid with a viscosity of 10 Pa s is stretched isothermally in the axial direction at a steady extensional rate of 1 s^{-1}. The outside diameter of the tube is initially 10 mm [$R_2(0) = 5$ mm], while the inside diameter starts out at 6 mm [$R_1(0) = 3$ mm]. The fluid is incompressible and uniform. The pressure is atmospheric inside and out.

(a) Unlike a solid rod, the tube could stretch by the outside surface moving in, the inside surface moving out, some combination of these, or some other motion all together. Which of these seems most likely, and why?

(b) Calculate all components of the extra stress tensor for the deformation you have chosen.

4-4. Ribbon candy is essentially a highly concentrated solution of polysaccharides, along with various colorants. Structure is developed by subjecting the mixture to repeated planar extensional deformations. This processing also causes the candy to become more opaque.

(a) If the candy is Newtonian with a viscosity of 10 kPa s, and is stretched at a maximum rate of 10 s^{-1}, what will be the pressure inside the strip?

(b) If the water activity in the candy is 0.1, and the temperature is 35 °C, will the conditions inside the ribbon be sufficient to boil the water? (In this problem, do *not* neglect atmospheric pressure!)

(c) (Open end) What else might lead to whitening during stretching?

4-5. A tapered die is often used to shape a thin rod or monofilament. Imagine that the walls of the die are lubricated such that the shear stresses are negligible, and the flow rate through the die is Q. Figure 4-6 shows a schematic of the setup.

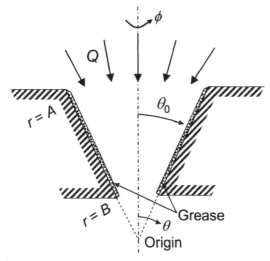

Figure 4-6. Sketch of lubricated conical die. The lubrication, in practice, lasts for but a few seconds before being swept away by the flowing polymer.[5] Methods for applying lubricant continuously have been described.[6,7] (Reprinted with permission of John Wiley & Sons, Inc.)

(a) Using spherical coordinates, describe the dependence of velocity V on position, i.e., fill in the relations for $\mathbf{V} = (V_r, V_\theta, V_\phi)$, where $\mathbf{V}$ is the velocity vector.

(b) Develop the equation of motion (again in spherical coordinates) and express in terms of the components of the total stress tensor, $\boldsymbol{\sigma}$.

(c) Assuming the fluid is exhibiting Newtonian behavior with viscosity η, describe the relevant extra stress components.

(d) Plot the variation of the total radial normal stress σ_{rr} and the pressure p with radius, and comment on these profiles.

4-6. (Open end) In 1974, two respected polymer rheologists published a short letter describing a disturbing observation.[8] The observation was that narrowly distributed PS samples of the same nominal molecular weight, but from difference sources, exhibited melt viscosities that differed by a factor of almost two. Both materials were made by anionic polymerization and both were checked carefully using GPC calibrated with PS standards from the same source.

(a) Assuming that the data are sound, hypothesize two possible sources of the problem.

(b) What modern methods could be applied to finding the most likely source of the problem?

REFERENCES

1. N. G. Remediakis, R. A. Weiss and M. T. Shaw, "Phase Structure Changes in a Sheared Blend of High-Molecular-Weight Polybutadiene and Polyisoprene Elastomers," *Rubber Chem. Technol.* **70,** 71–89 (1997).

2. M. T. Shaw and W. J. MacKnight, *Introduction to Polymer Viscoelasticity*, 3rd edition, Wiley, New York, 2005.

3. M. T. Shaw and Z. Liu, "Single-point determination of nonlinear rheological data from parallel-plate torsional flow," *Appl. Rheol.* **16,** 70–79 (2006).

4. K. B. Migler, "Droplet vorticity alignment in model polymer blends," *J. Rheol.*, 44, 277–290 (2000).

5. M. T. Shaw, "Flow of polymer melts through a well-lubricated, conical die," *J. Appl. Polym. Sci.*, **19**, 2811–2816 (1975).

6. C. W. Macosko, M. A. O'Cansey and H. H. Winter, "Steady planar extension with lubricated dies," *J. Non-Newt. Fluid Mechan.*, **11**, 301–316 (1982).

7. D. C. Venerus, T.-Y. Shiu, T. Kashyap and J. Hosttetler, "Continuous lubricated squeezing flow: A novel technique for equibiaxial elongational viscosity measurements on polymer melts," *J. Rheol.*, **54**, 1083–1095 (2010).

8. R. C. Penwell and W. W. Graessley, "Difference in viscosity among narrow-distribution polystyrenes of comparable molecular weight," *J. Polym. Sci., Polym. Phys. Ed.*, **12**, 213–216 (1974).

5

Generalized Newtonian Fluids—A Small but Important Step Toward a Description of Real Behavior for Polymers

One of the most prominent rheological characteristics of polymer melts or solutions is their non-Newtonian behavior. In most cases, this behavior is *pseudoplastic* in nature, which means the viscosity decreases with shear rate. The explanation for this behavior is, very roughly, that the velocity gradient tends to stretch out the polymer chains, which allows them to flow more freely past each other. Viscosities can decrease by orders of magnitude over a readily achievable shear-rate range, which is of huge practical importance. Thus, it is important to account for this behavior if realistic predictions are to be expected.

In this chapter, the fact that most non-Newtonian polymer fluids also exhibit elastic response (normal stresses) will be summarily ignored. Thus the application of the generalized Newtonian fluid (GNF) expressions should be confined to flows involving primarily steady shear, and where information about orientation is not needed. Any changes in deformation rates along the direction of flow must be very gradual for GNF descriptions to provide a reasonable estimate of the state of stress and the consequent pressure changes.

A. REASONS FOR INVENTING GENERALIZED NEWTONIAN FLUIDS—BEHAVIOR OF POLYMER MELTS

As pointed out above, most single-phase polymer melts and solutions tend to exhibit pseudoplastic behavior. Where does this term come from? Its origin clearly has something to do with *plastic* behavior, so let's compare these two by examining schematic *flow curves* for various materials. A flow curve is generally depicted as a graph of shear stress vs. shear rate, using either linear or log scales. Each type of scale has advantages.

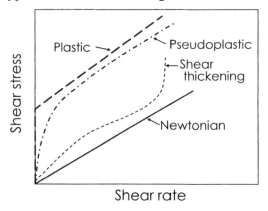

Figure 5-1. Depiction of the steady flow behavior of various materials using linear scales. Time is not supposed to be a factor for these materials, but often is.

Figure 5-1 shows schematic flow curves for Newtonian, plastic, pseudoplastic and shear-thickening fluids. The origin of pseudoplastic should thus be clear; it tends to soften the behavior of the strictly plastic response. Proper plastic materials feature a *yield stress*, that is, a stress that needs to be applied to get any deformation whatsoever. More information on plastic response will be provided in later chapters, especially Chapter 12.

1. A simple definition for simple shear

Simple shear has been defined precisely in Chapter 3, so we simply want to remind ourselves that simple shear can be visualized as a uniform sliding action similar to the sliding of a deck of cards. As with the cards, no distortion of the *shearing planes*[*] in the fluid is allowed. Experimentally, the shearing action in a fluid is controlled by parallel surfaces, one of which is moving tangentially at a steady velocity. The *shear rate* is the velocity gradient in the thickness direction. If all goes well, the velocity gradient will be the velocity of the moving plane divided by the sample thickness, which is another way of saying

---

[*] Shear planes, shearing planes, shearing surfaces and plane of shear all refer to the 1,3 plane.

that the fluid adheres to the surfaces, and the flow is homogeneous. See Figure 3-1 for a depiction of simple shear.

The SI units for shear rate are s^{-1}, often written as 1/s. Frequency has identical units, but there is no cyclic aspect of steady shearing.[†] The appropriate symbol, as we have seen, is $\dot{\gamma}$.

In practice, the achievable shear rate range between the moving and stationary plates in a rheometer is often restricted to 10^{-3} to 10 s^{-1}, although instrument manufacturers often claim more on each end. For example, one manufacturer of rotational rheometers specifies nearly 11 decades of rotational rate, which translates to 11 decades of shear rate for a given fixture geometry! The practical limitation at the low end, aside from the patience of the investigator, is the ability of the load cell to detect a significant stress. At the high end, non-uniform flows are very likely, similar to those depicted in Figure 4-2 but usually not as dramatic.

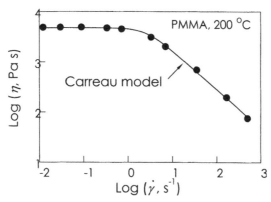

Figure 5-2. Typical response of polymer melt viscosity as the shear rate is changed for a special PMMA with a narrow molecular weight distribution. The data are adequately described by the Carreau model (see Table 5-1). This depiction uses the typical log-log scales.[‡]

2. Typical non-Newtonian behavior

Figure 5-2 shows a typical dependence of polymer melt viscosity on shear rate for a polymer, in this case, poly(methyl methacrylate), which has a relatively

---

[†] While not cyclic, one can visualize a rotational aspect to the flow similar to a roller bearing between the plates.

[‡] The scales shown in Figure 5-2 and many similar figures in this book (and frequently in the literature) commit the mathematical sin of taking the log of a dimensional quantity. The traditional nonlinear log scales appear to avoid this sin, but not really; they simply use an analog method of taking logs of the same dimensional numbers, like a slide rule. While not something to lose a lot of sleep over, the reader should remember that "log (η, Pa s)" is shortcut notation for "log [(η, Pa s)/1 Pa s]."

narrow molecular-weight distribution. Broadening the distribution will, logically enough, broaden the response (see Figure 5-4 below).

Even for commercially useful polymers, the viscosity range can be huge. Figure 5-3 shows the range of viscosities for commercial linear polyethylene, with the resins at the bottom end being describable as wax, while those at the high end are known as ultrahigh molecular weight resins (UHMWPE).

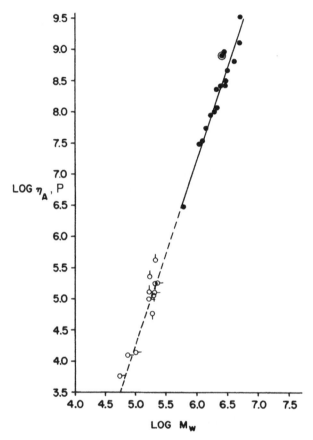

Figure 5-3. Viscosities at 200 °C of commercial polyethylene resins, showing the very wide range. Being old, the viscosity data were recorded in Poise (P), where 1 P = 0.1 Pa s. [Reprinted with permission of John Wiley & Sons, Inc. from Shaw (1977).[1]]

Unlike the special narrow-distribution PMMA described in Figure 5-2, commercial polyethylene often has a very broad molecular-weight distribution. An example of the viscosity response of a classical low-density polyethylene (LDPE) is shown in Figure 5-4. The transition from Newtonian behavior to a power-law dependence takes about 1 decade of shear rate for the PMMA. For the LDPE, the transition takes roughly 3 decades.

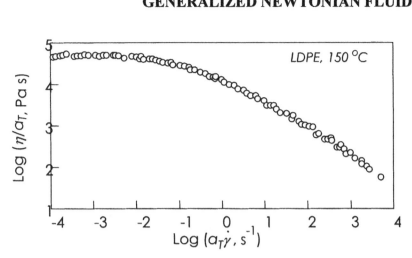

Figure 5-4. Depiction of the response of a LDPE melt over a shear-rate range of nearly 8 decades. To achieve this remarkably large range, several instruments as well as time-temperature superposition were used. The reference temperature is 150 °C. [Data from Laun (1978).[2] Used with permission of Springer-Verlag. Copyright © 1978.]

3. Example GNF model—the Cross model

More specific information can be obtained via the description of the viscosity behavior by a GNF model. In Figure 5-5 the Cross model has been used. The equations for a number of popular models have been assembled in Table 5-1. Using this result as an example, we will relate the parameters of the Cross model to common characteristics of the viscosity response.

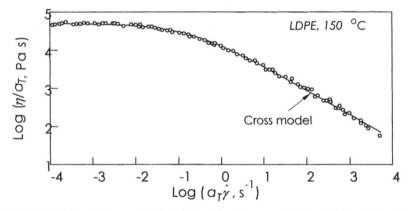

Figure 5-5. LDPE data of Figure 5-4 fitted with the Cross model (see Table 5-1). The size of the data points has been reduced to show more clearly the fit of this GNF model. [Data from Laun (1978).[2] Used with permission of Springer-Verlag. Copyright © 1978.]

The equation for the Cross model is

$$\eta(\dot{\gamma}) = \frac{\eta_0}{1 + (\dot{\gamma}/\dot{\gamma}_0)^{1-n}} \tag{5-1}$$

The three parameters for this equation are:
- η_0, the zero-shear-rate viscosity
- $\dot{\gamma}_0$, the characteristic shear rate
- n, the power-law index

For the LDPE data above, these parameters have the values $10^{4.7}$ Pa s, $10^{-0.73}$ s^{-1} and 0.36, respectively.[§] The characteristic shear rate is often reported as a time constant equal to $1/\dot{\gamma}_0$, but this suggests that the GNF has time dependence, which it does not. The material, however, does indeed have time dependence, so $1/\dot{\gamma}_0$ might still be quite useful as a characteristic time.

4. The power-law model

The power-law index n brings to mind the power-law itself, another example of a GNF model. The power-law model is given by

$$\eta = m\dot{\gamma}^{n-1} \tag{5-2}$$

where n is the power law index and m is the consistency index. Note that the slope of this line on the log-log plot will be $n-1$, not n. If $n = 1$, the fluid is Newtonian, a custom that is generally observed for all GNF models.

Note that as much as we might like, we cannot do anything about the crazy dimensions of m, which are Pa s^n when expressed in SI units. For example, we might try to write

$$\eta = \eta_0(\dot{\gamma}/\dot{\gamma}_0)^{n-1} \tag{5-3}$$

This expression has three parameters, whereas only two are required to describe a straight line (on log-log scales, of course). Thus we could arbitrarily set the $\dot{\gamma}_0$ equal to 1 s^{-1}, so now η_0 is the viscosity at a shear rate of 1 s^{-1}. Perhaps then we should call it η_1. Is this done? Occasionally, but not very often, although it is related to the mathematical issue of taking logs of dimensional numbers (see footnote for Figure 5-2).

[§] The form of the two dimensional parameters is a remnant of the search technique for the optimum values of the parameters. The search is done on the log of a dimensional parameter to keep the parameter itself always positive.

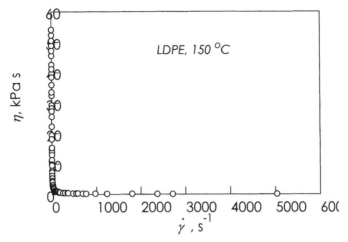

Figure 5-6. Data of Figure 5-4 replotted using arithmetic scales. The units of the ordinate have been modified from Pa s to kPa s to reduce the number of digits displayed. It is quite evident from this graph why log-log scales are preferred. [Data from Laun (1978).[2] Used with permission of Springer-Verlag. Copyright © 1978.]

It is instructive to visualize the asymptotic behavior of the model at low shear rates and at high. This is done for the LDPE example in Figure 5-7.

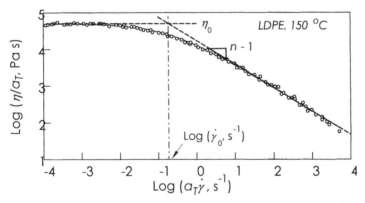

Figure 5-7. The dotted lines show the asymptotic behavior of the Cross model at low and high shear rates. It is no surprise that the intersection of these two is at the characteristic shear rate, $\dot{\gamma}_0$. [Data from Laun (1978).[2] Used with permission of Springer-Verlag. Copyright © 1978.]

C. GENERALIZING THE GNF TO THREE DIMENSIONS

So far, we have considered the pseudoplastic behavior of shear viscosity, that is, the viscosity measured in simple shear. We know that in simple shear, the shear rate is the same as the formal "rate of deformation" in rectangular coordinates; thus, we have not bothered to subscript the symbol $\dot{\gamma}$. We also learned earlier that in simple shear the shear rate is related to the magnitude of the rate-of-deformation tensor $\dot{\gamma}$. This magnitude becomes very important to the extension of the GNF to three dimensions.

1. Invariants of the rate-of-deformation tensor

The magnitude of a vector is the length of the vector. If the vector is rotated, the components change, but not the length. The length is thus *invariant* with displacement, rotation, or nature of the coordinate system. Why is this important? Because the length is real, whereas the components depend on an arbitrary coordinate system. This concept will now be extended to tensors.

$$\begin{pmatrix} 0 & \dot{\gamma}_{21} & 0 \\ \dot{\gamma}_{12} & 0 & 0 \\ 0 & 0 & 0 \end{pmatrix} \cdot \begin{pmatrix} 0 & \dot{\gamma}_{21} & 0 \\ \dot{\gamma}_{12} & 0 & 0 \\ 0 & 0 & 0 \end{pmatrix} = \begin{pmatrix} 0\cdot 0 + \dot{\gamma}_{21}\dot{\gamma}_{12} + 0\cdot 0 & 0\cdot\dot{\gamma}_{21} + \dot{\gamma}_{21}\cdot 0 + 0\cdot 0 & 0\cdot 0 + \dot{\gamma}_{21}\cdot 0 + 0\cdot 0 \\ \dot{\gamma}_{12}\cdot 0 + 0\cdot\dot{\gamma}_{12} + 0\cdot 0 & \dot{\gamma}_{12}\dot{\gamma}_{21} + 0\cdot 0 + 0\cdot 0 & \dot{\gamma}_{12}\cdot 0 + 0\cdot 0 + 0\cdot 0 \\ 0\cdot 0 + 0\cdot\dot{\gamma}_{12} + 0\cdot 0 & 0\cdot\dot{\gamma}_{21} + 0\cdot 0 + 0\cdot 0 & 0\cdot 0 + 0\cdot 0 + 0\cdot 0 \end{pmatrix}$$

$$= \begin{pmatrix} \dot{\gamma}^2 & 0 & 0 \\ 0 & \dot{\gamma}^2 & 0 \\ 0 & 0 & 0 \end{pmatrix} \quad \text{since} \quad \dot{\gamma}_{12} = \dot{\gamma}_{21} = \dot{\gamma}$$

Figure 5-8. Review of the dot product of two matrices, using the rate-of-deformation tensor in simple shear as an example. Note that the symbol $\dot{\gamma}$ is being used as the shear rate in simple shear; however, as shown below, it is also the magnitude of the shear rate in simple shear.

While a vector has but one magnitude or invariant, a tensor has three. As one might expect, this makes life a bit complicated. To add to the confusion, we find on reading the literature that the definitions of the invariants are not universal. How can this possibly be? The reason is that a combination of any two variants is also an invariant, so one can customize the invariants to suit some purpose. However, our focus is simplicity, so we will stick with the following:

1. The first invariant I_1 of $\dot{\gamma}$ shall be the trace of $\dot{\gamma}$, i.e., the sum of the diagonal elements $\dot{\gamma}_{ii}$. This definition is especially useful for σ because its trace is equal to the negative pressure (ssc).[**] For $\dot{\gamma}$, the trace is zero for incompressible materials:

$$I_1 = \text{tr } \dot{\gamma} \tag{5-4}$$

[**] See Appendix 3-3 for definitions of (ssc) and (fsc).

2. The second invariant shall be the trace of $\dot{\gamma} \cdot \dot{\gamma}$. What does this mean? This multiplication is the simple product of the two arrays, and is the familiar row × column. We will do an example below:

$$I_2 = \text{tr}(\dot{\gamma} \cdot \dot{\gamma}) \qquad (5\text{-}5)$$

3. The third invariant shall, thus, be the trace of $\dot{\gamma} \cdot \dot{\gamma} \cdot \dot{\gamma}$:

$$I_3 = \text{tr}(\dot{\gamma} \cdot \dot{\gamma} \cdot \dot{\gamma}) \qquad (5\text{-}6)$$

Sure enough, the definitions are easy enough to remember, so there must be a catch, and indeed there is. The example below will make this clear.

Example 5-1: *Find the invariants of the rate-of-deformation tensor $\dot{\gamma}$ for simple shear.*

The rate-of-deformation tensor for simple shear has been given in equation (3-10), which is reproduced below:

$$\dot{\gamma} = \begin{pmatrix} 0 & \partial V_1/\partial x_2 & 0 \\ \partial V_2/\partial x_1 & 0 & 0 \\ 0 & 0 & 0 \end{pmatrix} = \begin{pmatrix} 0 & \dot{\gamma} & 0 \\ \dot{\gamma} & 0 & 0 \\ 0 & 0 & 0 \end{pmatrix} \qquad (3\text{-}10)$$

The first invariant is just the trace of $\dot{\gamma}$, which is $0 + 0 + 0 = 0$. While this invariant doesn't cause any trouble, it also doesn't provide any information other than what we knew already.

The second invariant, following equation (5-5), involves finding the "dot" product of $\dot{\gamma}$ with itself. This operation is reviewed in Figure 5-8. The trace of the resulting matrix is $2\dot{\gamma}^2$.

The third invariant is found using equation (5-6). At first it might appear as though the there might be a difference depending on the order in which the multiplication is performed, i.e., $\dot{\gamma} \cdot (\dot{\gamma} \cdot \dot{\gamma})$ or $(\dot{\gamma} \cdot \dot{\gamma}) \cdot \dot{\gamma}$. Thanks to the symmetry of the matrices, there is no difference. The product is

$$\begin{pmatrix} 0 & \dot{\gamma} & 0 \\ \dot{\gamma} & 0 & 0 \\ 0 & 0 & 0 \end{pmatrix} \cdot \begin{pmatrix} \dot{\gamma}^2 & 0 & 0 \\ 0 & \dot{\gamma}^2 & 0 \\ 0 & 0 & 0 \end{pmatrix} = \begin{pmatrix} 0 & \dot{\gamma}^3 & 0 \\ \dot{\gamma}^3 & 0 & 0 \\ 0 & 0 & 0 \end{pmatrix} \qquad (5\text{-}7)$$

As can be seen, the third invariant is zero. Thus, like the first, it will not play a role in the properties of the GNF.

2. Connecting I_2 with shear rate

We are expecting the magnitude of the rate-of-deformation tensor to have something to do with the shear rate, yet the only invariant with non-zero magnitude in simple shear is $I_2 = 2\dot{\gamma}^2$. This is most annoying, and forces a somewhat messy fix. Thus we proclaim that for <u>any</u> deformation, the "shear rate" magnitude to be used in the formulas for the GNF shall be

$$\dot{\gamma} = \sqrt{\frac{I_2}{2}} \qquad (5\text{-}8)$$

where I_2 is given by equation (5-5). While this works perfectly for shear, what will the "shear rate" be for other flows? Let's see about simple extension.

Example 5-2: *Find the shear rate for uniaxial extension at an extension rate of $\dot{\varepsilon}$.*

The rate-of-deformation tensor of simple extension has been examined in Chapter 3, and is summarized for rectangular coordinates below:

$$\dot{\boldsymbol{\gamma}} = \dot{\varepsilon}\begin{pmatrix} 2 & 0 & 0 \\ 0 & -1 & 0 \\ 0 & 0 & -1 \end{pmatrix} \qquad (5\text{-}9)$$

The 1,1 term, which is equal to $2\dot{\varepsilon}$, is in the stretching direction, while the other two terms describe the shrinkage in the width and thickness directions. As expected for incompressible flow, the trace is zero.

The dot product can be found in the usual manner:

$$\dot{\varepsilon}\begin{pmatrix} 2 & 0 & 0 \\ 0 & -1 & 0 \\ 0 & 0 & -1 \end{pmatrix} \cdot \dot{\varepsilon}\begin{pmatrix} 2 & 0 & 0 \\ 0 & -1 & 0 \\ 0 & 0 & -1 \end{pmatrix} = \dot{\varepsilon}^2\begin{pmatrix} 4 & 0 & 0 \\ 0 & 1 & 0 \\ 0 & 0 & 1 \end{pmatrix} \qquad (5\text{-}10)$$

The resulting matrix gives the second invariant and "shear rate" as follows:

$$I_2 = tr(\dot{\boldsymbol{\gamma}} \cdot \dot{\boldsymbol{\gamma}}) = 6\dot{\varepsilon}^2 \qquad (5\text{-}11)$$

$$\dot{\gamma} = \sqrt{\frac{I_2}{2}} = \sqrt{3}\dot{\varepsilon} \qquad (5\text{-}12)$$

Thus the effective rate to be used in GNF models is not $\dot{\varepsilon}$ but $\sqrt{3}\dot{\varepsilon}$, a considerably higher value.

With this is mind, we are prepared to write a general, three-dimensional description of the GNF by recognizing that the viscosity of the Newtonian fluid can be influenced by the invariants of the rate-of-deformation tensor. Thus

$$\boldsymbol{\tau} = \eta(I_1, I_2, I_3)\dot{\boldsymbol{\gamma}} \tag{5-13}$$

Note that the viscosity, as modified by the invariants, multiplies each element of the rate-of-deformation tensor to get the corresponding element of the extra-stress tensor. Follow closely the example below to see how this works.

Example 5-3: *Predict the extensional stress at a stretch rate of $\dot{\varepsilon} = 1\ s^{-1}$ for the polyethylene depicted in Figure5-5. Use the Cross model parameters derived from the shear data.*

Using equation (5-9), the components of rate-of-deformation tensor are simply

$$\dot{\boldsymbol{\gamma}} = (1\,\mathrm{s}^{-1}) \begin{pmatrix} 2 & 0 & 0 \\ 0 & -1 & 0 \\ 0 & 0 & -1 \end{pmatrix} \tag{5-14}$$

Assuming the Cross model adequately describes the extensional viscosity of the melt[††] and further assuming that the role of I_3 is negligible, the viscosity can be calculated using equation (5-1) and the parameters

$$\eta(\dot{\gamma}) = \frac{\eta_0}{1 + (\dot{\gamma}/\dot{\gamma}_0)^{1-n}} = \frac{10^{4.7}}{1 + (\sqrt{3}/10^{-0.73})^{1-0.36}}\ \mathrm{Pa\ s} \tag{5-15}$$

Note the use of $\sqrt{3}\ \mathrm{s}^{-1}$ for the rate-of-deformation magnitude, according to equation (5-12). The result is a viscosity of 9700 Pa s. According to the GNF definition in equation (5-13), the extra stress tensor can be found as follows:

$$\begin{pmatrix} \tau_{11} & 0 & 0 \\ 0 & \tau_{22} & 0 \\ 0 & 0 & \tau_{33} \end{pmatrix} = (9700) \begin{pmatrix} 2 & 0 & 0 \\ 0 & -1 & 0 \\ 0 & 0 & -1 \end{pmatrix}\ \mathrm{Pa} \quad.$$

The tensile stress σ_T is given by $(\sigma_{11} - \sigma_{22}) = (\tau_{11} - \tau_{22}) = 9700 \times 3 = \underline{29\ \mathrm{kPa}}$.

[††] In fact, this assumption proves to be woefully inadequate, as will be shown in subsequent chapters.

D.   INVENTING RELATIONSHIPS FOR VISCOSITY VS. SHEAR RATE.

It is great fun to think up other GNF relationships for melts and solutions, but there are some guidelines to keep in mind. First of all, your invention may have already been found, but was written in a somewhat different form. For example, the original inventor may have expressed the result in terms of the log of viscosity instead of viscosity itself. Or perhaps the original inventor combined parameters in a fashion that is different from your result. So it is wise to examine your equation from all angles. Check first against those listed in Table 5-1, keeping in mind that this list is not exhaustive. Finally, though, your invention may have some serious physical or mathematical shortcomings. An extreme is the prediction of negative viscosities for some shear rates. To avoid these problems, we explore desirable and necessary characteristics below.

1. Desirable characteristics

The general behavior of a GNF function is important. It normally should describe a Newtonian plateau at low shear rates and shear-thinning behavior at high rates. At very high shear rates, it might be appropriate to have a finite limiting viscosity. However, this may have little practical importance, as viscosities at very high shear rates are nearly impossible to measure reliably. While many GNF functions approach power-law behavior at high shear rates, this feature is not necessary. Of course, an attractive feature of a GNF equation is the ability to describe both broad and narrow transitions by varying the value of a single parameter.

A more subtle argument can be made for smooth behavior at vanishing stresses. Consider simple shear, where the magnitude of the rate-of-deformation tensor is the magnitude of the shear rate itself. We can achieve the same value of $\dot{\gamma}$ by applying either positive or negative shear rate. Plotting shear stress vs. shear rate should result in a flow curve that goes through zero in a straight line. Symmetry also suggests that the Taylor expansion of the flow curve should have only odd powers, i.e.,

$$\sigma = a\dot{\gamma} + b\dot{\gamma}^3 + \cdots \qquad (5\text{-}16)$$

where b is negative for pseudoplastic materials. Thus, this behavior at low shear rates would be a desirable characteristic of our new equation.

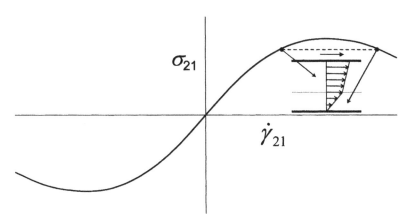

Figure 5-9. Cubic flow curve showing proper symmetry, but the possibility of two flow regimes at higher shear rates.

Another subtlety is the behavior at high shear rates. We mentioned that the viscosity should not become negative, and might approach a non-zero limiting value. However, we might also examine the flow curve at high shear rates for other problems. While opinions differ, we perhaps should be uncomfortable with a shear stress that decreases with shear rate, as this suggests that the flow could separate into a high-shear-rate region and a low-shear-rate region without violating any equations of motion (Figure 5-9). This means that the slope of the viscosity vs. shear rate on log-log scales should not be lower than −1.0, which represents a constant shear stress.

The analysis of slip in capillary flow is introduced in Chapter 7.

Example 5-4: *The bell-shaped curve below looks like it may be a reasonable choice for the behavior of a GNF. The equation used to draw this curve (Figure 5-10) is $\eta_R = \exp(-\dot{\gamma}_R^2)$, where $\eta_R = \eta/\eta_0$ and $\dot{\gamma}_R = \dot{\gamma}/\dot{\gamma}_0$. Is this an appropriate choice for a GNF model?*

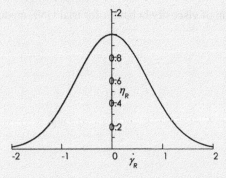

Figure 5-10. Proposed GNF viscosity function depicted on linear scales, where $\eta_R = \eta/\eta_0$ and $\dot{\gamma}_R = \dot{\gamma}/\dot{\gamma}_0$.

We can see from the graph and the equation that there are no negative viscosity values. Expanding the equation at low values of shear rate gives

$$\eta_R = 1 - \dot{\gamma}_R^2 + \gamma_R^4 / 2 ...$$

This shows quadratic behavior at low shear rates, a desirable result. The final test is the log-log slope at high shear rates. Finding the natural log of both sides gives

$$\ln \eta_R = -\dot{\gamma}_R^2$$

The derivative of this expression is

$$\frac{d \ln \eta_R}{d \dot{\gamma}_R} = -2 \dot{\gamma}_R$$

Finding the log-log slope involves multiplying both sides by $\dot{\gamma}_R$ to yield the final result:

$$\frac{d \ln \eta_R}{d \ln \dot{\gamma}_R} = -2 \dot{\gamma}_R^2$$

This is definitely not good news, as the log-log slope grows increasingly negative as the shear rate increases. The log-log plot is shown below.

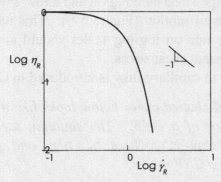

Figure 5-11. Illustration of viscosity behavior for trial GNF model shown in Figure 5-10.

Table 5-1. A selection of empirical Generalized Newtonian Fluid (GNF) equations for viscosity of polymer melts and solutions.[a]

# Parameters	Code	Name	n	η_R
2	BU	Bueche-Harding	1/4	$1/[1 + \dot\gamma_R^{\,0.75}]$
	FY	Ferry[b]	1/2	$1/[1 + \eta_R \dot\gamma_R]$
	DH	DeHaven	1/3	$1/[1 + (\eta_R \dot\gamma_R)^2]$
	SP	Spencer-Dillon	0	$\exp(-\eta_R \dot\gamma_R)$
	EY	Eyring	0	$\sinh^{-1}(\dot\gamma_R)/\dot\gamma_R$
3	CA	Carreau	n	$1/[1 + \dot\gamma_R^2]^{(1-n)/2}$
	CR	Cross	n	$1/[1 + \dot\gamma_R^{1-n}]$
	EL	Ellis	n	$1/[1 + (\eta_R \dot\gamma_R)^{(1-n)/n}]$
	MR	Mieras	n	$1/[1 + (\eta_R \dot\gamma_R)^2]^{(1-n)/2n}$
	SU	Sutterby	n	$[\sinh^{-1}(\dot\gamma_R)/\dot\gamma_R]^{1-n}$
	QD	Quadratic	-	$\exp[-a(\ln \dot\gamma_R)^2]$
4	SA	Sabia	n	$1/[1 + (\dot\gamma_R)^{(1-n)/a}]^{a-\eta_R}$
	VN	Vinogradov	n	$1/[1 + a(\dot\gamma_R)^{(1-n)/2} + (\dot\gamma_R)^{1-n}]$
	GR	Generalized rate	$1-ab$	$1/[1 + \dot\gamma_R^a]^b$
	GS	Generalized stress	$1/(1+ab)$	$1/[1 + (\eta_R \dot\gamma_R)^a]^b$

General Notes: η_R is the reduced viscosity $\eta(\dot\gamma)/\eta_0$, where η_0 is the zero-shear-rate viscosity. $\dot\gamma_R$ is the reduced shear rate $\dot\gamma/\dot\gamma_0$, where $\dot\gamma_0$ is the characteristic shear rate. σ_R is the reduce shear stress and is also written as $\eta_R \dot\gamma_R$.
[a] (Adapted from Elbirli and Shaw.[3] Copyright © 1978 by The Society of Rheology, Inc. All rights reserved.)
[b] The Ferry equation can be solved explicitly for η_R; it's $\eta_R = \left(-1 + \sqrt{1 + 4\dot\gamma_R}\right)/2\dot\gamma_R$.

2. Introduction to the deficiencies of the GNF concept

As mentioned previously, the GNF concept, while widely used for modeling viscosity data, has severe limitations as a general model for polymeric fluids. The GNF, for example, has no time dependence in spite of a time-like parameter that scales the shear-rate values.

It should be pointed out, however, that the time constant obtained from the modeling of steady viscosity alone does correlate with time constants obtained with linear viscoelastic measurements, e.g., the maximum relaxation time. This is not unexpected. In fact, there is a "rule-of-thumb" that connects the two in an easy-to-remember fashion. This is popularly known as the Gleissle mirror relationship,[4] which states that the nonlinear viscosity vs. shear-rate response is a mirror image of the linear response of the same fluid in a steady-shear stress-growth experiment. Of course, the two should be drawn on equal log-log scales. In equation form, this relationship can be stated as

$$\eta(\dot{\gamma})/\eta_0 \approx \eta^+(t)/\eta^+(\infty)\Big|_{t=1/\dot{\gamma}} \qquad (5\text{-}17)$$

where $\eta^+(t)$ is the stressing viscosity. As $\eta^+(\infty)$ should be the same as η_0 in the linear region, we can simply state that the two are viscosities are approximately equal. As the slope of $\eta^+(t)$ vs. t is simply the relaxation modulus, the connection between characteristic times and curvature in the viscosity function becomes obvious. It should also be evident that the Gleissle mirror relationship is the same as the Tobolsky-Chapoy approximation[5] published many years earlier. This relationship is

$$\eta(\dot{\gamma}) \approx \int_0^{\gamma_B/\dot{\gamma}} G(t)dt \qquad (5\text{-}18)$$

where γ_B is a parameter referred to as the breaking strain. For equivalence with the Gleissle mirror relationship, $\gamma_B = 1$.

E. SHORT PRIMER ON FINDING GNF PARAMETERS FROM DATA

1. Introduction

The title of this section includes the two modifiers "short" and "primer," but does not include the one we would like to see: "easy." There are no easy ways to fit data with GNF models, so software is required. With this statement of fact, we would like to see adjectives such as "cheap" and "user-friendly," and here there is some help, especially for students, in the form of academic discounts on software. But first a word about the problem itself.

2. The mathematics of fitting data with models

If our GNF equation were simple, such as the power-law model, we can see easily how to solve for the parameters if we have two data points,

$$\eta_1, \dot{\gamma}_1 \text{ and } \eta_2, \dot{\gamma}_2 .$$

The steps we would follow would be:

Start with power-law model:

$$\eta(\dot{\gamma}) = m\dot{\gamma}^{n-1} \tag{a}$$

Take logs of both sides:

$$\ln \eta(\dot{\gamma}) = \ln m + (n-1)\ln \dot{\gamma} \tag{b}$$

Write two equations using the two points:

$$\ln \eta_1 = \ln m + (n-1)\ln \dot{\gamma}_1 \tag{c}$$

$$\ln \eta_2 = \ln m + (n-1)\ln \dot{\gamma}_2 \tag{d}$$

Solve equations (c) and (d) for m and n:

$$n = 1 - \frac{\ln \eta_1 - \ln \eta_2}{\ln \dot{\gamma}_1 - \ln \dot{\gamma}_2} \tag{e}$$

$$\ln m = \ln \eta_1 - \left(\frac{\ln \eta_1 - \ln \eta_2}{\ln \dot{\gamma}_1 - \ln \dot{\gamma}_2} \right) \ln \dot{\gamma}_1 \tag{f}$$

However, suppose we have 10 data points instead of two? We would be able to write ten equations like equations (c) and (d), and only two parameters to find. The problem would then be *over-determined*; more equations than parameters. Another approach is needed.

The approach is really simple in concept, but tricky in application. The method is to guess at values for m and n, and use these values along with equation (a) to calculate 10 values of the viscosity. These will be then compared with the corresponding experimental values to see if the agreement is good. By good, we usually examine the sum of the squares of the differences between the calculated and observed viscosities. Then m and n are varied until the sum of the squares is minimized. This is known as the "least squares" method.

The result of this exercise is the "best fit" curve to the data. Oddly enough, in spite of being the "best" representation of the data, the curve does not go exactly through any of the data. The vertical distance between the curve and each point is called the residual. All the residuals, taken together, should be normally distributed[‡‡] with a mean of around zero, and some standard deviation, hopefully small. Taken one at a time in order, there should be no order, i.e., the signs and magnitudes of the residuals should not form any pattern. The signs should alternate in a random, coin-tossing fashion, while the magnitudes should not correlate at all with the point position in the sequence.

3. Error in the parameters

The parameters in the power-law model are m and n. The parameters of some other models are named in Table 5-1. The value of the parameter is a characteristic of the polymer and can at least be correlated, and sometimes derived, from the molecular structure of the polymer. An example is the well-known correlation between the zero-shear-rate viscosity η_0 and the weight-average molecular weight M_W of the polymer, i.e.,

$$\eta_0 = kM_W^{3.4} \tag{5-19}$$

Parameters are also used to compare two polymers. For example if Polymer A has a measured η_0 of 4.1 kPa s, and Polymer B has an η_0 of 3.9 kPa s, are the two viscosities significantly different? Realizing that these two values are subject to error, it is helpful to know the magnitude of this error and how it compares to the difference between the two parameters. Most programs will provide this error in the form of the standard error, or as the 95% confidence interval.[§§]

[‡‡] Normally distributed values of x follow the normal probability density function $f(x_i) = (1/\sigma\sqrt{2\pi})\exp[-(x_i - \mu)^2/2\sigma^2]$. Here, x_i is the ith observation, and μ and σ are the estimated average and standard deviation of the population from which the residuals were "drawn."

[§§] This problem is a bit more complex than explained here. The reason is that the viscosity values for the different shear rates are not independent observations, as they are normally gathered by "sweeping" shear rate from low to high or high to low. Thus the errors associated with the parameters are deceptively lower than what would result from repeating each measurement with fresh samples.

4. Tricks of the trade

Sometimes viscosity measurements are taken over a large range covering several decades of viscosity values. For example, the data in Figure 5-4 covers 2½ decades of viscosity. If we attempt to fit the viscosities as listed, the high values at low shear rates will have a huge influence on the results, whereas the low values at high shear rates will have almost no effect on the results. One "trick" to avoid this problem is to fit to the logarithms of the values instead of the values themselves. This procedure also avoids handling of large numbers associated with the viscosities of most polymer melts.   Large numbers, combined with a bad guess for the parameters and a large data set, can lead to numerical difficulties.

A common difficulty is when the search algorithm chooses a value for a parameter that is incompatible with a mathematical operation. For example, consider the Cross equation, shown previously in equation (5-1), but reproduced below:

$$\eta(\dot\gamma) = \frac{\eta_0}{1+(\dot\gamma/\dot\gamma_0)^{1-n}} \qquad (5\text{-}1)$$

In searching for the best fit, suppose the parameter n is currently 0.3, which means the exponent $1 - n$ is 0.7. Then, for some reason, the program chooses a value of -0.1 s^{-1} for $\dot\gamma_0$. This forces the value in the parentheses to also become negative because the shear rates are all positive. The result will be a numerical error, and the program will quit.

The trick to avoid this problem is to code the equation such that the search is done on logs of all dimensional parameters. Thus the Cross model could be coded as

$$y=a-\text{Log10}(1.+10.^{\wedge}(c*(x-b))) \qquad (5\text{-}20)$$

where:
$\quad$ y = log η
$\quad$ x = log $\dot\gamma$
are the dependent and independent variables, and
$\quad$ a = log η_0
$\quad$ b = log $\dot\gamma_0$
$\quad$ c = 1 $- n$
are the parameters.   This form is resistant to out-of-range arguments of functions, such as negative arguments for the log function Log10.

F. SUMMARY OF GNF CHARACTERISTICS

The Generalized Newtonian Fluid (GNF) model is a limited but useful step in the direction of characterizing nonlinear behavior in polymer solutions and melts. There are countless applications in the literature using the more popular empiricisms. In most of these, the objective may be quite specific: find the zero-shear-rate viscosity given a set of steady-flow data. We have seen that the model has no time dependence, which is the reason for the "Newtonian" label. However, for most sets of data, a time-like parameter can be found, and this parameter correlates strongly with characteristic times found using linear viscoelastic measurements.

APPENDIX 5-1: FITTING DATA WITH EXCEL®

The spreadsheet program Excel® published by Microsoft is readily available to most students. Can it be used to fit equations to data, e.g., GNF models to viscosity data? The answer is mixed: yes it can, although the process is less obvious than with many mathematics and graphics packages such as MathCad®, Mathematica®, Origin®, Polymath®, PSI Plot®, Sigma Plot®, etc. The process for doing the fitting is outlined here. The process for obtaining the standard error and confidence intervals for the parameters is even more involved, and will not be attempted.[***] To practice this method, the student is encouraged to attempt Problem 5-12.

The first step is to modify Excel® by using the "Solver" Add-In. To install this Add-In, go to Tools/Add-Ins and select Solver. The installation will add Solver in your Tools menu.[†††] At this point, set up your problem by putting the viscosity and shear rate data in columns and initial parameter values in separate cells. For example, the parameters of a three-parameter GNF equation might be in cells A1, A2, and A3, while the shear rate and viscosity might occupy columns B and C, respectively (Figure 5-12). Then use the selected GNF function to calculate viscosity values using the initial parameter values. The calculated values might occupy column D. Next, subtract these values from the

---

[***] A Monte Carlo method can be set up, but very tedious. Using the variance of the residuals as a measure of pure error, generate N other sets of "data" by adding random errors to the calculated values. The large N, the better. Fit the chosen function to each of these data sets to generate N sets of parameters. The standard deviation of these is an estimator for the standard error of the original parameters. Multiplying these by the appropriate t value will give the confidence intervals.

[†††] Some versions of Excel® may not come with the Solver Add-In.

experimental ones. These are the *residuals*. Put the residuals in column E, as they may useful for a *residual plot* to illustrate the quality of the fit. Square each of the residual values and put these in next column. Sum the squares into a separate cell, say G1. Select the cell containing the sum and go to Tools/Solver. On the window that opens, select Min and start the operation by hitting Solve. The values of the parameters will change to minimize the sum of the squares. When the search for the minimum is finished, the final values of the parameters will appear in place of the initial guesses. (It may be wise to reproduce in a safe place the initial values for modification in case the search runs into trouble, or to check for other possible minima.)

#	A	B	C	D	E	F	G
1	3.7	-1.91	3.69	3.682	0.0082	0.0001	1.48
2	0	-1.52	3.69	3.672	0.0182	0.0003	
3	0.9	-1.05	3.69	3.643	0.0467	0.0022	
4		-0.47	3.67	3.551	0.1191	0.0142	
5		-0.14	3.65	3.447	0.2026	0.0410	
6		0.52	3.49	3.095	0.3952	0.1562	
7		0.84	3.31	2.864	0.4462	0.1991	
8		1.55	2.84	2.278	0.5621	0.3160	
9		2.22	2.29	1.688	0.6023	0.3628	
10		2.7	1.88	1.258	0.6216	0.3864	

Figure 5-12. Illustration of (real) spread-sheet sequence described above for fitting a GNF model to viscosity (Column C) vs. shear rate (Column B) data. The data are in the form of logs to the base 10. Equation (5-20) was used to generate the calculated values (Column D) using the initial parameter values (Column A). Column E carries the residuals, whereas Column F shows the squares of the residuals. The value in Column G is the sum of the squares, i.e., sum of the values in Column F. This is the number that Solver will minimize.

If there are problems, first check all the formulas. Make sure all the signs are correct. If everything checks out, try modifying the search conditions in the Solver Options box. This is reached by Tools/Solver/Options. Increasing the number of iterations might help if the search is slow, or consider changing the items in the last row. For example, central derivatives might be more accurate than forward derivatives. Good initial guesses can be critical to the success of the search. One approach is to try a model with fewer parameters, e.g., the Bueche-Harding model. Then use the optimized parameter values as initial values for models with more parameters, e.g., the Cross model. If all the optimized parameter values for the Cross model are significant, then use these as starting values in an attempt to fit a four-parameter model such as the Generalized Rate model.

PROBLEMS

5-1. (Challenging) Consider the following set of GNF equations: Ferry, Cross, Carreau, Eyring and Ellis. Which of these, if any, can be solved explicitly for:

(a) $\sigma(\dot{\gamma};\dot{\gamma}_0,\eta_0,n)$ (the usual "flow curve.")

(b) $\dot{\gamma}(\sigma;\dot{\gamma}_0,\eta_0,n)$ (often appearing in older literature)

(c) $\eta(\sigma;\dot{\gamma}_0,\eta_0,n)$ (often quite independent of temperature)

In these equations, the three parameters to the right of the semicolon are sufficient to describe the entire flow curve.

In your explicit expressions, all terms in the parentheses must be on the right-hand side of the equation—none on the left. The *parameters*, which are listed following the semicolon, are characteristic shear rate $\dot{\gamma}_0$, zero-shear-rate viscosity η_0, and power-law index n (for the three-parameter equations). The *argument* is the variable preceding the semicolon.

5-2. The data for Figure 5-2 are listed below. Plot these data using linear-linear scales, and log-linear scales. For the latter, use the log scale for the shear rate and the linear scale for the viscosity. Describe the general shape of these curves.

Log ($\dot{\gamma}$, s^{-1})	Log (η, Pa s)
-1.91	3.69
-1.52	3.69
-1.05	3.69
-0.47	3.67
-0.14	3.65
0.52	3.49
0.84	3.31
1.55	2.84
2.22	2.29
2.70	1.88

5-3. An important skill for student and practitioner alike is the digitizing of data from figures in old journal articles for the purpose of re-analyzing the data. If the article is sufficiently old, the figure will need to be scanned; otherwise an electronic image of the graph can be accessed from the electronic version, which is usually in PDF format. Using the shear stress vs. shear rate data in Figure 1 from Shaw,[6] replot the indicated set as viscosity vs. shear rate. Use digitizing software to extract the data.

5-4. Basic mechanics suggests that the behavior of the viscosity function at very low shear rate should be quadratic, that is

$$\eta = \eta_0[1-(\dot{\gamma}/\dot{\gamma}_0)^2 + \alpha(\dot{\gamma}/\dot{\gamma}_0)^4 + \cdots] \qquad (5\text{-}21)$$

(a) (Computer) Try to fit this equation using only up through the quadratic term to the PMMA data of Problem 5-2. What are the problems?

(b) Show that the Carreau model will behave in this fashion at very low shear rates.

(c) Does the Cross model follow quadratic behavior at low shear rates? If not, what does it do? Plot some results to illustrate the behavior. (Hint: use linear scales.)

(d) For part (b), explore what happens if the third term is included. Is α positive or negative?

5-5. Show that the rate-of-deformation tensor in simple (uniaxial) extension at constant extension rate has a finite third invariant I_3. Derive the expression for I_3 in terms of the extension rate $\dot{\varepsilon}$.

5-6. A variation of the Carreau GNF model is

$$\eta(\dot{\gamma}) = \frac{\eta_0}{[1 + \dot{\gamma}/\dot{\gamma}_0]^{1-n}} \tag{5-22}$$

(a) Compare the behavior of this model with the Carreau model shown in Table 5-1 for low shear rates. Make clear plots on both log-log and linear-linear scales.

(b) Plot each using reduce variables, i.e., η/η_0 and $\dot{\gamma}/\dot{\gamma}_0$, and compare the breadth of the transitions from Newtonian to power-law behavior.

(c) For both equations, compare the values of $\dot{\gamma}_0$ with the intersection of the lines describing the power-law region at high shear rates with the Newtonian plateau at low shear rates.

5-7. For convenience, practitioners often describe limited ranges of polymer melt flow curves ($\log \sigma_{21}$ vs. $\log \dot{\gamma}$) with a quadratic equation, i.e.,

$$\log \sigma = a_0 + a_1 \log \dot{\gamma} + a_2 (\log \dot{\gamma})^2 \tag{5-23}$$

Discuss the problems with this equation for describing normal polymer behavior at low and high shear rates. Graphs may be useful.

5-8. It was stated without proof that the Gleissle mirror relationship and the Tobolsky-Chapoy approximation are different expressions of the same empirical correlation between nonlinear steady-shear viscosity and linear viscoelastic response. Demonstrate this equivalence using some of the leads provided in Chapter 5, Section D-1.

5-9. If there is indeed a relationship between the linear relaxation modulus and non-Newtonian viscosity, then one can use familiar relaxation functions to generate new GNF models. Suppose the stress-relaxation behavior follows the one-element Maxwell model, i.e., $G(t) = G_0 \exp(-t/\tau)$.

(a) What will the corresponding GNF model look like according to the Tobolsky-Chapoy approximation?

(b) Relate the parameters in the Maxwell model (G_0 and τ) to those in your GNF equation (η_0 and $\dot{\gamma}_0$)

(c) What desirable or undesirable attributes will this GNF have?

5-10.   Narkis, Hopkins and Tobolsky[7] examined the relaxation modulus of a series of narrowly distributed PS samples and found the relaxation modulus was described quite well by the stretched-exponential (KWW) function

$$G(t) = G_0 e^{-(t/\tau)^{1/2}} \tag{5-24}$$

Find the GNF model corresponding to this function and plot the result using the reduced variables η/η_0 and $\dot{\gamma}/\dot{\gamma}_0$. Plot this model and comment on its suitability. [Hint: Use the substitution $u = (t/\tau)^{1/2}$ to help with the integration.]

5-11. Examine equation (5-20) and show that is indeed a reformulation of the Cross model. If an appropriate program is available, compare the parameters resulting from fits of equations (5-20) and (5-1) to the data in Problem 5-2. Comment on why the results are different.

5-12. With limited access to specialized software, one can use Excel® to fit nonlinear equations to data, as described in Appendix 5-1. Try this with the data in Problem 5-2 and the

(a)   Cross Model

(b)   Carreau Model

(c)   Sutterby Model

Which of these models has the lowest sum of squares of the residuals? Which gives the highest number of runs (sequences of residuals of same sign)?

5-13. The Ferry model as written in Table 5-1 is implicit in viscosity, but can be solved for η explicitly; the result is shown in the footnote. Verify the result shown. Plot the result using log-log scales and reduced variables. Compare the resulting curve with that generated using the Cross model with $n = \frac{1}{2}$.

5-14. Repeat Problem 5-13, but find $\eta(\sigma)$ instead.   Name a characteristic stress parameter appropriately, and explain clearly what this parameter means.

5-15. The DeHaven model is similar to the Ferry model, but with a quadratic dependence on stress in the denominator.

(a)   Can this model be solved explicitly for viscosity? Show you reasoning.

(b)   Compare the behavior of the DeHaven and Ferry models using the reduced variables η/η_0 and σ/σ_0, where $\sigma_0 = \eta_0\dot{\gamma}_0$

5-16. (Small project) For polymer solutions, most GNF equations can be modified by adding a limiting viscosity η_∞, e.g.,

$$\frac{\eta - \eta_\infty}{\eta_0 - \eta_\infty} = \frac{1}{1 + (\dot{\gamma}/\dot{\gamma}_0)^{1-n}} \tag{5-25}$$

for the Cross model, where η_∞ is an additional parameter, perhaps close to the solvent viscosity η_m. Note that the $0 \to 1$ function on the right-hand side defines the unaccomplished change in the viscosity on going from η_0 to η_∞, rather than from η_0 to η_m, which would require the LHS to be $(\eta - \eta_m)/(\eta_0 - \eta_m)$. The latter suggests that the viscosity behavior at high rates is totally independent of the polymer, and depends only on the solvent.

(a) Attempt to find a research report that explores the question of the solvent-dependence of η_∞. Discuss the reported findings, including the discrepancy between η_∞ and η_m.

(b) Another issue is the dependence of the viscosity drop $\eta_0 - \eta_\infty$ on solvent nature. Suggest a research program that would find experimental evidence for the notion that if the polymer chain expansion were the same in two different solvents, the shear-rate-dependent part of the viscosity would be indistinguishable.[‡‡‡]

(c) Assuming that the coil expansion is dependent on the polymer-solvent solubility parameter, find a few solvent pairs that might provide similar coil expansion for polystyrene. Compare their viscosities under these conditions. (Hint: tables of these properties are available in *Polymer Handbook*.)

(d) Intrinsic viscosity $[\eta]$ is a rough measure of coil expansion. Find theoretical or empirical relationships that relate coil expansion explicitly to intrinsic viscosity.

(e) Following on (c), find literature on the effect of shear rate on intrinsic viscosity.

5-17. Often when viscosity data is plotted as log η vs. log σ, the viscosity appears to tumble precipitously at high stress. Yet the form for the Ellis fluid in Table 5-1 suggests that on a log η vs. log $\dot{\gamma}$ plot, this behavior is quite normal, i.e., power-law at high shear rates.
(a) Show that the form in Table 5-1 is correct for values of $n = 0$, ½ and 1.
(b) What is the final slope of log η vs. log σ if $n = 0$?

5-18. In Table 4.5-1 of Bird et al.,[8] the Ellis equation is written as

$$\frac{\eta_0}{\eta} = 1 + \left(\frac{\tau}{\tau_{1/2}} \right)^{\alpha - 1}$$

where τ is the shear stress (σ_{21}) and $\tau_{1/2}$ is the characteristic stress (σ_0). Note that the exponent is $\alpha - 1$, not $n - 1$. Using the result in Table 5-1, find a relationship between α and n. What is the advantage of the form shown above?

[‡‡‡] Finding two factors to have indistinguishable effects is, of course, logically impossible. Instead, one must decide beforehand what fractional difference might be considered negligible and design the experiment to have enough statistical power to avoid falsely deciding that the difference is negligible.

5-19. A useful property of a GNF equation is to have a correspondence between the characteristic shear rate and the shear rate at which the extrapolation of the power-law region intersects the zero-shear viscosity.

(a)   Why might this property be useful?

(b)   Which, if any, of the three-parameter models in Table 5-1 do *not* have this correspondence?

(c) (Open end) Pick a four-parameter model and investigate the relationship between the above-described intersection and the model parameters.

5-20. (Open end) The Cross model is perhaps the most-widely used of the three-parameter models, as it has a simple form and generally describes the viscosity function of polymers quite well.   However, there is one curious aspect of its form, which is reproduced below:

$$\eta(\dot{\gamma}) = \frac{\eta_0}{1 + (\dot{\gamma}/\dot{\gamma}_0)^{1-n}} \qquad (5-1)$$

While this equation is well behaved at low shear rates, approaching η_0 as expected, an approach to Newtonian behavior as $n \to 1.0$ produces the seemingly incorrect result that $\eta = \eta_0/2$.

(a)   Investigate this issue using analytical and/or numerical methods.

(b)   Compare your result with the Carreau-type equation shown in Problem 5-6, equation (5-22).

REFERENCES

1. M. T. Shaw, "Melt characterization of ultra high molecular weight polyethylene using squeeze flow," *Polym. Sci. Engr.*, **17**, 266–268 (1977).

2. H. M. Laun, "Description of the non-linear shear behaviour of a low density polyethylene melt by means of an experimentally determined strain dependent memory function," *Rheol. Acta*, **17**, 1–15 (1978).

3. B. Elbirli and M. T. Shaw, "Time constants from shear viscosity data," *J. Rheol.*, **22**, 561–570 (1978).

4. W. Gleissle in *Rheology Volume 2, Fluids* (G. Astarita, G. Marrucci and L. Nicolais, eds.), Plenum, New York, 1980, pp. 457–462.

5. A. V. Tobolsky and L. L. Chapoy, "Viscosity as a function of shear rate," *J. Polym. Sci., Polym. Lett. Ed.*. **6**, 493–497 (1968).

6. M. T. Shaw, "Detection of multiple flow regimes in capillary flow at low stress," *J. Rheol.*, **51**, 1303–1318 (2007).

7. M. Narkis, I. L. Hopkins and A. V. Tobolsky, "Studies on the stress relaxation of polystyrenes in the rubbery-flow region," *Polym. Eng. Sci.*, **10**, 66–69 (1970).

8. R. B. Bird, R. C. Armstrong and O. Hassager *Dynamics of Polymeric Liquids*, Vol. 1, 2nd ed., Wiley-Interscience, New York, 1987.

6

Normal Stresses— Ordinary Behavior for Polymeric Fluids

A. INTRODUCTION

In Chapter 5, we examined in some detail the phenomenon of pseudoplastic behavior of the viscosity function, as expressed by the Generalized Newtonian Fluid (GNF) models. Most polymer melts and concentrated solutions exhibit such behavior. Toward the end of the same chapter it was suggested that the characteristic time of these models might be related to their elastic behavior.[*] The primary question to be addressed in Chapter 6 is the following: are there any other signs of non-Newtonian behavior in steady shear? By its very nature, steady shear does not involve time; nevertheless, another characteristic of the flow appears that is not at all Newtonian or even GNF in nature. This characteristic is the presence of *normal stresses*. While commonly observed for solids of all sorts in shear, normal stresses are rare for nonpolymeric fluids, although foams and some particle suspensions also can exhibit modest normal stresses.

---

[*] See Chapter 8 for a more precise connection of these two.

B. WHAT ARE NORMAL STRESSES?

When regarding the subject of normal stresses, our first inclination is to examine the phenomena and relate these to the stress tensor. As learned in Chapter 2, there are two stress tensors: (1) the total stress σ, and extra or material-generated stress τ. The two differ only in that the diagonal elements of the former include the isotropic pressure. However, the pressure can be eliminated by subtracting any two diagonal terms. The two components most commonly differenced are σ_{11} and σ_{22}, giving the first-normal-stress-difference N_1. Thus

$$N_1 = \sigma_{11} - \sigma_{22} \qquad \qquad \text{(ssc)}^\dagger \quad \text{(6-1a)}$$

$$N_1 = -(\sigma_{11} - \sigma_{22}) \qquad \qquad \text{(fsc)} \quad \text{(6-1b)}$$

By definition, the Newtonian fluid will generate no normal stresses in simple shear because there is no deformation in either the 1 or 2 direction. However, we can surely generate normal stresses by simply stretching the Newtonian fluid. In planar extension, just to keep the flow in a plane, the rate-of-deformation tensor is

$$\dot{\gamma} = \begin{pmatrix} 2 & 0 & 0 \\ 0 & -2 & 0 \\ 0 & 0 & 0 \end{pmatrix} \dot{\varepsilon} \qquad \qquad \text{(6-2)}$$

where the 1 and 2 positions correspond to the stretch and compression directions, respectively, while $\dot{\varepsilon}$ is the velocity gradient in the stretch direction. For a Newtonian fluid with viscosity η, we can see that the first normal-stress difference N_1 is

$$N_1 = \sigma_{11} - \sigma_{22} = \tau_{11} - \tau_{22} = 4\eta\dot{\varepsilon} \qquad \qquad \text{(ssc)} \quad \text{(6-3a)}$$

$$N_1 = -(\sigma_{11} - \sigma_{22}) = -(\tau_{11} - \tau_{22}) = 4\eta\dot{\varepsilon} \qquad \qquad \text{(fsc)} \quad \text{(6-3b)}$$

So, a Newtonian fluid can generate a finite N_1 in this flow. But what about simple shear? For the answer to this, we turn to the Mohr's circle construct (see Figure 2-11). This construct says that for plane stress, as is the case for Newtonian fluids in shear, the maximum normal-stress difference is at 45° to the shearing direction, and the magnitude of this stress difference will be simply $2\sigma_{21}$, where σ_{21} is the shear stress. This can be confirmed by

---

† See Appendix 3-3 for definitions of (ssc) and (fsc).

subtracting equation (2-17b) from equation (2-17a), and noting that $\sigma_{11} - \sigma_{22}$ is zero for simple shear for Newtonian fluids.

Thus, a Newtonian fluid in simple shear generates normal stresses at an angle to the shearing direction, but this is a purely geometrical issue: N_1, as defined, remains zero. However, polymer melts and solutions will, in general, produce a positive N_1, that is a positive (tensile) stress σ_{11} in the flow direction as if the fluid were being stretched in the flow direction (ssc). How do we know there is such a stress, as it is rather inconvenient to measure σ_{11} between plates that are moving relative to one another? The answer is that we don't try; we instead measure the stress pushing the plates apart. The stress difference $\sigma_{11} - \sigma_{22}$ remains the same, except σ_{11} becomes zero (no pull or push), whereas σ_{22} becomes negative (ssc) due to the compression applied to keep the plates from moving apart. This action is illustrated in the Figure 6-1 with a rubber band stretched tight and placed into a fixed gap between two lubricated plates. If the ends of the rubber band are now released, the plates will keep the material from recovering by pressing against the sides of the band. The plates will thus be applying a negative stress (ssc) to the band.

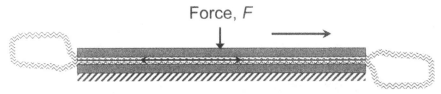

Figure 6-1. Cartoon of lubricated plates being pushed apart by a tensile stress in the flow direction produced by a stretched rubber band. When the ends are released, a compressive normal force F is required to keep the gap constant.

C. ORIGIN OF NORMAL STRESSES IN SIMPLE SHEAR

As with the rubber-band illustration in Figure 6-1, we expect that the flow in simple shear will result in stretching and alignment of the polymer molecules in the solution or melt. The alignment tends to be a bit more in the flow direction than in the gradient direction because of a torque applied by the velocity gradient to the stretched molecule. If the molecules are long and stretched out, their alignment tends to be very much in the flow direction.

How do we know this? One demonstration is to place fluorescent groups along a polymer chain that is very long. The molecule is carefully placed in a stationary shearing field, which can be achieved by moving both top and bottom plates, but in opposite directions. Using a high-resolution optical microscope[‡] operating with a dark field and ultraviolet illumination, one can observe the stretching of the chain by looking at the points of light produced by the fluorescing groups. Of course, one cannot see the molecule *per se*, but the groups can be detected as they stretch out. There are other ways of looking at the alignment, including neutron scattering from deuterium-tagged molecules located in an analogous untagged matrix. Optical birefringence is another and will be discussed in Chapter 7 in some detail.

Given that the typical polymer fluid exhibits normal stresses, it would be convenient to have equations for such that are as easy to use as the GNF models. As might be expected, there are such equations, but to make them useful in three dimensions, they are complicated and their presentation and discussion will be postponed to Chapter 8. By sticking to simple shear, we can say a few things about normal stresses. For one thing, we can surmise that normal stresses will not depend on the direction of shearing. Shear stresses, on the other hand, change sign when the shear rate changes sign. This is illustrated schematically in Figure 6-2.

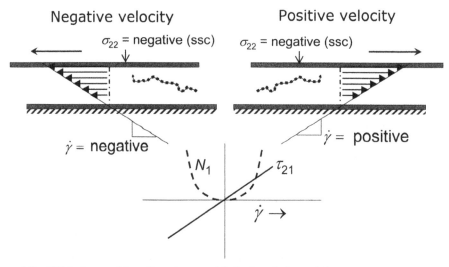

Figure 6-2. With the positive direction established as left to right, the shear stress becomes negative when the plate is moved steadily from right to left (left panel). The first normal-stress difference, however, retains its sign. Thus we expect N_1 to depend on $\dot{\gamma}^2$ at low shear rates. The dotted line depicts a polymer molecule in the flow field; it is stretched in the flow direction regardless of the sign of the velocity gradient.

Because of the symmetry of the first normal-stress difference, we expect that a positive material property should be (and is) defined in the following fashion:

[‡] Normally required for this experiment is a microscope equipped with confocal laser scanning optics to "optically slice" at a particular plane containing the target molecule.

$$\Psi_1 = N_1 / \dot{\gamma}^2 \qquad (6\text{-}4)$$

Thus, at low shear rates, we expect the shear stress τ_{21} to rise as $\dot{\gamma}^1$ and the normal stress to increase as $\dot{\gamma}^2$. Being really bold, we could propose that over a wider range of shear rates

$$N_1 \propto \tau_{21}^2 \qquad (6\text{-}5)$$

which actually seems to work quite well for melts of linear polymers with fairly narrow molecular weight distributions. Figure 6-3 shows a result for polystyrene; the slopes are quite close to 2.0. To make this relationship somewhat more general, we might propose a power-law relationship for N_1 similar to that for shear stress. Thus

$$\tau_{21} = m\dot{\gamma}_{21}^n \qquad \text{(ssc)} \quad (6\text{-}6)$$

from equation (5-2), and by analogy

$$N_1 = m'\dot{\gamma}^{n'} \qquad \text{c)} \quad (6\text{-}7)$$

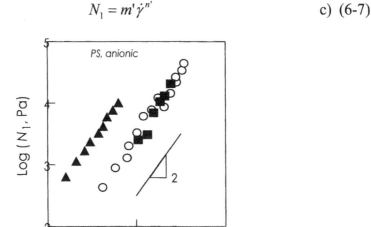

Figure 6-3. Relationship between normal and shear stress for anionic PS samples of various molecular weights. The triangles are for a blend of two anionic PS samples giving a M_w similar to the sample represented by the open circles. [Adapted with permission of John Wiley & Sons, Inc. from Oda et al. (1978).[1]]

where the subscripts on the shear rate $\dot{\gamma}$ have been dropped. Thus, if these relationships were correct, the normal stress would scale with shear stress in the following fashion:

$$N_1 = m'\dot{\gamma}^{n'} = m'\left(\tau_{21} / m\right)^{n'/n} \propto \tau_{21}^{\,n'/n} \qquad \text{(ssc)} \quad (6\text{-}8)$$

This relationship demonstrates how the slope of graphs such as those shown in Figure 6-3 can be close to 2 even at high shear rates. The only requirement is that $n'/n \sim 2$.

Oddly enough, in spite of the power-law nature of the relationships shown in equations (6-6) and (6-7), a time constant τ can be formulated.[2] The relationship is

$$\tau = \left(\frac{m'}{2m} \right)^{1/(n'-n)} \tag{6-9}$$

The derivation of this relationship involves analogs between dynamic oscillatory and steady-flow properties. See Problem 6-8 and Figure 6-11.[§]

Needless to say, while the simple relationships shown above may be helpful for some problems, they cannot be easily generalized to three dimensions; thus, they are specific for flows that are at least close to simple shear.

D. THE SECOND NORMAL-STRESS DIFFERENCE

The second normal-stress difference is defined as

$$N_2 = \sigma_{22} - \sigma_{33} = \tau_{22} - \tau_{33} \qquad \text{(ssc)} \quad \text{(6-10a)}$$

$$N_2 = -(\sigma_{22} - \sigma_{33}) = -(\tau_{22} - \tau_{33}) \qquad \text{(fsc)} \quad \text{(6-10a)}$$

The analogous material coefficient is

$$\Psi_2 \equiv N_2 / \dot{\gamma}^2 \tag{6-11}$$

N_2 can be viewed as a measure of the relative stretching of polymer molecules in the 2 direction, that is, in the direction of the velocity gradient versus that in the 3 or neutral direction. Figure 6-4 illustrates this schematically, where we are peering along the flow direction. While intuition suggests that N_2 might be positive, as the principal-stress direction for Newtonian fluids leans in the gradient direction, the observations available for polymer fluids indicate a negative N_2. Thus there is a very slight "log-rolling" effect between the two plates that stretches the molecules out in the 3 direction.

[§] The even behavior of normal stresses in oscillatory shear means that the frequency of normal-stress oscillation is twice that of the shear stress, a mechanical frequency-doubling device! See also Figure 6-11 and Problems 6-6 and 8-21.

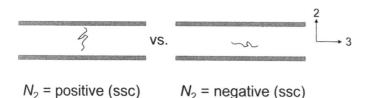

N_2 = positive (ssc) N_2 = negative (ssc)

Figure 6-4. Illustration of a slight distortion of a polymer chain in the 2 (gradient) direction (left panel) versus the 3 (neutral) direction (right panel), giving positive and negative N_2, respectively (ssc). The flow direction between the plates is into the page, which is also the direction of the major alignment of the chain.

As hinted above, N_2 is tough to measure. Not only is its magnitude small, the measurements are anything but direct. These problems will be described in more detail in Chapter 7. One might argue if N_2 is so small, it must not be very important. To a certain extent, this is true; however, it does enter into some processing flows. If nothing else, accurate values of N_2 are useful for checking the validity of theories concerning the generation of normal stresses in steady shearing flows.

E. NORMAL-STRESS COEFFICIENTS AND EMPIRICAL FINDINGS

Over the years, ample rheological data on polymer solutions and melts have been taken using steady simple-shear flows of one sort or another. Thus, it is not surprising that much has been said and debated about normal stresses and their relationships with shear stresses and other measurements. In steady simple shear, the fluid is fully characterized by the three simple material functions $\eta(\dot{\gamma})$, $\Psi_1(\dot{\gamma})$, and $\Psi_2(\dot{\gamma})$. Nothing more can be said about the material. This is a very appealing concept, especially as the normal stresses have been associated with all kinds of elastic phenomena, some of which show up in shear-free flows (i.e., extension).

Naturally enough, there is also a compelling interest in connecting the steady-flow material functions to linear viscoelastic properties. We know the dynamics are completely different, yet it seems that changing the frequency in a dynamic oscillatory flow is somehow analogous to changing the shear rate in a steady flow. Amazingly, this works better than one should expect. We will also explore the connections suggested by continuum and molecular theory in Chapter 8 and 9.

Based on observations alone, a number of useful empiricisms have been developed. These are described below:

1. The Cox-Merz rule

One of the earliest correlations noticed by rheologists[3] was the close relationship between the magnitude of the complex dynamic viscosity $|\eta^*|(\omega)$ and the steady-flow viscosity $\eta(\dot{\gamma})$ at roughly $\omega = \dot{\gamma}$. As this relationship seems to work quite well for many simple polymer melts and solutions, many have imbued the Cox-Merz relationship with a status approaching equivalency. One reason for this is that $|\eta^*|(\omega)$ is much easier to measure than $\eta(\dot{\gamma})$. But it's easy to see how it can fail miserably. Take, for example, a gel. Its viscosity is infinite, but the magnitude of the complex viscosity is very finite.

Figure 6-5 shows an example where the Cox-Merz rule is working well, and a graph illustrating mediocre agreement.

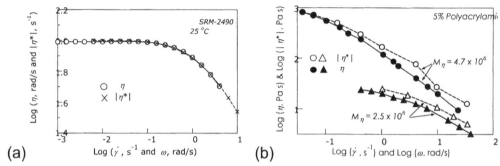

(a)   (b)

Figure 6-5. Examples of test of the Cox-Merz relation for viscosity. (a) SRM-2490, a solution of PIB (polyisobutylene) in 2,6,10,14-tetramethylpentadecane shows good agreement.   (b) A solution of polyacrylamide, shows mediocre agreement.   Generally fluids with complex, delicate structures are likely to show poor agreement.   [Data of Kulicke and Porter (1980).[4] Used with permission of Springer-Verlag, © 1980]

2. Normal stress and linear viscoelastic properties

The "Cox-Merz" of normal stress is the approximation

$$\Psi_1(\dot{\gamma}) \approx 2G'/\omega^2 \Big|_{\dot{\gamma}=\omega} \tag{6-12}$$

where G' is the in-phase component of the complex dynamic modulus G^*. By "in phase" we mean in phase with the strain applied to the sample. Thus, G' reflects the elastic character of the polymer solution or melt. This relationship between nonlinear steady flow response and linear viscoelastic response has had less testing than the Cox-Merz rule, and for good reason. Although G' can be measured quite handily up to frequencies ω of around 100 rad/s, it is very difficult to find reliable N_1 data at shear rates of 100 s^{-1}, or even close to this rate, due to flow instabilities. Similarly, at low frequencies and shear rates, both N_1 and G' become very small and thus difficult to measure accurately. Figure 6-6 features an example of the correlation between normal stress and the dynamic modulus; however, this interrelation does not hold very well for some other mixtures.

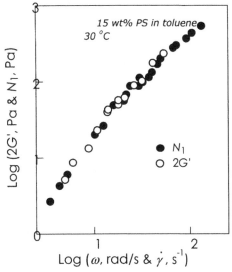

Figure 6-6. Normal stresses compared with the linear elastic response of a solution of polystyrene in xylene. Not all solutions studied showed such close agreement. (Adapted with permission from K. Osaki, M. Tamura, T. Kotaka and M. Kurata,[5] *J. Phys. Chem.*, **69**, 3642–3645. Copyright 1965, American Chemical Society.)

3. Other correlations

Mentioned already are the Gleissle mirror relationship and the equivalent Tobolsky-Chapoy equation relating the nonlinear, steady-shear viscosity to the shear stress-relaxation modulus $G(t)$. Can these be extended to normal stresses in a simple fashion? Perhaps indirectly through the relationship investigated by Middleman, who describes an excellent correlation (Figure 6-6) for many polymer solutions via the relationship

$$\frac{\Psi_1(\dot{\gamma})}{\Psi_{1,0}} \approx \left(\frac{\eta(\dot{\gamma})}{\eta_0}\right)^2 \tag{6-13}$$

Thus, a combination of equations (6-13) and (5-18) should give an approximation to the ratio on the left-hand side of equation (6-13). [Note that equation (6-13) is a restatement of equation (6-5).] An approximation to $\Psi_{1,0}$ is $\eta_0/\dot{\gamma}_0$, although this tends to be a bit on the low side.

Proposed some years ago by R. B. Bird et al.[6] is an approximation for normal stresses in terms of the viscosity function. The equation is

$$\Psi_1(\dot{\gamma}) \approx \frac{4}{\pi} \int_0^\infty \frac{\eta(\dot{\gamma}) - \eta(\dot{\gamma}')}{(\dot{\gamma}')^2 - (\dot{\gamma})^2} d\dot{\gamma}' \qquad (6\text{-}14)$$

which unfortunately is a bit difficult to apply as it requires the viscosity not only at the shear rate in question, but over the entire shear-rate range. A suggested maneuver is to fit the viscosity data with one of the empirical equations mentioned in Chapter 5, and then set up the integration using that function. This can be done with a spread sheet (Problem 6-4). Other empiricisms and approximations can be found in Table 6-1.

Table 6-1. Summary of interrelations and approximations

Formula	#	Notes and References
Exact:		
$$\eta_0 = \int_0^\infty G(t)\,dt$$	A	Not often used, as the contributions from low and high values of t are difficult to measure.
$$\eta_0 = \lim_{\omega \to 0} \frac{G''(\omega)}{\omega} = \lim_{\omega \to 0} \eta'(\omega)$$	B	This and the next are useful because the dynamic properties may be easier to measure at low frequencies.
$$\psi_1^0 = \lim_{\omega \to 0} \frac{2G'}{\omega^2} = \lim_{\omega \to 0} \frac{2\eta''}{\omega}$$	C	
Approximate:		
$$\eta(\dot{\gamma}) \approx \eta'(a\omega) = G''(a\omega)\,/\,a\omega; \quad a \approx 1.$$	1	See S. Middleman, *The Flow of High Polymers*, Wiley-Interscience, New York, 1968, p. 188.
$$\eta(\dot{\gamma}) \approx |\eta^*|(\omega)\big|_{\dot{\gamma}=\omega}$$	2	The Cox-Merz rule.[3]
$$\eta(\dot{\gamma}) \approx \int_0^{\gamma_B/\dot{\gamma}} G(t')\,dt'$$	3	The Tobolsky-Chapoy approximation.[7] This is equivalent to the Gleissle mirror relationship below when $\gamma_B = 1.0$
$$\eta(\dot{\gamma}) \approx \eta^+(t)\big|_{t=1/\dot{\gamma}}$$	4	Gleissle mirror relationship.[8]
$$\Psi_1(\dot{\gamma}) \approx 2G'(a\omega)\,/\,(a\omega)^2 \quad a \sim 1.$$	5	Derivable from the Spriggs model. See equation (1) above.

Formula	#	Notes and References
$$\Psi_1(\dot\gamma) \approx \frac{4}{\pi} \int_0^\infty \frac{\eta(\dot\gamma) - \eta(\dot\gamma')}{(\dot\gamma')^2 - (\dot\gamma)^2} d\dot\gamma'$$	6	Bird et al.[6]
$$\Psi_1(\dot\gamma) \approx \frac{2\eta''(\omega)}{\omega} \left[1 + \left(\frac{\eta''}{\eta'}\right)^2 \right]^{0.7} \Bigg|_{\omega = \dot\gamma}$$	7	Laun's rule.[9]
$$\Psi_1(\dot\gamma) \approx \Psi_1^+(t)\Big|_{t=k/\dot\gamma}$$	8	$k \approx 3$. See Reference 8.
$$\Psi_1(\dot\gamma) \approx \frac{2}{\dot\gamma} \int_0^\infty \sigma_{21}^-(t)\,dt$$ $$= \int_{-\infty}^\infty t\sigma_{21}^-(t)\,d\ln t = 2\int_0^\infty \eta^-(t;\dot\gamma)\,dt$$	9	See Reference 8.
$$\frac{\Psi_1(\dot\gamma)}{\Psi_{1,0}} \approx \left(\frac{\eta(\dot\gamma)}{\eta_0}\right)^2$$	10	Derivable from the White-Metzner model as well as a prediction of the rigid-dumbbell model.
$$\Psi_{1,0} \approx \eta_0 \tau_0$$	11	The characteristic time τ_0 is not specified.

F. TRANSIENT RHEOLOGICAL FUNCTIONS

Heretofore, we have been discussing steady-state-viscosity and normal-stress responses of polymer solutions and melts. Is there anything to be learned by examining the details encountered in initiating a flow starting with a perfectly quiescent melt or solution? There certainly should be. Visualize for a moment the quiescent melt with highly entangled polymer chains. On application of a fixed shear rate, the molecules distort and fight back. Then, as the strain increases, they begin to disentangle in response to the stress and aided by the alignment of the chains. At this point, the "viscosity" falls below the linear values, i.e., the plot of stress divided by strain rate vs. time. Yes, some important information might emerge from this experiment.

1. Transient viscosity function

In accord with the feeling that the word "viscosity" should be reserved for steady-state values, we call the viscosity during the transient the *stressing viscosity*, and give it a special symbol η^+. The full definition is shown below:

$$\eta^+(t;\dot{\gamma}) \equiv \sigma_{21}(t;\dot{\gamma})/\dot{\gamma} \qquad \text{(ssc)} \quad (6\text{-}15)$$

In this equation, we are expecting the shear rate to be raised instantly from 0 to $\dot{\gamma}$. The shear stress σ_{21} then climbs from zero to reach eventually its steady value corresponding to the applied shear rate. Meanwhile, we record the sometimes complex response.

If the shear rate is very slow, the response is fully linear and related to the relaxation modulus $G(t)$ by the Boltzmann equation as follows:

$$\eta^+(t) = \frac{1}{\dot{\gamma}} \int_0^t G(t-t')\dot{\gamma}(t')dt' = \int_0^t G(t-t')dt' \qquad (6\text{-}16)$$

As can be seen, the "stressing viscosity" is not dependent on shear rate in the linear regime and all values of η^+ will fall on the same curve.

In the nonlinear regime, the stressing viscosity starts out following equation (6-16) before falling to its steady value. In a sense, the quiescent melt is "broken" up at the higher stresses, and becomes a different material. A goal of rheology is to describe this process completely.

2. Transient normal-stress growth

The analogous functions for transient normal-stress growth are:

$$\Psi_1^+(t;\dot{\gamma}) \equiv N_1(t;\dot{\gamma})/\dot{\gamma}^2 \qquad (6\text{-}17)$$

$$\Psi_2^+(t;\dot{\gamma}) \equiv N_2(t;\dot{\gamma})/\dot{\gamma}^2 \qquad (6\text{-}18)$$

Like the steady-state normal-stress functions, these functions are also difficult to measure accurately. But one characteristic is very, very clear: the normal stress transient is much slower than that for the shear stress. Again, rheologists would like to describe this behavior.

3. The stress-decay functions

The name "stress-decay function" is easy to confuse with the linear stress-relaxation function, and indeed there are similarities. But first of all, we need to describe the experiment. We first allow the sample to reach its fully steady flow state at a shear rate $\dot{\gamma}$, and then instantaneously return the shear rate to zero. The shear and normal stresses will relax toward zero, but the way they relax will depend upon the initial shear rate. Thus we name the functions

$$\eta^-(t;\dot{\gamma}) \equiv \sigma_{21}(t;\dot{\gamma})/\dot{\gamma} \qquad\qquad \text{(ssc)} \quad (6\text{-}19)$$

$$\Psi_1^-(t;\dot{\gamma}) \equiv N_1(t;\dot{\gamma})/\dot{\gamma}^2 \qquad\qquad (6\text{-}20)$$

$$\Psi_2^- \equiv N_2(t;\dot{\gamma})/\dot{\gamma}^2 \qquad\qquad (6\text{-}21)$$

which are completely analogous to the stress-growth functions, but not the same. However, as one might expect, the transients for the normal stresses cover a wider time span than do the shear-stress transients.

4. Approximations involving the transient functions

The most quoted of the approximations involving the transient functions is Gleissle mirror relationship, discussed above. An analogous approximation, also proposed by Gleissle,[8] for normal stresses is

$$N_1(\dot{\gamma}) \approx N_1^+(t;\dot{\gamma})\big|_{t\sim 3/\dot{\gamma}} \qquad\qquad (6\text{-}22)$$

Note that the two are compared not at $t = 1/\dot{\gamma}$, but at approximately three times this value.

5. Summary of equivalents and approximations

Table 6-1 summarizes the equivalencies and approximations mentioned in this chapter, plus some others that are perhaps less widely used.

D. TEMPERATURE EFFECTS AND SUPERPOSITION OF STEADY-FLOW DATA

The influence of temperature on the steady-flow properties of polymer melts and solutions is of practical and fundamental importance. As with linear viscoelastic properties,[10] rheologists have attempted to generalize the influence of temperature on steady-flow properties by superposing the data with a combination of shifts along the shear-rate and stress axes to generate a master curve. The temperature dependence of these shifts has been described by various semi-empirical expressions including the Arrhenius equation and the WLF modification of the Doolittle expression. In some cases, these shifts have been related to structural features of the polymer, including branching and characteristics of the molecular-weight distribution.[11]

1. Ideas from simple liquids

As temperature is increased, the molecules move more rapidly, which increases the specific volume of the fluid. The additional kinetic energy suggests a type of activation for flow in a fashion similar to chemical reactions. The activated state would occur when a molecule in the fluid has moved into a position between two low-energy sites into a high-energy position, and is about to move into the lower velocity site, along with its momentum.[**] But Hildebrand argued that the total flux of momentum via the diffusion of actual molecules would not be controlled by the population of the activated state (the Arrhenius concept from reaction-rate theory), but by the density of vacant sites, i.e., the free volume in the fluid. The two-parameter version of Hildebrand's equation is

$$\eta = \frac{B}{v_{sp}/v_{sp,0} - 1} \tag{6-23}$$

where v_{sp} is the specific volume, and B and $v_{sp,0}$ are parameters. The Doolittle equation, which was developed earlier, is

$$\eta = \alpha \exp\left(\frac{\beta}{v_{sp}/v_{sp,0} - 1}\right) \tag{6-24}$$

The three parameters are α, β and $v_{sp,0}$. Note that both of these equations predict a very rapid increase in viscosity as the specific volume v_{sp} drops toward $v_{sp,0}$. At this point, the fluid is jammed.

Equation (6-24) is the starting point for the Williams, Landel, Ferry (WLF) equation for polymers above, but near the glass transition temperature T_g. The key assumption in the development of the WLF equation is a constant coefficient of expansion above the T_g. The resulting equation is

$$\eta(T) = \eta_R 10^{\left(-\frac{c_1(T-T_R)}{c_2 + T - T_R}\right)} \quad \text{or} \quad \log\left(\frac{\eta(T)}{\eta(T_R)}\right) = -\frac{c_1(T - T_R)}{c_2 + T - T_R} \tag{6-25}$$

While there appear to be four parameters η_R, c_1, c_2 and T_R in this equation, there are really only three, as T_R can be set arbitrarily. The hope was that this

---

[**] Liquids are liquids because of significant interaction between the molecules due to van der Waals forces of various types as well as stronger interactions such as hydrogen bonding. Thus the low-energy site for a diffusing molecule would be such that the distances and orientation of the molecule relative to its neighbors is favorable. The high-energy transition would involve less favorable interactions, e.g., excessively close proximity to the neighboring molecules.

equation would apply to all simple glassy (amorphous) polymers if T_R were set equal to T_g. Furthermore, if all amorphous polymers behaved in the same fashion near T_g, one might expect the parameters in equation (6-25) to be independent of the chemical nature of the polymer; however, this has not been realized. Appendix 6-1 lists some transforms of the WLF equation, including the widely used Vogel-Fulcher-Tammann version.

The equivalent to equation (6-25), but based on the Hildebrand equation, is

$$\eta(T) = \frac{\eta_R}{1 + a_1(T - T_R)} \tag{6-26}$$

where again T_R can be set arbitrarily leaving as parameters η_R, and a_1. The viscosity at T_R is η_R.

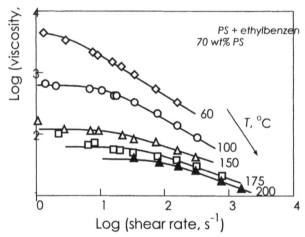

Figure 6-7. Viscosity data for PS solutions in ethylbenzene, with temperature as a parameter. (Adapted from R. A. Mendelson.[12] Copyright © 1980 by The Society of Rheology, Inc. All rights reserved.)

Other equations for temperature dependence abound. Aside from the Arrhenius equation, one of the more popular is the Fox-Loshaek equation[13]

$$\eta(T) = \eta_R \exp\left[a_1 \left(\frac{1}{T^{a_2}} - \frac{1}{T_R^{a_2}} \right) \right] \tag{6-27a}$$

which is somewhat of a modified Arrhenius equation. The parameter a_1 has very complex dimensions; thus a more agreeable form is

$$\eta(T) = \eta_R \exp\left\{ a_1' \left[\left(\frac{T_R}{T} \right)^{a_2} - 1 \right] \right\} \tag{6-27b}$$

where the parameter a'_1 is dimensionless. Note that the reference temperature, T_R in these equations can be set arbitrarily; in fact, it must be set arbitrarily or the equation will have too many parameters.[††] As the viscosity of polymers can change drastically with temperature, it is usually appropriate to fit the equation of choice to the data using logarithms of the viscosity.

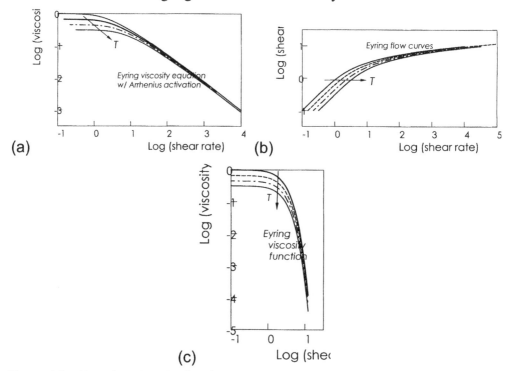

Figure 6-8. (a) Eyring viscosity function with equal temperature shifts on both viscosity and characteristic time. (b) Same data, but plotted as a flow curve. Note that only a shift along the horizontal axis is required to superpose the flow curves. (c) Viscosity vs. shear stress requires only a vertical shift. Data are often plotted in this fashion to find the activation energy for viscosity.

2. Empiricisms developed for polymers

From the study of linear viscoelasticity, the reader should be familiar with the concept of time-temperature superposition, whereby one supposes that a time-

---

[††] An equation $y = f(x; a, b, c, ...)$ has too many parameters (a, b, c, etc.) if it is possible to combine two or more of the parameters such that the combination can be replaced by a single new parameter, eliminating one (or more) of the original parameters. This was seen earlier with the power-law viscosity model, which we might try to express with three parameters as $y = a(x/b)^c$. Rearranging the right-hand-side gives $(a/b^c)x^c = a'x^c$, where the new parameter $a' = a/b^c$ has eliminated b. As the equation gets more complicated, overparameterization is more difficult to spot.

dependent material property such $G(t; T)$ at a given temperature T is similar in shape to $G(t; T_R)$ at a reference temperature T_R. Thus the two responses [e.g., log $G(t)$ vs. log t curves on log-log plot] can be shifted along the log time axis, with perhaps a slight adjustment along the log G axis (vertical shift). The analogous frequency-temperature superposition is also widely practiced. If the linear viscoelastic properties behave such that superposition is possible, the material is considered to be *thermorheologically simple*. Many polymers are thermorheologically complex.

As there seems to be an urge to search for universality in polymer behavior, it is not surprising that superposition of nonlinear material properties has been explored. Using the equations introduced above, we can, of course, superpose the viscosity of Newtonian fluids and possibly the zero-shear-rate viscosity of non-Newtonian polymer melts and solutions. Needless to say, things get a bit more complicated.

Let's consider some data for a single polymer run at different temperatures. The viscosity functions are depicted in Figure 6-8a, whereas the flow curves (stress vs. shear rate) are shown Figure 6-8b. Superficially, it looks like the flow curves can be superposed by horizontal shifts alone, while the viscosity curves will require both vertical and horizontal shifts. How is this to be interpreted? To avoid being fooled by scatter in the data, let's deal with some "data" generated with the help of empirical viscosity equations, such as those in Table 5-1.

We will explore first the Eyring equation, which has two parameters and an n value of 0, i.e., at high shear rates the slope of the viscosity function on a log-log plot approaches −1.0, eventually. We will assume the viscosity is reduced with temperature according to the Arrhenius relationship, equation (4-3), and the characteristic time is reduced in exactly the same fashion. The resulting viscosity functions are displayed in Figure 6-8a, while the flow curves are plotted in Figure 6-8b. To superpose those in Figure 6-8a will require first a vertical shift to match the zero-shear-rate viscosities, followed by a horizontal shift to match the curves at higher shear rates. For this data, the two shifts are identical.

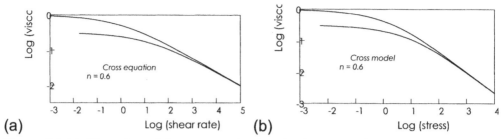

(a)   (b)

Figure 6-9 (a) Illustration of data generated using the Cross equation with $n = 0.6$. These curves cannot be superposed with equal shifts along the vertical and horizontal axes. (b) Same viscosity data, but plotted vs. shear stress. It is evident that comparing the viscosities at high stress would give a very bad estimate of the viscosity activation energy.

Figure 6-8c shows the same viscosity data plotted against log stress. With the Eyring equation, the drop in viscosity with stress is quite severe; in fact, it approaches a vertical drop, representing a limiting stress beyond which the fluid cannot go. A limiting stress appears in the log-log plot of viscosity vs. shear rate as a line with a slope of -1. The reason for a plot like Figure 6-8c is that vertical shifts can be used to get the flow activation energy. However, one has to be cautious about this, as we shall see.

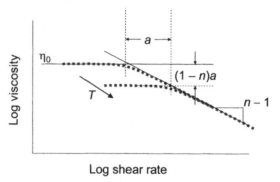

Figure 6-10. Schematic illustrating shifts required for superposition along the vertical and horizontal axes for a polymer that exhibits temperature-independent power-law behavior at high deformation rates.

With values of the limiting power-law exponent less than 1.0, but greater than 0.0, the shifts along the shear-rate axis must be greater than those for the viscosity axis. This is illustrated by examining data generated with the Cross equation using a high value of $n = 0.6$ to demonstrate this point more graphically (Figure 6-9a). Again, a base curve is generated (solid line) and then another representing a higher temperature. The shifts are chosen such that the two merge at high shear rates, a feature of many observations. As is evident, a larger shift, by a factor of $1/(1-n)$, will be required in the horizontal direction than the vertical to superpose these two curves. This is illustrated schematically in Figure 6-10.   Note that the shifts are on log scales; the multiplicative factors from Figure 6-10 will be 10^a and $10^{(1-n)a}$ for the viscosity and shear-rate values, respectively.

How do the viscosity data from Figure 6-9 look when plotted vs. shear stress? Figure 6-9b shows that a vertical shift only will not superpose the data. Thus the activation energy will not be independent of stress.

In summary, empirical information concerning temperature effects can be derived from the steady-flow viscosity data gathered at different temperatures. In particular, the indications are that the characteristic time derived from the shape of the flow curve has greater temperature sensitivity than the zero-shear-rate viscosity, at least for many polymer melts. The reasons for this behavior need to be explored.

In any case, whenever data are superposed using T-t superposition, it is obligatory to indicate this fact clearly on the resulting graphs be using axis labels such as $a_T \dot\gamma$ and by designating the reference temperature. As a reader, how does one tell if a graph has superposed data? Signs include a very broad variable range. Very few instruments with a single fixture and load cell can gather decent data over more than four decades of stress or rate. Another sign is clumped data. Isothermal data are usually gathered at equal spacing along the log stress or log shear rate axis, whereas superposing of equally spaced data invariably results in clumps of points.

APPENDIX 6-1: TRANSFORMS OF THE WLF EQUATION

The WLF expression was originally developed as a universal description of viscosity changes above the glass-transition temperature, T_g.   Thus, the reference temperature in equation (6-25) was considered to be T_g to achieve universality. The proposed universal constants established by examination of T-t shifts for several amorphous polymers were listed as $c_1 = 17.4$ and $c_2 = 51.6$ K.[‡‡] Equation (6-25) is written below with T_g as the reference temperature:

$$\log a_T = -\frac{c_1(T - T_g)}{c_2 + T - T_g} \tag{a}$$

Interestingly, the same form is valid for any reference temperature and the shifts for an arbitrary reference temperature T' can be written in terms of those using T_g as the reference temperature. The relationships are

---

[‡‡] Equation (6-25) and equation (a) in this section are often written with natural log instead of log to the base 10. Either way is fine, except the value of c_1 will change by the factor 2.303.

$$c_1' = \frac{c_1}{1 + \dfrac{T' - T_g}{c_2}} \tag{b}$$

$$c_2' = c_2 + T_g - T' \tag{c}$$

In these equations, the unprimed c_1 and c_2 are those for T_g, and the primed symbols are for the new temperature T'. These work for moving between any two temperatures, not just between T_g and another temperature. The general form is then

$$\log a_T = -\frac{c_1'(T - T')}{c_2' + T - T'} \tag{d}$$

where T' is any reference temperature.

A particularly important transform of equation (a) is one in which the reference temperature causes c_2 to disappear! One can see from equation (d) that the correct temperature T_v to chose can be found by equating $c_2 + T - T_g$ to $T - T_v$ to give $T_v = T_g - c_2$. In other words, T_v will be considerably lower than T_g. This temperature is known as the Vogel temperature. The result of the shift will be the equation

$$\log a_T = \frac{-c_v(T - T_g)}{T - T_v} \tag{e}$$

A free-volume interpretation of the Vogel temperature is the temperature to which the liquid should be cooled to eliminate all the free volume in the liquid. Of course, this is just a concept; it is not possible to check this experimentally.

By eliminating T_g in favor of $T_v + c_2$ in the numerator, the equation

$$\log \eta = a + \frac{b}{T - T_v} \tag{f}$$

results, where a, b and T_v are parameters. This form is coined the Vogel, Fulcher and Tammann (VFT) equation.

PROBLEMS

6-1. The data below are from the thesis of J. Starita, who related the mixing characteristics of polyethylene (shown below) and polystyrene to their rheological properties.[14] The data were gathered using a parallel-disk fixture on a rotational rheometer, a common method for low shear rates (see Chapter 7). The reported normal stress is a combination of the first and second normal-stress differences, sometimes called the third normal-stress difference N_3.[§§]

Rim shear rate, s^{-1}	Corrected viscosity, lbf s/in.^{2}	$\tau_{11}-2\tau_{22}+\tau_{33}$, lbf/in.^{2}
0.00747	1.80340	********
0.01194	1.45330	********
0.01510	1.38430	********
0.02390	1.17060	0.03570
0.02429	1.20650	0.03320
0.03730	1.02170	0.04910
0.03848	1.03950	0.04090
0.05970	1.03360	0.06430
0.06006	0.94890	********
0.09190	0.77640	0.11050
0.09600	0.75160	0.08800
0.11940	0.67890	0.13940
0.14770	0.63780	0.15480
0.15100	0.68900	0.17670
0.23320	0.52390	0.29240
0.24290	0.53100	0.34860
0.37330	0.46100	0.46760
0.38480	0.46660	0.56000
0.54690	0.40540	0.68140
0.60050	0.40190	0.77620
0.96010	0.35070	1.09940
0.98400	0.35900	0.97110
1.19400	0.34700	1.23960
1.47800	0.30910	1.55660
1.51000	0.31570	1.33560
2.39200	0.25780	1.71260
2.42900	0.26230	2.20350
3.73300	0.23520	2.44340
3.84800	0.22200	2.45960
4.80100	0.20060	2.79110
5.96900	0.18030	2.93680
7.68000	0.14850	3.40560
9.18500	0.13310	3.34710
12.01000	0.09210	4.09310
15.10000	********	3.82320
23.92000	********	4.84050
37.33000	********	5.82140
59.69000	********	6.77950
91.85001	********	7.93330

(a) What combination of N_1 and N_2 is displayed in the last column?

[§§] Although easily measured, N_3 is by no means universally or even widely recognized as a material characteristic as it involves two functions that in principle are independent of each other. However, it has been argued that N_3 can be used to test constitutive equations as easily as the more recognized material functions. Of course, if N_2 is small, $N_3 \approx N_1$.

(b)   Use this data to check the scaling exponent between normal and shear stresses. What is the value of the scaling exponent?

(c)   Does the scaling exponent differ significantly from 2.0?

(d)   Plot the normal and shear stress data using SI units, as opposed to the original English units, and include the regression line in the plot.

6-2. (Open end) With certain liquid-crystal polymers (LCPs), negative N_1 values have been reported.[15]   Find an example of such data and report on the proposed reasons for this behavior.

6-3. (Short project) Using suitable observations, answer the question: Does shearing an elastomer produce a positive N_1?   The suggested method is to twist a latex rubber tube and measure the length of the tube under constant axial load.   Keep the strain low enough so the tube does not buckle, yet high enough to produce an effect.   Now the hard part: the tube needs to be twisted about its axis in such a fashion that no additional axial loads are applied during the twisting.

6-4. (Computer) Using the viscosity data in Problem 6-1, predict the reduced normal-stress function and then the normal stress itself using the relationship

$$\Psi_{1,0} \sim \eta_0 \tau_0 \tag{6-28}$$

where τ_0 is a time constant.   Compare your result with the approximation shown in equation (6-13) and, of course, with the actual data.   (Hint: assume $\tau_0 = 1/\dot{\gamma}_0$, where $\dot{\gamma}_0$ is a characteristic shear rate.)

6-5. (Literature search) The steady-state recoverable creep compliance, a linear viscoelastic property, is a direct indication of elastic response in polymers, and thus should be related to normal stresses in steady shear (a nonlinear property).   Investigate this connection, if any.

6-6. (Computer) There is often a decent analog between the behavior of gels with strain and the behavior of fluids with strain rate.   With this in mind, check the scaling according to the quadratic relations between shear and normal stresses suggested by equation (6-5) using the data from Figure 6-11.

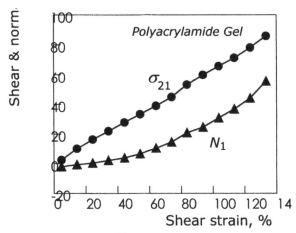

Figure 6-11. Figure from Janmey et al.[16] showing the stress-strain response for a gel made from polyacrylamide. The triangles signify the normal stress, while the circles show the shear stress. (Reproduced with permission of Nature Publishing Group. Copyright © 2007.)

6-7. This problem involves the Gleissle mirror relationship, but starts off with a bit of linear viscoelasticity.

(a)   Find expressions for the linear stressing viscosity function $\eta^{+}(t)$ for a fluid with a shear stress-relaxation modulus given by $G(t) = G(0)/[1 + (t/\tau)^n]$ using values of $n = 1/2$, 1, 3/2, 2, etc. (Hint: If you do not have access to software for symbolic manipulation such as Maple®, Mathematica® etc., use a web-based integration tool. An example of a web-base symbolic tool is the one currently supported by Wolfram, Inc. at http://integrals.wolfram.com/index.jsp.)

(b)   Plot the results in dimensionless form [$\eta^{+}(t)/\tau G(0)$ vs. t/τ] showing the influence of n on the shape of the curves.

(c)   What are the limits on n for a fluid?

(d)   Plot the viscosity function predicted by the Gleissle mirror relationship for $n = 2$.

6-8. Derive the expression shown in equation (6-9) for the time constant suggested by power-law behavior of the shear- and normal-stress responses with shear rate. (Hint: review the time constant for a Maxwell viscoelastic model in terms of the response of the dynamic functions with frequency.)

6-9. The data below are for a commercial polystyrene with a number- and weight-average molecular weights of 122 and 265 kDa, respectively.[7] The data were gathered at 200 °C. Note that the viscosity units are poise (P).

Shear Rate, s^{-1}	Log (viscosity, P)
0	5.47
0.303	5.36
0.810	5.25
1.69	5.13
9.58	4.75

Shear Rate, s^{-1}	Log (viscosity, P)
20.8	4.53
55.6	4.27
117	4.02
260	3.75
520	3.52
1300	3.21

(a)   Describe these data using the Cross equation. Check the quality of the fit by plotting the data and the fitted equation using log-log scales.

(b)   Assuming the Cross equation is correct over the entire shear rate range, use the Tobolsky-Chapoy approximation (Table 6-1) to derive an expression for $G(t)$, setting $\gamma_B = 1.0$. Check the validity of your result at the limits of $\dot\gamma = 0$ and ∞, corresponding to $t = \infty$ and 0, respectively. Hint: Extract $G(t)$ by differentiating the integral using Leibniz's rule. Leibniz's rule is given by the expression:

$$\frac{d}{dx}\int_{a(x)}^{b(x)} f(x,x')dx' = \int_{a(x)}^{b(x)} \frac{d}{dx} f(x;x')dx' + f(b)\frac{db(x)}{dx} - f(a)\frac{da(x)}{dx} \qquad (6\text{-}29)$$

6-10. Show that the parameter T_R in equation (6-27b) can be eliminated to return the equation to a three-parameter viscosity equation.

6-11. Explore the behavior of the Eyring model with temperature by assuming the value of $\dot\gamma_0$ is increased by the same factor as η_0 is reduced when the temperature is increased.

(a)   Will this keep the limiting stress the same?

(b)   Will the change in viscosity with temperature be independent of shear stress?

6-12. The data below are for a standard oil used for calibrating glass capillary viscometers. The density is provided to allow the calibration to be presented as kinematic or dynamic viscosity, but it also provides an opportunity for checking on the validity of the Doolittle and Hildebrand equations for viscosity.

T, °C	η, mPa s	ρ, g/cm^3
20.00	8.588	0.8266
25.00	7.237	0.8231
37.78	4.927	0.8146
40.00	4.639	0.8130
50.00	3.604	0.8063
60.00	2.880	0.7998
80.00	1.966	0.7863
98.89	1.457	0.7736
100.00	1.434	0.7730

(a)   Check on the variation of the coefficient of thermal expansion $\alpha = d\ln v_{sp}/dT$ over the range 20 to 100 °C. Report the value of α and any evidence for variation with T.

(b)   Examine the fits to the Doolittle and Hildebrand equations using the density values listed.

6-13. The data below are for Standard Reference Material (SRM) 2490, a solution of PIB dissolved in 2,6,10,14-tetramethylpentadecane. The testing temperature was 25 °C. A certificate describing this SRM in more (much more) detail may be retrieved from www.nist.gov by following links to Standard Reference Materials. (You may also buy some of this solution, but the cost is very high.)

Shear rate, s^{-1}	Viscosity, Pa s	Shear rate, s^{-1}	Viscosity, Pa s	N_1, Pa
0.001000	97.9	0.1585	96.1	2.4
0.001585	98.1	0.2512	93.7	5.5
0.002512	98.3	0.3981	90.0	12.9
0.003981	97.9	0.6310	84.6	26.5
0.006310	98.4	1.000	77.6	50.1
0.01000	98.1	1.585	69.2	87.6
0.01585	98.7	2.512	59.98	148.2
0.02512	98.8	3.981	50.56	236.8
0.03981	98.6	6.310	41.44	377
0.06310	98.4	10.00	33.04	585
0.1000	97.5	15.85	25.60	880
		25.12	19.36	1280
		39.81	14.26	1800
		63.10	10.22	2462
		100.0	7.22	3319

(a)   Plot the shear and normal stresses vs. shear rate using log-log scales. Compare the slopes of these two curves at low shear rates.

(b)   Estimate values of η_0 and $\Psi_{1,0}$ using GNF-type equations described in Chapter 5.

(c)   Check the approximation given by the equation

$$\frac{\Psi_1(\dot\gamma)}{\Psi_{1,0}} \approx \left(\frac{\eta(\dot\gamma)}{\eta_0} \right)^2 \qquad (6\text{-}30)$$

appearing in Table 6-1.

(d)   The concentration of polymer in SRM-2490 is reported to be 0.114 mass fraction. Estimate the molecular weight M of the PIB using the Bueche relationship

$$\Psi_{1,0} = \frac{12\eta_0^2 M}{\pi^2 cRT} \qquad (6\text{-}31)$$

where c is the mass concentration (e.g., kg/m^3) of polymer.

6-14. Dynamic viscoelastic data are available for the SRM described in Problem 6-13. These data are reproduced in the table below.

ω, rad/s	G'	G''	$|\eta^*|$, Pa s
0.00631	0.6180	99.84	
0.01000	0.9827	99.78	
0.01585	0.02062	1.562	99.86
0.02512	0.05525	2.477	99.8
0.03981	0.1379	3.922	99.63
0.0631	0.3343	6.189	99.01
0.1000	0.7819	9.704	97.9
0.1585	1.738	15.07	96.1
0.2512	3.711	23.06	93.3
0.3981	7.565	34.59	89.24
0.631	14.56	50.63	83.77
1.000	26.45	71.98	76.94
1.585	45.34	99.10	68.98
2.512	73.51	131.9	60.3
3.981	113.1	169.7	51.38
6.31	166.0	211.2	42.69
10.00	232.7	254.8	34.63
15.85	314.8	298.8	27.47
25.12	411.0	341.5	21.33
39.81	518.6	381.2	16.25
63.1	637.4	421.7	12.15
100.0	764.7	459.1	8.923
158.5	897.8	494.8	6.444
251.2	1026	529.7	4.603
398.1	1124	565.1	3.295

(a) Plot $|\eta^*|$ and η vs. ω and $\dot{\gamma}$, respectively, using log-log scales. Does SRM-2490 obey the Cox-Merz rule? If not, where are the largest discrepancies?

(b) Check the agreement of this data with Laun's rule, approximation #7 in Table 6-1.

(c) Check the agreement with the approximation #5 in Table 6-1

REFERENCES

1. K. Oda, J. L. White and E. S. Clark, "Correlation of normal stresses in polystyrene melts and its implications," *Polym. Eng. Sci.*, **18**, 25–28 (1978).

2. P. J. Leider and R. B. Bird, "Squeezing flow between parallel disks. I. Theoretical analysis," *Ind. Eng. Chem. Fund.*, **13**, 336–341 (1974).

3. W. P. Cox and E. H. Merz, "Correlation of dynamic and steady flow viscosity," *J. Polym. Sci.*, **28**, 619–622 (1958).

4. W.-M. Kulicke and R. S. Porter, "Relation between steady shear flow and dynamic rheology," *Rheol. Acta*, **19**, 601–605 (1980)

5. K. Osaki, M. Tamura, T. Kotaka and M. Kurata, "Normal stresses and dynamic moduli in polymer solutions," *J. Phys. Chem.*, **69**, 3642–3645 (1965).

6. R. B. Bird, O. Hassager and S. I. Abdel-Khalik, "Co-rotational rheological models and the Goddard expansion," *AIChE. J.*, **20**, 1041–1066 (1974).

7. A. V. Tobolsky and L. L. Chapoy, "Viscosity as a function of shear rate," *J. Polym. Sci., Polym. Lett. Ed.*, **6**, 493–497 (1968).

8. W. Gleissle, in *Rheology, Vol. 2. Fluids* (G. Astarita, G. Marrucci and L. Nicolais, eds.), Plenum, New York, 1980, pp. 457–462.

9. H. M. Laun, "Prediction of elastic strains of polymer melts in shear and elongation," *J. Rheol.*, **30**, 459–502 (1986).

10. M. T. Shaw and W. J. McKnight, *Introduction to Polymer Viscoelasticity*, Wiley-Interscience, New York, 2005.

11. P. Wood-Adams and S. Costeux, "Thermorheological behavior of polyethylene: Effects of microstructure and long chain branching," *Macromolecules*, **34**, 6281–6290 (2001).

12. R. A. Mendelson, "Concentrated solution viscosity behavior at elevated temperatures—Polystyrene in ethylbenzene," *J. Rheol.*, **24**(6), 765–781 (1980).

13. T. G. Fox and S. Loshaek, "Influence of molecular weight and degree of crosslinking on the specific volume and glass transition of polymers," *J. Polym. Sci.*, **15**, 371–390 (1955).

14. J. M. Starita, *Microstructure of Melt-Blended Polymer Systems*, Ph.D. Thesis, Princeton University, 1970.

15. G. Kiss and R. S. Porter, "Rheology of concentrated solutions of poly(γ-benzyl-glutamate)," *J. Polym. Sci., Polym. Symp.*, **65**, 913–211 (1978).

16. P. A. Janmey, M. E. McCormick, S. Rammensee, J. L. Leight, P. C. Georges and F. C. MacKintosh, "Negative normal stress in semiflexible biopolymer gels," *Nature Mater.*, **6**(1), 48–51 (2007).

7

Experimental Methods

Rheology can be a useful tool for characterizing the structure of polymer melts, solutions, suspensions and gels, and is used widely in technology because of its close connection with the applications of these and other materials. As pointed out in Chapter 1, rheology is a bulk measurement that averages over spatial extents of, usually, millimeters, but sometimes down to the upper nanometer range. Thus, conventional rheological techniques cannot reveal the specific molecular-scale interactions that are ultimately responsible for the material's behavior. However, much can be inferred by combining bulk and surface chemical analysis of the components with rheological measurements. For example, the rheological behavior of poly(ethylene-co-acrylic acid) is far different than that of poly(ethylene-co-vinyl acetate). It doesn't take long to surmise that the carboxyl groups may be dimerizing to stick the molecules together, impeding flow. Infrared spectroscopy confirms this. Neutralizing a small fraction of the carboxyl groups has an even more profound effect, and small-angle X-ray scattering confirms the presence of ionic clusters. Thus, the rheology, combined with a bit of logic, often results in a fairly complete and accurate picture.

Of course, we need methods for determining the rheological properties of a huge variety of polymeric compositions.[*] By "properties" we mean the fundamental material functions discussed in the previous chapters. Certainly, we do not want a measured material function to depend on the instrument or procedure used for the measurement. That said, instrument-dependent

---

[*] Most of the procedures described can also be used to characterize non-polymeric fluids, pastes and emulsions. For tests common to quality control and certification, see Chapter 11.

measurements do have their place, especially in quality control (see Chapter 11), as they are often quick, easy and quite precise.[†]

Fundamental to any rheological measurement is the ability to measure force, distance and time. As described in Chapter 1, even the gravity-driven capillary experiment for determining the viscosity of polymer solutions needs these basic measurements to derive the absolute viscosity.[‡] The application or measurement of force can involve load cells or actuators working through solid surfaces, body forces (e.g., gravity), or hydrostatic pressure. Deformation and deformation rate are deduced from the sample geometry combined with the displacement of the material, usually as inferred from motion of confining surfaces or bulk flow rates. Displacement can be axial (motion along a straight line) or rotational (motion around an axis). Flows may be created by dragging the material (*drag flow*), or by using a pressure gradient (*pressure-driven flow*). With the former, there is no pressure gradient in the flow direction, whereas with the latter, the pressure gradients along the flow direction can be huge, a possible disadvantage. Some more complex flows combine the two.

A. MEASUREMENT OF VISCOSITY

The viscosity of a polymeric solution, suspension or melt is the most basic of all the rheological measurements. Viscosities of all sorts of substances have been measured for centuries using a vast array of techniques and geometries. These days, however, everyone must have a computer-controlled instrument to run the test and gather the data. This fact has focused the major instrument makers' efforts on few rather than many geometries and techniques, although smaller companies constantly introduce clever innovations that make life interesting for the analytical rheologist.

One of the earliest methods of measuring viscosity was to observe the flow of the material through a tube. Why a tube? Why not an orifice? Or a funnel? These questions will be explored below.

[†] We will be using the words "precise" and "accurate" in this chapter. Precision has to do with the variability of repeat measurements, whereas accuracy refers to how close the average (or other characteristic statistic) of the observations is to the real value. The precision of an average can be improved by additional observations, whereas accuracy is improved by better calibration of the instrument using standards, e.g., standard reference materials supplied by NIST. Thus, precision and accuracy are quite different, and certainly not interchangeable.

[‡] As explained in Chapter 1, we do not need to know anything about the geometry to derive the intrinsic viscosity $[\eta]$, which has units of inverse concentration, if the flow is laminar and the material exhibits Newtonian behavior during the measurement.

1. Capillary flow—Newtonian fluid

The capillary is also known as a tube, pipe, etc. They are all the same. The important aspect is the geometry: a long hollow cylinder with the polymer melt or solution flowing axially due to a pressure gradient in the axial (z) direction. The reason "capillary" is preferred is because the typical laboratory would like to test small samples. To get to high shear rates without running out of polymer, the tubes must be small, as we shall see.

"Long" was also used. What this means is that the aspect ratio—length divided by diameter—must be large, again for reasons that will become clear.

The flow of Newtonian fluids through a capillary was solved long ago, presumably by Poiseuille, so flow in the capillary geometry is also called Poiseuille flow. We can also guess where the viscosity unit "Poise" came from. The development (see Appendix 7-1) seems straightforward in hindsight; one simply considers the flow as a series of concentric layers, adds the flow rate in all the shells to get the total flow rate Q, which is

$$Q = \frac{\pi R^4 P}{8\eta L} \tag{7-1a}$$

Here, R and L are the capillary radius and length, respectively; P is the pressure drop;[§] and η is the viscosity of the Newtonian fluid. Of course, we can easily solve equation (7-1a) for the viscosity in terms of the flow rate to give

$$\eta = \frac{\pi R^4 P}{8QL} \tag{7-1b}$$

This equation shows us a couple of the important aspects of the capillary experiment. One is that the result is highly sensitive to the capillary radius, which may be somewhat difficult to determine with high accuracy.[**] A mere 1% error in the radius will produce a 4% error in the viscosity. The other point is that the volumetric flow rate must be measured in some fashion. In typical commercial instruments, the flow rate is inferred from the velocity of a piston in a cylinder. Lacking this, the material passing through the capillary is

---

[§] The symbol P (upper case) will often be used for pressure drop as a simplification for $\Delta P = |p_2 - p_1|$, where p_1 and p_2 are pressures measured along the flow direction.

[**] Common methods included wire go/no-go gauges, optical measurement at the ends, and filling the capillary with a fluid of known density. All have their problems. The specified radius may change due to wear, corrosion, or deposits.

gathered for a measured time period and weighed. In this case, the polymer density at the flow temperature must also be known.

The third and perhaps most important point is the ratio P/L. What we really want to express, though, is the pressure gradient, $-\partial p / \partial z$, where z is the axial coordinate, and p represents the pressure along the capillary. The local pressure gradient is what drives the fluid, whereas the apparent gradient P/L over the entire capillary may have problems due to flow patterns in the capillary entrance and exit. We use the partial derivative just in case the pressure also depends on radius.

Examining Appendix 7-1 will reveal a number of other assumptions in deriving the Poiseuille equation.[††] One is that the velocity of the fluid at the wall of the capillary is zero. Speaking of the wall, we also must believe that the wall does not influence the viscosity. This seems like a small point, and probably is for solutions and melts; however, for suspensions, the effect of the wall can be quite pronounced. Also the viscosity must be independent of pressure, which is high at the entry and low near the exit. We further assume that a steady-state velocity profile is established quickly at the fluid enters the capillary. This is not a huge issue for Newtonian fluids at low flow rates, but can be for non-Newtonian melts.

There are some tricks that can help to reduce the effects of problems; these work-arounds will be discussed next. As the flow in the capillary is hidden from view, the problems are too often "out of sight, out of mind." Unfortunately, at high stress levels, we will never know what we are measuring, as the capillary is one of the few methods that can reach the high shear rates and stresses where the problems occur. But certainly one can "read the signs" and report accordingly. One sign is the condition of the exiting strand; if highly distorted, it is likely the flow in some section of the duct is not stable.

2. The capillary—non-Newtonian fluids

The analysis of the capillary-flow experiment gets complicated for non-Newtonian fluids because we must deduce the velocity at each radial position to get the total flow. We do not have to worry about the stress level; for steady flow, it is automatically linear with radius, i.e.,

$$\tau_{rz} = -\frac{\partial p}{\partial z}\frac{r}{2} \qquad\qquad (ssc)^{‡‡} \ (7\text{-}2)$$

[††] Also known as the Hagen-Poiseuille equation.

[‡‡] See Appendix 3-3 for definitions of (ssc) and (fsc).

where τ_{rz} is the shear stress and r is the radial position. Note the shear stress is zero at the center of the capillary, and rises to a maximum at the capillary wall. We will call this maximum value τ_w, where the subscript w stands for the wall. Thus

$$\tau_w = -\frac{\partial p}{\partial z}\frac{R}{2} \approx \frac{R\Delta P}{2L} \qquad \text{(ssc) (7-3)}$$

where the approximation on the right-hand side of the equation is good for long capillaries, i.e., those with a high *aspect ratio*. Note that the properties of the fluid do not enter here at all.

But, how do we know all this? The quick and easy answer is from the equations of motion, but it is also wise to have a more physical feel for this important point. This can be accomplished by using what is called a shell momentum balance, which is a fancy name for a force balance.[§§] This balance is depicted in Figure 7-1.

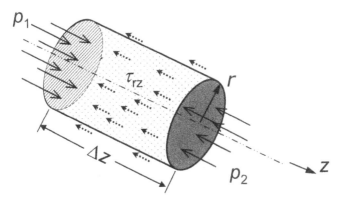

Figure 7-1. Shell balance on a piece of material in a tube with an applied pressure differential, where $p_1 > p_2$. The dotted arrows represent shear drag τ_{rz} on the surface of the cylinder. It is important to note that the material need not be a fluid and doesn't need to be flowing; however, it must not be accelerating. Note that τ_{rz} is negative (ssc) due to the negative dP/dz; however, τ_w (shear stress at wall) will be considered positive, as a convenience.

We shall agree that in the absence of acceleration or other body forces, the sum of the z-direction forces on a piece of the fluid must be zero. The pressure p_1 is higher than p_2, but both act on the same area, πr^2, to give a z-direction force of $\pi r^2 (p_1 - p_2)$, or $\pi r^2 \Delta p$. This force must be resisted by the shear stresses along

---

[§§] The use of a free body force balance can be tricky if the velocity of material into the element is different from that leaving, i.e., the material is compressible. In this case, the use of the full equation of motion is recommended.

the surface of the fluid element. The surface area is $2\pi r \Delta z$; thus, the total force is $2\pi r \Delta z \, \tau_{rz}$. Combining this result with the pressure force gives equation (7-3). As can be seen, nothing is mentioned concerning the properties of the material or how the material interacts with the wall.

The difficult part of this simple relationship is that we need the pressure gradient where the velocity profile has stabilized to have any hope of relating τ_w to the fluid viscosity. It is not easy to measure the pressure gradient or velocity profile in a capillary. Thus a trick is used; the problem areas near the entrance and the exit are subtracted out by comparing a short and a long capillary. This process is very clear if the two capillaries have the same diameter and are both long enough so the entrance and exit problems are identical. We shall lump these problems into one term, P_e, where, conveniently, the subscript e stands for both exit and entrance. Obviously, the two capillaries should be used at exactly the same flow rate. In this case, the shear stresses should be exactly the same in each capillary. For the long capillary, which we will subscript with a 2, the pressure drop is

$$\Delta P_2 = P_e + 2(L_2/R)\tau_w \qquad (7\text{-}4a)$$

while for the short capillary (number 1) the pressure drop is

$$\Delta P_1 = P_e + 2(L_1/R)\tau_w \qquad (7\text{-}4b)$$

Subtracting these two equations and solving for τ_w gives

$$\tau_w = \frac{R(\Delta P_2 - \Delta P_1)}{2(L_2 - L_1)} \qquad (7\text{-}5)$$

The P_e term has disappeared. This process is illustrated graphically in Figure 7-2.

Finding the shear stress by extrapolating away the end effects with capillaries of different lengths is a fairly obvious fix. More innovative and exciting was the idea proposed by Bagley[1] that the capillaries need not be of the same diameter as long as the exit and entrance are geometrically similar. By this we usually mean that regardless of the capillary diameter, the flow of polymer is from a very large reservoir into a relatively small capillary with a constant, well-defined entrance angle. Usually this angle is 180°, which is also known as a flat entry angle.[***] The exits of all the capillaries must also be

---

[***] Alternatively, and quite often, one sees the entry angle defined as the angle between the capillary axis and the conical surface. Thus a flat entry would be 90°, not 180°.

similar.   Thus, big capillaries look like small ones if the latter are simply magnified.

The argument Bagley made is as follows: The only way the polymer molecule, which is typically less than 50 nm, can "size" the capillary is through the stress it experiences, which is determined to a large extent by a characteristic time of the flow versus the characteristic time of the fluid.   There are, unfortunately, two characteristic times for the flow.   One is the average residence time in the capillary, while the other is the reciprocal of the shear rate in the capillary.   As one might expect, the ratios of these to the fluid time have been given names.   These are the Deborah number (**De**) and the Weissenberg number (**Wi**).   While definitions vary, we will use the following:[2]

$$\mathbf{De} \equiv \text{characteristic fluid time/flow time} \tag{7-6}$$

$$\mathbf{Wi} \equiv \text{characteristic fluid time} \times \text{rate of deformation} \tag{7-7}$$

Both of these are designed to get larger as the flow rate increases.

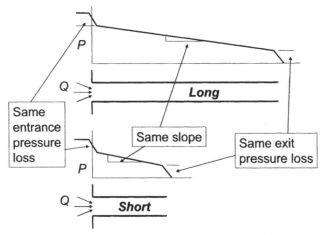

Figure 7-2.   Depiction of the subtraction of two capillaries of different length, but equal diameter.

Before finishing the argument of Bagley, which leads to the Bagley correction and the Bagley plot, we need to return briefly to the Poiseuille equation, shown earlier in equation (7-1) and repeated below but explicit in pressure:

$$\Delta P = \frac{8\eta QL}{\pi R^4} \tag{7-1c}$$

By multiplying both sides by R and dividing by $2L$, we arrive at

$$\frac{R\Delta P}{2L} = \eta \frac{4Q}{\pi R^3} \tag{7-8}$$

which is easily recognized as a form of Newton's law $\tau = \eta \dot{\gamma}$ (ssc). As the left-hand side is an estimate of the shear stress at the wall, τ_w, then the $4Q/\pi R^3$ must be the shear rate at the wall. Thus, for Newtonian fluids,

$$\dot{\gamma}_w = \frac{4Q}{\pi R^3} \tag{7-9}$$

This is a good relationship to commit to memory, as it is the starting point for many discussions about the flow in capillaries.

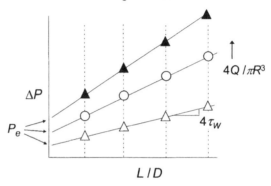

Figure 7-3. Schematic of a Bagley plot using data from four capillaries. The slope of the line is four times the shear stress, while the intercept, P_e, is a combination of the entrance and exit pressure losses, mostly the former.

If the fluid is not Newtonian, then equation (7-9) will be not be correct. To point this out, writers have assigned additional subscripts such as N for Newtonian or A for apparent. We will use the former, i.e., $\dot{\gamma}_{W,N}$. For polymer melts and solutions that are pseudoplastic (the most common case), $\dot{\gamma}_W$ is higher than the Newtonian value $\dot{\gamma}_{W,N}$. However, the two are uniquely related and functionally proportional.

From the proceeding discussion, we can easily see that the Bagley correction should apply for two capillaries having different diameters, but compared at the same values of $4Q/\pi R^3$, because the pressure drop through the entrance region can depend only on the value of $\dot{\gamma}_{W,N}$. The importance of this is the entrance problems in the capillary experiment can be removed with capillaries of arbitrary diameters as long as their aspect ratios L/D are different. Similarly, if the aspect ratios of two capillaries are the same, their pressure drops at fixed

$4Q/\pi R^3$ should be identical.[†††] The huge advantage of this development is that it allows the analysis to include data from many different capillaries, each used over its optimum range.[‡‡‡]

Figure 7-3 presents a schematic of the Bagley plot, which deserves some comment, especially when compared to the actual data in Figure 7-30 in Problem 7-4. It is easy to show that the shear stress is given by the equation

$$\tau_W = \frac{1}{4}\frac{d(\Delta P)}{d(L/D)} = (\text{slope})/4 \qquad\qquad \text{(ssc)} \quad (7\text{-}10)$$

where τ_W is the shear stress at the wall and ΔP is positive as shown in Figure 7-3. The Bagley procedure presumably frees this value of problems associated with the entrance and exit flows. Also, to a limited extent, it increases the precision of the result because the process of finding the slope reduces errors such as those resulting from difficulties in determining the diameter of the capillary. In effect, it is an averaging process.

The reality is usually somewhat different. While the schematic depicts a straight line, often the slope changes with capillary length. In a very real sense, the local slope $dp/d(L/D)$ can be regarded as the stress in the extra length of the capillary, which is an argument for taking the local slope at the higher L/D values. However, the data at higher L/D are associated with higher pressure, which can raise the viscosity of the fluid, thus artificially increasing the slope. The higher pressures in the longer capillaries can reduce wall slip, leading also to a higher slope. The pressure effect raises the slope above the desired value, while the reduction in wall slip raises the slope toward the true value. Unfortunately, it takes more work to distinguish the two.

Examining carefully some real data in Problem 7-4 reveals a problem at low capillary length: the points systematically fall below the line established by the data gathered using higher L/D capillaries. Why should very short capillaries (orifices) lead to low pressure drops? One explanation is that the part of the deformation experienced by the melt going into the capillary is stored temporarily as elastic or recoverable energy. This stored work dissipates as the melt moves down the capillary. However, if the capillary is very short, the

---

[†††] If they are not, then one explanation is the presence of slip at the wall. Another is the presence of a contribution due the breakdown of the assumption that the reservoir diameter D_b is much larger than the capillary diameter D_c. Thus $\Delta P_e = f(\mathbf{Wi}, D_c/D_b)$.

[‡‡‡] For instruments with fixed piston speed, the capillaries provided are often sized to provide constant $4Q/\pi R^3$ by matching the capillary diameter with the piston speed. If such a capillary set is not available, then the data must be interpolated.

stored energy is recovered, and the pressure is restored to some extent. It's as if the stretched polymer pulls itself out of the orifice.

A *velocity-profile correction* is also required for non-Newtonian fluids. If the fluid is Newtonian, the viscosity is given directly by the Poiseuille equation (7-1b). This equation doesn't work for non-Newtonian fluids. Neither does the equation (7-9) for the shear rate at the wall. However, the shear stress at the wall is still given by equation (7-3).

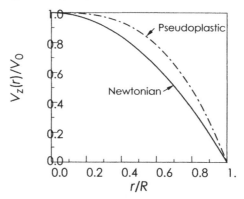

Figure 7-4. Velocity profiles in tube flow for a Newtonian and pseudoplastic fluid. The pseudoplastic fluid, in this case, is a power-law fluid with $n = 0.5$. V_0 is the velocity at the centerline.

For pseudoplastic fluids, e.g., most polymer solutions and melts, the stress present at the wall can be sufficient to reduce the viscosity at the wall to a level which is considerably lower than the viscosity at the center line. As a result, the shear rate at the wall will be higher than predicted by equation (7-9). The velocity profile will then acquire a blunted shape, as shown in Figure 7-4.

Before examining the details of a correction for this distorted velocity profile, let's find out what the velocity profile is for a Newtonian fluid. The example below shows how to do this in two ways.

Example 7-1: *Show that a quadratic velocity profile is consistent with a Newtonian fluid.*

The essential characteristic of a Newtonian fluid for this problem is a shear viscosity that is independent of shear rate. Figure 7-4 shows a plot of the velocity profile in capillary for a Newtonian fluid. The equation used to plot the curve is

$$\frac{V_z(r)}{V_0} = 1 - \left(\frac{r}{R}\right)^2 \tag{7-11}$$

where V_0 is the flow rate at the center line and R is the capillary radius. To find out if this profile is consistent with a Newtonian fluid, we need to calculate the viscosity and show that it is independent of radial position, r. We know that the shear stress is linear with radius, i.e., $\tau = \tau_W (r/R)$ regardless of the fluid, so all that is needed is the shear rate as a function of radial position. Checking shear rate using the tables in Appendix 3-1 shows that relevant component of the rate of deformation is

$$\dot{\gamma}_{zr} = \dot{\gamma}_{rz} = \frac{\partial v_z}{\partial r} + \frac{\partial v_r}{\partial z} \tag{7-12}$$

Because there is no flow in the radial direction, the second term on the right-hand side is zero. Thus, we can proceed directly to the shear-rate magnitude. First

$$\dot{\gamma}_{rz} = \frac{\partial v_z}{\partial r} = -2V_0 \left(\frac{r}{R} \right) \frac{1}{R} \tag{7-13}$$

Recalling that

$$\dot{\gamma} = \sqrt{\frac{I_2}{2}} \tag{5-8}$$

where I_2 is the second invariant of the rate-of-deformation tensor, gives the now-obvious result that

$$\dot{\gamma}(r) = 2V_0 \frac{r}{R^2} \tag{7-14}$$

The viscosity at each radial position is then

$$\eta(r) = \frac{\tau(r)}{\dot{\gamma}(r)} = \frac{\tau_W r / R}{2V_0 r / R^2} = \frac{R\tau_w}{2V_0} \tag{ssc} \tag{7-15}$$

The variable r has disappeared, showing that the viscosity is independent of radial position. Thus, the quadratic velocity profile of equation (7-11) is consistent with the steady, developed flow of a Newtonian fluid in a tube.

In the example below, we start from first principles to show that the velocity profile is quadratic.

Example 7-2: *Find the velocity profile for axial flow of a Newtonian fluid in a tube.*

In Example 7-1 we have shown that the quadratic velocity profile of equation 7-11 is consistent with the flow of a Newtonian fluid in a capillary. However, we might worry about the possibility that some other profile might work as well. To counter this concern, we start with equation $\tau = \tau_W \, r/R$, and get the shear rate at each radial position by dividing the shear stress by the viscosity, which is constant. Thus,

$$\dot{\gamma}_{rz} = \tau_W r / R\eta \qquad (7\text{-}16)$$

As shown in Example 7-1, the shear rate is simply the velocity gradient, $-dV_z(r)/dr$, where we have used the total differential because V_z doesn't vary with anything except r. The negative sign can be seen from equation (7-13) above. Equating this with the right-hand side of equation (7-16) yields the differential equation

$$\frac{dV_z(r)}{dr} = -\frac{\tau_w r}{R\eta} \qquad \text{(ssc)} \quad (7\text{-}17)$$

Separating variables gives

$$dV_z(r) = -\frac{\tau_w}{R\eta} r \, dr \qquad \text{(ssc)} \quad (7\text{-}18)$$

which can be integrated from $r = 0$, where $V_z = V_0$ to $r = R$ where $V_z = 0$, if there is no slip. Completing the integration and entering these limits gives

$$0 - V_0 = -\frac{\tau_w}{2R\eta}(R^2 - 0) \qquad \text{(ssc)} \quad (7\text{-}19)$$

Solving for V_0 gives

$$V_0 = \frac{R\tau_w}{2\eta} \qquad \text{(ssc)} \quad (7\text{-}20)$$

Repeating all this but stopping the integration at r gives the analogous

$$V_z(r) - V_0 = -\frac{\tau_w}{2R\eta}(r^2 - 0) \qquad \text{(ssc)} \quad (7\text{-}21)$$

Substituting for V_0 yields the final result

$$V_z(r) = \frac{R\tau_w}{2\eta} - \frac{\tau_w r^2}{2R\eta} = \frac{R\tau_w}{2\eta}\left(1 - \frac{r^2}{R^2}\right) \qquad \text{(ssc) (7-22)}$$

Perhaps the above exercise was a bit more detailed than needed, but it does set the stage for the development of the velocity-profile correction, which allows for the measurement of non-Newtonian viscosity in capillary flow.

One aspect of the Poiseuille law remains to be shown and that is the relationship between flow rate Q and wall shear rate $\dot{\gamma}_W$, as given in equation (7-9). The reason we need this information is that the velocity profile is difficult to get relative to the total flow rate. The method for doing this is illustrated in Example 7-3.

Example 7-3: *Given the maximum (centerline) velocity V_0, find the flow rate Q for a Newtonian fluid flowing in a capillary.*

The starting point is the Newtonian velocity profile shown in equation (7-11), and reproduced below

$$\frac{V_z(r)}{V_0} = 1 - \left(\frac{r}{R}\right)^2 \qquad (7\text{-}11)$$

Imagine that the flow in the capillary as a telescoping set of thin-walled, concentric tubes of fluid, each moving at a velocity $V_z(r)$ down the capillary. The flow rate contribution of each of these will be the cross-sectional area of thin tube times its velocity. The total flow rate will be a sum of all of these. If the telescoping tubes are differentially thick, then we can write

$$Q = \int_0^R V_z(r)2\pi r\,dr \qquad (7\text{-}23)$$

This relationship is generally true, i.e., it doesn't depend on the nature of the fluid. However, the velocity profile does.

Plugging in the velocity profile from equation (7-11) gives

$$Q = \int_0^R V_0\left(1 - \frac{r^2}{R^2}\right)2\pi r\,dr = 2\pi V_0 \int_0^R \left(1 - \frac{r^2}{R^2}\right)r\,dr \qquad (7\text{-}24)$$

We are now ready to integrate by splitting the integral into the two terms

$$Q = 2\pi V_0 \left[\int_0^R r\,dr - \frac{1}{R^2} \int_0^R r^3\,dr \right] \tag{7-25}$$

and integrating each to give

$$Q = 2\pi V_0 \left[\frac{r^2}{2} \Big|_0^R - \frac{1}{R^2} \frac{r^4}{4} \Big|_0^R \right] = 2\pi V_0 \left[\frac{R^2}{2} - \frac{R^2}{4} \right] \tag{7-26}$$

Combining these terms gives the result

$$Q = \pi R^2 V_0 / 2 \tag{7-27}$$

At this point we need to refer back to the relationship between V_0 and $\dot{\gamma}_W$ shown in equation (7-14). At the wall the result is

$$\dot{\gamma}(R) \equiv \dot{\gamma}_W = \frac{2V_0}{R} \tag{7-28}$$

Substituting in V_0 from equation (7-27) gives the final result

$$\dot{\gamma}_W = \frac{2}{R} \left(\frac{2Q}{\pi R^2} \right) = \frac{4Q}{\pi R^3} \tag{7-29}$$

This is essentially what Poiseuille went through many years ago.

The derivation for non-Newtonian fluids is a bit more complex, and is left to Appendix 7-2. The result is the Weissenberg-Rabinowitsch-Mooney equation,[§§§] which we will refer to as the WRM equation. The WRM equation is

$$\dot{\gamma}_W = \frac{4Q}{\pi R^3} \left[\frac{3}{4} + \frac{1}{4} \frac{d\ln\left(4Q/\pi R^3\right)}{d\ln\tau_W} \right] \tag{7-30}$$

Other than "no slip at the wall," this result should apply to the flow of all homogeneous fluids. Note that if the differential has a value of 1.0, we recover the Newtonian result for $\dot{\gamma}_W$. Also note that this term is equivalent to

---

[§§§] See, e.g., Reference 48. One also sees this equation called simply the Rabinowitsch equation (correction) and the Weissenberg-Rabinowitsch equation.

$d\ln Q/d\ln P$, where Q is the flow rate, and P is the corrected pressure drop through the capillary.

Now let's examine the complications of the WRM equation. First of all, because of the derivative, its application requires a set of flow data to obtain even one value of the wall shear rate. Then, as the observations of flow rate and pressure are generally discrete, we need to figure out how to take the derivative of the flow curve. Of course, we can plot the data using log scales and find the slope with a ruler, but this seems a bit quaint these days. Regardless of how we find the derivative, the results near the ends of the data set will be less reliable than those near the center. Appendix 7-1 contains some general advice for finding derivatives, as well as hints for using spread sheets for this operation. As will be seen, this is a common problem in the rheometry of non-Newtonian materials.

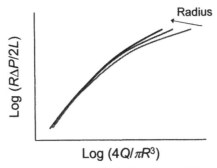

Figure 7-5. Schematic illustration of an early indication of the presence of slip in capillary flow at the higher stress levels. Uncorrected pressure drop can also be used because, at constant $4Q/\pi R^3$, the wall shear stresses should be the same if the capillaries have the same L/D.

After successfully applying the Bagley end correction and the WRM velocity profile correction, one may or may not start to worry about slip at the wall, especially if the stresses are high. General signs of flow problems include pressure oscillations, irregular shape of the extrudate, and sudden drop in the viscosity as the shear rate is increased. A more certain diagnosis is to examine shear stress vs. $4Q/\pi R^3$ curves for capillaries that have the same aspect ratio L/D, but different diameter; the data from these two should fall close together. If, however, the data gathered with the small-diameter capillary give higher values of $4Q/\pi R^3$ at equal shear stress, then slip is a strong possibility. Slip is a near certainty if the match is good at low shear stress but the discrepancy appears at high shear stress. This is illustrated schematically in Figure 7-5.

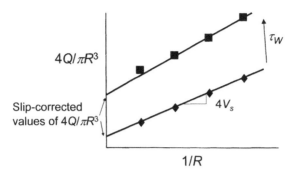

Figure 7-6. Illustration of a Mooney plot based on four capillaries of different radii. The slight curvature in the plot at the higher stress level could be the result of viscous heating, which is a bigger problem in the larger diameter capillaries.

If the slip velocity V_s depends only on wall shear stress τ_W, then some progress can be made in using the flow data gathered under slip conditions to find the viscosity of the fluid. The procedure is to create a Mooney plot, which is illustrated schematically in Figure 7-6. The derivation of the Mooney equation

$$\frac{4Q}{\pi R^3} = \frac{4Q}{\pi R^3}\bigg|_{corr} + \frac{4V_s}{R} \quad (\tau_W = \text{const.}) \tag{7-31}$$

is given in Appendix 7-3. The Mooney plot is thus the observed values of $4Q/\pi R^3$ vs. $1/R$ at constant τ_W.

Does the slip velocity depend only on shear stress? Certainly, the data must be very plentiful and very precise to see other effects. One of these signs is a dependence of slip velocity on capillary size, which is indicated by curvature in the Mooney plot (Figure 7-6). One can also argue that there might be a pressure dependence of slip. Thus higher L/D capillaries run at the same shear stress might show lower slip velocities than shorter capillaries. This effect may be particularly noticeable if the polymer contains dissolve gas, which can lead to slip at the wall.

3. Sandwich or sliding-plate geometries

Viscosity is often defined in terms of shearing a sample between two plates and measuring the measuring the force F_x required to drag the moveable plate along at velocity V_x. Given the sample area A (shape doesn't matter) and plate gap H, the viscosity is given by

$$\eta = \frac{F_x}{A} \frac{1}{V_x/H} \tag{7-32}$$

The factor F_x/A on the right is easily recognized as the shear stress, whereas the V_x/H is the shear rate. Couldn't be much simpler. However, is this actually done, in practice?

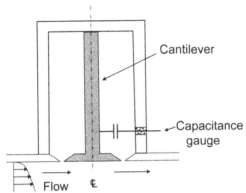

Figure 7-7. Schematic of a direct shear-stress transducer. The shear stress generated by the flow causes a slight movement of the cantilever, which is detected by a sensitive capacitance gauge. The polymer may flow slightly past the knife edges into the housing, but this does not affect the steady-state position of the cantilever.

The answer is a qualified "yes" in that most instruments of this type are custom made. One, the Sliding Plate Rheometer,[3] has been offered commercially as an accessory for the MTS servo-hydraulic load frame. A feature of this device is a direct shear-stress transducer (Figure 7-7), which removes the need to account for friction in the mechanism holding the sliding plate. An excellent example of a custom-made rheometer is described in papers written by Mather and his associates.[4] The big advantage of this geometry, which has been exploited in both of these designs, is the ability to examine the flow optically. Another advantage is the ability to apply uniform electric fields to the sample by using conductive plates.[5] An alternative design is to sandwich the sliding plate between two stationary surfaces and with two samples, one on each side. This has the advantage that the center plate does not need to be supported by bearings; thus, it is free of friction.

There are disadvantages to the sliding plate geometry. The most worrisome is the limited strain that can be achieved without running out of travel. If the travel is limited to 10 cm and the gap is 1 mm, the maximum shear strain is only 100. While this level of strain may be adequate for transient measurements, it may not be enough to allow the sample to reach steady flow conditions. Another issue is the necessity of maintaining a very constant gap, as any deflection of the plates will change the normal forces drastically. The reason for this is the sample is very wide compared to its thickness, so if the plates move together or apart, the deformation is essentially hydrostatic. This will have relatively little effect on the shear stresses.

4. Rotational geometries—the parallel-plate fixture

The obvious fix to the low-strain issue with the sliding-plate geometry is to rotate the plates instead of sliding them in a straight line. Thus we move forward to geometries that can be accommodated on a rotational machine. There are three that maintain their popularity: parallel plates, or parallel disks; cone and plate; and Couette, or cup and bob. Each has advantages and disadvantages.

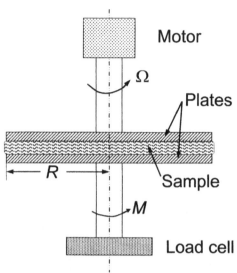

Figure 7-8. Schematic of the parallel-plate geometry. Either the top or bottom fixture may be rotated, and the torsional load cell may be attached to either the rotating or fixed plate.

We will examine first the parallel-plate geometry. The reasons for using this geometry are both numerous and compelling. Some are listed below:

- Easy to fill and gap the fixtures
- Making custom fixtures, including disposable plates, is straightforward
- Samples can be stamped from a sheet and loaded into the fixtures without trapping air
- Can run crosslinked materials in the form of disks
- Has been used for multi-dimensional optical measurements
- Has been used for application of magnetic and electric fields to the sample. The fixed thickness of the sample means electric fields are uniform.

The principal disadvantage is the very evident fact that the flow is not uniform. The reason for this is that the tangential velocity increases from the center on out to the rim, whereas the spacing between the plates is constant. Thus

$$\dot{\gamma}(r) = \frac{r\Omega}{H} \qquad (7\text{-}33)$$

where Ω is the rotational rate in radians per second (rad/s) and H is the gap between the plates. The shear stress also varies with radius, but according to the viscosity function for the material in the gap. On application of a rotational rate Ω (rad/s), one observes the torque, M. To understand the torque calculation, imagine the sample as a series of rings, each exerting a moment on the plate according to its radial location and the stress on its surface. Thus

$$M = \int_0^R \tau(r)\, r\, 2\pi r dr \qquad (7\text{-}34)$$

where the area of the ring is the $2\pi r dr$ part of the integrand. The value of $\tau(r)$ is not known, but we need to get at it to find the viscosity.[****]

The situation presented in equation (7-34) is reminiscent of that for the capillary, and indeed there are very similar except that we measure torque instead of flow rate, and the shear rate is known and linear with radius instead of the shear stress, as with the capillary. It is no surprise, then, that the process for solving the problem is very similar. The result for the shear stress, τ_R, at the rim (radius $r = R$) is

$$\tau_R = \frac{2M}{\pi R^3}\left[\frac{3}{4} + \frac{1}{4}\frac{d\ln M}{d\ln \dot{\gamma}_R}\right] \qquad (7\text{-}35)$$

where M is the torque and $\dot{\gamma}_R$ is the shear rate at the rim. It is also no surprise that the term $2M/\pi R^3$ is the shear stress for a Newtonian fluid. The viscosity at the rim is just the shear stress divided by the rim shear rate. For a Newtonian fluid, the viscosity is

$$\eta = \frac{2MH}{\pi R^4 \Omega} = \frac{2M}{\pi R^3 \dot{\gamma}_R} \qquad (7\text{-}36)$$

Note the sensitivity of the viscosity to the radius of the plate, R. Thus accurate measurements require very careful alignment and filling of the fixtures.

As might be expected, several single-point methods have been suggested for those in a hurry.[49] One that is easy to remember is to calculate the viscosity,

---

[****] We are using the symbol τ to represent the relevant component of the extra stress tensor. The actual component in this case is $\tau_{z\theta}$. Also, we have not attached a sign to the moment M, which is properly a vector.

using equation (7-36) as if the fluid were Newtonian, and assign this viscosity to 3/4ths of the rim shear rate. We will call this the "3/4ths rule."

5. Rotational geometries—cone-and-plate fixtures

The cone-and-plate fixtures simplify matters considerably, because the flow is homogeneous, or at least it is fairly homogeneous if the cone angle α is small. The analysis of the flow in the gap is properly done using spherical coordinates, considering the surfaces of the cone and plate to be a constant-ϕ surfaces and the outside of the melt to be a constant-r surface. But a simpler approach is to use cylindrical coordinates, as with the parallel plates. The result is the same, as we expect. Thus, the (positive) torque is again given by

$$M = \int_0^R r\tau \, 2\pi r \, dr \qquad \text{(ssc)} \quad (7\text{-}37)$$

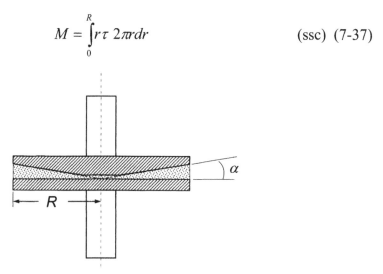

Figure 7-9. Diagram of cone-and-plate fixtures. The angle has been greatly exaggerated in this drawing; usually it is less than 0.1 rad.

As the stress is constant, the integral is easy to evaluate. Solving for stress gives

$$\tau = \frac{3M}{2\pi R^3} \qquad \text{(ssc)} \quad (7\text{-}38)$$

The viscosity is obtained by dividing through by the shear rate, which is evaluated as for the parallel plate by dividing the tangential velocity $r\Omega$ by the gap, which is αr. The resulting expression for the viscosity is consequently

$$\eta = \frac{3M\alpha}{2\pi R^3 \Omega} \qquad (7\text{-}39)$$

at a shear rate of

$$\dot{\gamma} = \frac{\Omega}{\alpha} \tag{7-40}$$

Note that the dependence of torque on radius is somewhat less than for the parallel-plate fixture. This is an advantage; filling and trimming are not quite as critical to the precision and accuracy of the result. However, there are some unique opportunities for other problems. For one, the cone and plate fixture must be spaced such that cone tip hits the plate. This doesn't actually happen, as the cone is truncated by a slight amount. Also, the cone angle must be known, and can be difficult to measure accurately.[6]

6. Rotational geometries—the Couette fixtures

Couette fixtures, also known as the "cup and bob" or "coaxial cylinder" viscometer, has a very long history involving many variations. It consists indeed of a cylindrical cup into which is immersed a cylindrical bob that is slightly smaller than the inside diameter of the cup (Figure 7-10). The test material is contained in the annular space between the two cylinders, and along the bottom space. It is readily apparent that, compared to the parallel plate fixture, more fluid will be required, but much more torque will be developed. Thus the Couette geometry is the right choice for measuring the viscosity of dilute polymer solutions or suspensions. With suspensions, it is the least susceptible of the geometries discussed to errors due to sedimentation. Unfortunately, Couette fixtures are difficult to load with viscous samples.

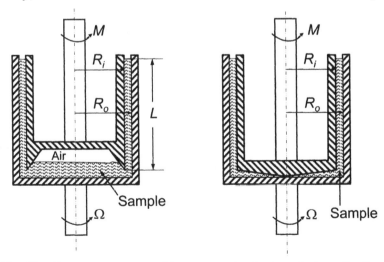

Figure 7-10. The Couette geometry with two popular bottom designs for the bob. This drawing shows the cup being rotated, with the torque measured by the stationary bob.

A variety of designs are used for the bottom of the bob. For non-Newtonian fluids, we would like to have either no torque at all from the bottom surface, or some geometry that can be analyzed. The two most common treatments are shown in Figure 7-10.

The fact that there is a finite gap between the bob and cup surfaces presents some problems in analyzing the torque M vs. rotational speed Ω to get the non-Newtonian viscosity function. The reason for this is that unlike the parallel-plate geometry, we cannot write down a simple equation for M in terms of Ω. Given a value of Ω, we have no idea what each concentric layer of fluid is doing, except that the sum of the shear rates in each layer, times its thickness, must equal the overall tangential velocity of the moving fixture. If end effects are ignored, we do know the stress in each layer because the torque on the inner cylinder must balance that on the outer. The relationship for the shear stress is thus

$$\tau(r) = \frac{M}{2\pi r^2 L} \qquad \text{(ssc)} \quad (7\text{-}41a)$$

where L is the height of the bob and M is positive.

If the gap is very narrow, then uniformity across the gap might be assumed. In this case, the estimate of the average shear stress is given safely by the equation

$$\tau = \frac{M}{2\pi \langle R \rangle^2 L} \qquad \text{(ssc)} \quad (7\text{-}41b)$$

In this equation, $\langle R \rangle$ is the average radius, i.e., $(R_i + R_o)/2$, where R_i and R_o are the inner and outer radii, respectively. The corresponding narrow-gap estimate for the shear rate is

$$\dot{\gamma} = \frac{\langle R \rangle \Omega}{R_o - R_i} \qquad (7\text{-}42)$$

This is equivalent to dividing the tangential velocity by the gap. If the gap is not so small, then one needs to use a better approximation, which unfortunately involves the derivative

$$n = \frac{d\ln M}{d\ln \Omega} \qquad (7\text{-}43)$$

with all its problems. Then, one must pick where the shear rate is wanted—at the bob surface, or near the cup. This choice influences only the shear stress that is used. Choosing the inner radius, we have

$$\dot{\gamma}_i = \frac{2\Omega}{n\left[1 - \left(\dfrac{R_i}{R_o}\right)^{2/n}\right]} \tag{7-44}$$

Correspondingly, we pick the inner radius to calculate the shear stress

$$\tau_i = \frac{M}{2\pi R_i^2 L} \qquad \text{(ssc)} \tag{7-45}$$

The viscosity *at the inner radius* is simply the ratio $\tau_i / \dot{\gamma}_i$ using these two results.

Suffice it to say, most modern rheometers have these or even more exact approximations built into their analysis software. However, it is always prudent to record the raw M-Ω data to make sure everything is working properly.

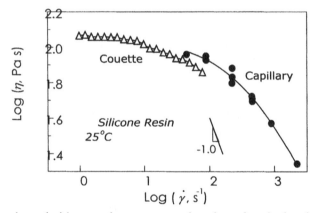

Figure 7-11. Flow irregularities can show up as a rather sharp drop in the viscosity as the shear rate increases for this Couette data, leading to a ~20% mismatch with the capillary data. Shown is data for a silicone resin with a molecular weight of around 100 kDa. (Unpublished data, S.-P. Sun.)

7. Problems at high shear rates

Given a stable sample, measurements at low shear rates are usually plagued only by lack of signal, that is, the torques or pressure drops are too low to be measured properly. At high rates, however, problems do arise, regardless of the geometry of the fixtures.

One such problem has been shown previously as Figure 4-2. This rolling fracture in a parallel-plate fixture is presumably the result of excessively high elastic stress. Incipient problems of this sort usually are evidenced by a rapidly dropping viscosity. An example of this behavior is shown in Figure 7-11. Other reasons for this behavior include viscous heating and slippage. Distinguishing between these can be quite involved.   Figure 7-12 shows a common flow irregularity found with pressure-driven flows.

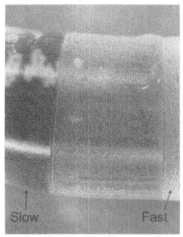

Figure 7-12. Surface "shark skin" roughness on a polybutadiene extrudate from a 1-cm-diameter tube, illustrating the presence of a flow irregularity that increases with extrusion rate. In this case, the source of the roughness may be the exit flow, as the strand did not show any roughness when the capillary was quenched and taken apart.

Flow irregularities in a sliding-plate rheometer have been reported by Koran and Dealy.[3] One is the expected flow resulting from a N_1 value that exceeds $-N_2 = \sigma_{33} - \sigma_{22}$ (ssc). This imbalance causes shrinkage in the flow direction and a concomitant expansion in transverse direction. More worrisome are the observations of local irregularities initiating at the plate surface. Different patterns can form depending upon the polymer, the nature of the plates, and the stress level. Some of these may have their origin at the edge of the sample. Figure 7-13 shows data from a sliding-plate rheometer showing the importance of the material used for the plates.

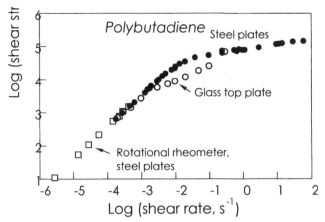

Figure 7-13. Evidence of slip using the sliding plate rheometer, and the dependence of slip on materials used to construct the plates.   (Adapted from Koran and Dealy.[3] Copyright © 1999 by The Society of Rheology, Inc. All rights reserved.)

In capillary flow, the picture is complicated by the presence of an entry flow feeding the capillary, and the exit. Another complication is the variation of wall normal stress, σ_{rr}, along the capillary surface. Again, flow irregularities of all sorts have been reported, with dependencies on the polymer, the wall material, and the stress level. Signs of flow irregularities include rough surfaces of the extrudate (Figure 7-12); pressure oscillations; and, occasionally, direct observation of velocity variations in the channel.

B. NORMAL STRESSES FROM SHEARING FLOWS

Of interest in the shear rheometry of polymers are the normal stresses developed by the polymer melt or solution, as these stresses are the direct result of deformation of polymer chains. While normal stresses are easily detected, their precise measurement can be challenging.

Examination of the literature will reveal many clever techniques for measuring normal stresses in both drag and pressure-driven flows. Many of these are included in Table 7-1. In spite of years of work on this problem, the gold-standard technique for measuring N_1 remains the direct use of normal force, F, using the cone-and-plate fixtures in a rotational rheometer. The relevant equation is

$$N_1 = \frac{2F}{\pi R^2}$$

(ssc) (7-46)

(Note that F is taken as positive if N_1 tends to push the plates apart.) We will examine this flow in detail, but first, we need to look at the somewhat similar parallel-plate fixture to see what not to do.

Table 7-1. Partial list of methods used to assess the magnitude of normal stresses in pressure-driven shearing flows.

Name	Geometry	Direct or indirect?	Comments	Ref.
Die swell	Capillary	Indirect	Analyses have been presented.	7
Exit pressure	Slit, capillary	Indirect	Analyses have been presented.	8
Birefringence	Slit	Indirect	Stress-optical coefficient must be known.	9, 10
Thrust	Capillary	Direct	High N_1 should reduce the jet thrust on a capillary over that expected from usual momentum balance.	11
Pitot tube	Duct	Direct [a]	Similar to thrust, but applies only to large ducts.	12
Radial pressure	Axial annular flow	Direct [a]	Gives N_2. Flush transducers are needed on curved surfaces.	13
Hole-pressure error.	Slit die, equipped with surface slots	Direct [a]	Has been commercialized. Pressure difference in transverse slot vs. flat surface gives N_1. Very sensitive transducers needed.	14

[a] "Direct" means that an analysis is available for determining the normal-stress values; however, most of these devices should be calibrated or checked against data taken using a cone-and-plate rheometer.

1. Normal stresses from the parallel-plate rheometer

As explained previously, the parallel-plate fixtures are often preferred for measurements in the linear range, and are workable, with due care, for measuring a portion of the non-Newtonian viscosity function, for example, up to shear stresses of around 5 kPa for poly(dimethyl siloxane).[15] Can normal stresses be determined by measuring the force that is needed to maintain the plate gap? Yes, but only to some extent. Let's take a look.

The "development" below might be considered to be a natural extension of the derivation of the relationship for shear stress expressed in equation (7-35); however, it is wrong. The reason it is wrong does, however, illustrate nicely the complications of geometries with curved coordinate axes and the necessity of using the equations of motion for these *curvilinear* systems.

The first thought would be that the total normal force should be simply the sum of the contributions of thin circular strips of fluid. Thus,

$$F = -\int_0^R \tau_{zz}(r)\, 2\pi r\, dr$$

By substituting $\dot{\gamma}(r)$ for the variable r, and doing the usual differentiation of the integral to isolate $\tau_{zz}(R)$ gives the "answer"

$$\tau_{zz}(R) = \frac{2F}{\pi R^2}\left[\frac{1}{2} + \frac{1}{4}\frac{d\ln F}{d\ln \dot{\gamma}_R}\right]$$

While this equation certainly looks good, it's wrong. Why? First of, we have forgotten that it is the total stress σ_{zz}, not τ_{zz}, that is being sensed by the force on the plate. But can we argue that this is a drag shear flow, so pressure should be essentially atmospheric. Again, this sounds good, but neglects the fact that the normal stress acts along flow lines that are curved. This curvature means that each circular strip acts like a rubber band and squeezes the material at lower radius. Thus, the pressure builds up at the lower values of radius. We are really much better off if we simply start with the proper equations of motion and have faith that these equations are correct.

So, where does the correct path start? It starts with a free-body force balance on the upper plate.[tttt] The balance between thrust F of the load cell on the plate vs. thrust exerted by the fluid yields the equation

$$F = -\int_0^R \sigma_{zz}(r)\, 2\pi r\, dr \qquad \text{(ssc) (7-47)}^{\ddagger\ddagger\ddagger\ddagger}$$

[tttt] The balance could be done on either plate, but we will assume the upper plate is connected to the normal-force load cell, with alignment maintained by an air bearing. This design prevents the air bearing from being damaged by spilled sample. Some older models did have the load cell in the bottom, e.g., the System 4 rheometer.

Now the equations of motion should be examined to relate σ_{zz} to the kinematics of the flow, which we know. Which equation do we use? As we need to know the radial dependence of σ_{zz}, it is reasonable to look for terms like $\partial\sigma_{zz}/\partial r$, which suggests the r component of the equation of motion. This equation can be found in Appendix 2-1. We notice no $\partial\sigma_{zz}/\partial r$ term in this equation, although the $\partial p/\partial r$ term could be converted using the relationship $\sigma_{zz} = \tau_{zz} - p$ (ssc). On the left-hand side,

$$\rho\left(\frac{\partial v_r}{\partial t} + v_r\frac{\partial v_r}{\partial r} + \frac{v_\theta}{r}\frac{\partial v_r}{\partial\theta} - \frac{v_\theta^2}{r} + v_z\frac{\partial v_r}{\partial z}\right) = \text{LHS} \quad \text{(ssc)} \quad (7\text{-}48)$$

Assuming only θ-direction flow, all the terms are zero except the "centrifugal force" term $\rho v_\theta^2/r$, which will be small at moderate rotation rates. For the right-hand side,

$$-\frac{\partial p}{\partial r} + \left(\frac{1}{r}\frac{\partial}{\partial r}(r\tau_{rr}) - \frac{\tau_{\theta\theta}}{r} + \frac{1}{r}\frac{\partial\tau_{r\theta}}{\partial\theta} + \frac{\partial\tau_{rz}}{\partial z}\right) + \rho g_r \quad \text{(ssc)} \quad (7\text{-}49)$$

Right away, we can see that the body force (gravity) term ρg_r is zero if the r direction is horizontal. Very convenient. Also, changes in the z and θ directions are zero (hopefully), so the terms $\partial\tau_{rz}/\partial z$ and $\partial\tau_{r\theta}/\partial\theta$ are out of the picture. These simplifications, plus the substitution for pressure, plus some combining of terms, gives

$$\frac{\partial\sigma_{zz}}{\partial r} = \frac{N_1}{r} + \frac{\partial N_2}{\partial r} \quad \text{(ssc)} \quad (7\text{-}50)$$

While fairly simple looking, this equation tells us two sad facts. One is that both N_1 and N_2 will contribute to the normal force. The other is that a single measurement of normal force will not give a single measure of the normal-stress differences—several measurements will be needed, along with differentiation. The latter worry has been already encountered with the evaluation of non-Newtonian viscosity using the parallel-plate fixtures. However, because normal forces will be measurable only over a rather limited range of rotation rates due to their high sensitivity to shear rate, we can expect to obtain reliable results over an even smaller range.

---

‡‡‡‡ The negative sign for (ssc) gives a positive force against the plate if the normal stress is negative, i.e., compressive. This is similar to stretching a rubber band and then attempting to hold it stretched by pressing in on its sides.

While not as obvious, to get the normal force F we need to integrate over radius (or $\dot{\gamma}$) to get σ_{zz} and then integrate this result again over radius according to equation (7-47) to get the normal force. The final result of all this is an equation for a combination of normal stresses, thus

$$N_1 - N_2 = \frac{F}{\pi R^2}\left[2 + \frac{d\ln F}{d\ln\dot{\gamma}_R}\right] \qquad \text{(ssc) (7-51)}$$

The normal stresses are, of course, those corresponding to the shear rate at the rim of the fixture, $\dot{\gamma}_R$. Note that if the trace of τ is zero, $N_1 - N_2 = -3\tau_{22}$, a direct determination of an extra-stress component! As the 2 direction is the z direction, we were somewhat right in looking at τ_{zz}, but the details make a big difference.

An obvious application of equation (7-51) is for finding N_2 if N_1 is known from cone-and-plate measurements. However, the extraction of N_2 involves the difference of two values that both have error and are very close. Thus N_2 is not only small, but the variance[§§§§] of this small quantity will be the sum of the variances in each of the two methods. The percent error will thus be huge.

A better way to find N_2 at low to moderate shear rates is to examine the stress profile in the cone-and-plate fixtures. This will be discussed in the next section.

2. Normal stresses using a cone-and-plate rheometer

As already mentioned in discussing the parallel-plate rheometer, the cone-and-plate geometry gives the first normal-stress difference N_1 directly, according to the equation

$$N_1 = \frac{2F}{\pi R^2} \qquad \text{(ssc) (7-46)}$$

where F is the normal force. While not strictly proper, a positive value of F means that the force is applied inward toward the sample, as depicted in Figure 6-1.

[§§§§] On adding or subtracting two numbers, the variances of the two numbers add. Variance is the square of the standard error of the mean of the set of observations used to get the result. The standard error of the mean is simply the standard deviation of the observations divided by the square root of the number of observations.

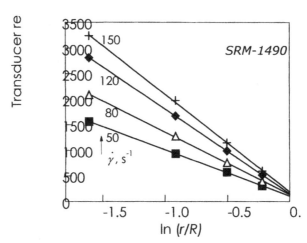

Figure 7-14. Example of pressure measurements in the cone-and-plate rheometer using four flush pressure transducers. The transducer readings are "gauge" pressure, equivalent to $-\sigma_{22}$, where σ_{22} is negative (ssc). SRM-1490 is a 10% PIB solution in normal hexadecane (cetane). (Adapted from Baek and Magda.[16] Copyright © 2003 by The Society of Rheology, Inc. All rights reserved.)

As for the second normal-stress difference N_2, one either needs to replace the cone with a flat plate and use equation (7-51) or use information about the pressure profile. To see how the latter technique works, we integrate the radial component of the equation of motion in spherical coordinates (Appendix 7-4), giving the relationship

$$P_G = -\sigma_{22} - P_A = -(N_1 + 2N_2)\ln(r/R) - N_2 \qquad \text{(ssc) (7-52)}$$

where P_G is the gauge reading from the transducers (assuming there are referenced to the atmosphere and calibrated with positive pressure), P_A is atmospheric pressure, and σ_{22} is the total normal stress that is applied to the sample by the plates. σ_{22} is compressive in nature, thus it is negative for (ssc). Note that σ_{22} is referenced to zero, which is why the P_A is included; in the absence of deformation, $\sigma_{22} = -P_A$. If all of this is confusing, just pretend that atmospheric pressure is zero. Remember, though, it is important to know at what pressure the transducer have been zeroed: zero absolute (vacuum), or atmospheric.

The intriguing aspect of equation (7-52) is that N_2 can be accessed from a plot of the transducer readings vs. $\ln(r/R)$. Such a plot is shown in Figure 7-14 for a polymer solution. The intercept at $r = R$ is N_2. In addition, the slope, $-(N_1 + 2N_2)$, when combined with the intercept, provides a check on the value of N_1 derived from the normal force F. Note that for the polymer solution depicted in Figure 7-14, the intercept is positive, meaning that N_2 must be negative.

Of course, one has to figure out how to measure the "pressure" as a function of radius to get such a plot. In principle, flush pressure transducers like those depicted in Figure 2-3 can be mounted in the plate at various radial positions; they do not need to be lined up. The more transducers, the more precise will be the answer. A more modern method is to use solid-state sensors distributed in an array[16] and individually addressable; this gives a huge data set that will give very high precision.***** Accuracy, as always, depends upon careful calibration and compensation for drift and temperature changes.

C. EXTENSIONAL RHEOLOGY

1. Uniaxial extension

The saga of experimental extensional rheology is long and intriguing, but somewhat messy. By long, we refer to an analysis from 1906 by Trouton,[17] who demonstrated that $\eta_E = 3\eta$ for Newtonian fluids. The real interest, however, started in the late 1960's and early 1970's with the advent of better measuring techniques that for the first time could subject a sample to either a fixed tensile stress (not a fixed force) or a fixed elongation rate. Notable early designs were those of Cogswell, but particularly those designed by Meissner and associates.

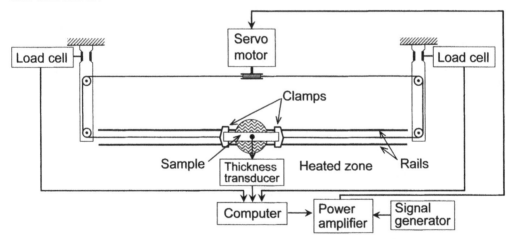

Figure 7-15. Schematic of device for stretching samples at fixed rate or fixed stress. With this geometry, the ends of the sample do not have to be displaced in an exponential fashion because the sample hardens as it moves out of the heated zone. The key component is the thickness transducer. [From Shaw and Lin,[19] *Current Topics in Polymer Science Vol. II*, (S. Inoue, L. A. Utracki and R. M. Ottenbrite, eds.), Hanser, Munich, 1987, with permission.]

***** Such a device was commercialized by Rheosense, Inc.

Why does one need to run the experiment at fixed rate $\dot{\varepsilon}$ or fixed stress, σ_T? The hope is, of course, that with these conditions, the observed stress or rate, respectively, will reach a steady value, leading to an extensional viscosity η_E given by

$$\eta_E = \frac{\sigma_T}{\dot{\varepsilon}} \qquad \text{(ssc) (7-53)}$$

In spite of diligent effort and some success, most investigators have found that it is virtually impossible to reach a steady state except at very low extension rates, where the response is essentially linear. At higher rates, the sample fails, either by thinning or by a more solid-like failure similar to a rubber band. Thus most of the attention was focused on the details of the transient function $\eta_E^+(t; \dot{\varepsilon})$, defined as

$$\eta_E^+(t; \dot{\varepsilon}) = \frac{\sigma_T^+(t; \dot{\varepsilon})}{\dot{\varepsilon}} \qquad \text{(ssc) (7-54)}$$

An equivalent could be defined for the constant-stress transient, but such experiments are more difficult to run and thus less popular.

The outcome of the experiment is usually simply a graph of $\eta_E^+(t; \dot{\varepsilon})$ vs. time t at various extension rates. An example is shown in Figure 7-16. Another outcome is the use of the data to build or test constitutive equations. The parameters for these equations might be derived by fitting the data, which means in principle that the result could be used to predict what will happen in shear. One of the challenges of rheology is that this seldom works very well.

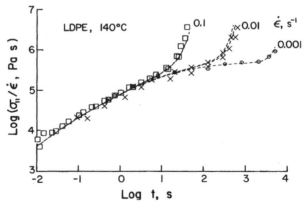

Figure 7-16. Uniaxial extension transient of a LDPE melt measured using the apparatus shown in Figure 7-15. (From Shaw and Lin,[19] *Current Topics in Polymer Science Vol. II*, (S. Inoue, L. A. Utracki and R. M. Ottenbrite, eds.), Hanser, Munich, 1987, with permission.)

If one is anxious to fit models to transient data, it might be appropriate to ask if it is absolutely necessary to maintain constant extension rate. One problem with transient rheometry is that it is difficult for real machinery to change instantly from an initial extension rate of zero to a high and constant extension rate. In principle, this would require infinite acceleration with the associate infinite forces or torques. Would it be possible and appropriate to start the experiment with a more gradual acceleration, and then convert this data to constant rate or constant stress using a model?

This question was addressed many years ago by White et al.,[18] who developed an extensional-flow rheometer that used a weight to stretch the sample. A somewhat more recent examination of the problem was described[19] wherein the sample was stretched by the device shown in Figure 7-15, but by opening the servo loop to apply nearly constant force to the sample. Various integral constitutive equations were then used to convert the data to constant rate, and the resulting profiles were virtually independent of the constitutive equation used for the conversion. The converted results were also very similar to the controlled-rate results.

For melts, the Sentmanat extensional rheometer (SER)[20] for existing rheometers has become very popular, because it uses the temperature and rate control of the rheometer, produces constant-rate and reasonably uniform deformations, and is relatively easy to use compared to many of the extensional devices described in the older literature. As the sample is not supported while warming up, it is best used for relatively viscous melts.

For measuring the extensional viscosity of solutions, many clever devices have been developed. A recent innovation is what can be viewed as a tack test, wherein the sample is stretched into a thread between two small disks. One can either keep stretching until failure; or, at a certain point, the stretching can be stopped abruptly. In the latter case, capillary forces cause the fluid in the thread to keep stretching—rapidly for low-viscosity fluids, slowly for higher viscosity. The fluid from the thread is added to the pools at the top and bottom of the sample. Force measurements and movies of the thread (to get the diameter) provide the data needed to derive the extensional viscosity.

Aside from the problems of establishing a constant axial velocity gradient and measuring small forces, there is the more fundamental issue of uniformity in the gauge region of the sample. Images gathered during stretching can provide a check on thickness uniformity along the axis. However, something more must be done to confirm that there is a negligible radial gradient of the axial velocity, that is, $\partial V_z/\partial r \approx 0$. This can be done using particle position tracking, or by examining the length of traces produced by a pulse of light. A less invasive method has been described by Paul et al.,[21] who examined the fluorescence from a thin plane of excited species uniformly dispersed in the melt. A more direct way of controlling the uniformity is to encase the sample with another polymer that is very stable when stretched.[22] In principle, the melt could even be loaded into a thin rubber tube.

Some of the other devices used to measure uniaxial extensional viscosity are listed in Table 7-2.

Table 7-2. References to additional extensional techniques

Name[a]	Description	Typical samples	Ref.[c]
Inverse (ductless) siphon	A thread is pulled up from a solution using a tube connected to a vacuum source. Force on tube is measured.	Solutions	23
Spinning	A stream from a tube or die is fed onto a rotating drum. Force on tube is measured as speed is changed.	Solutions and melts	24
Stagnation flow	Opposed jets feed solution against each other. Resulting flow is an outward-spreading sheet of fluid. Force pushing the two jets apart is measured.	Solutions	25
Tensile testing	Uses a tensile frame (usually servo hydraulic) to apply controlled deformation to a rod or strip of polymer. Many examples.	Melts	26
Weight drop[d]	Uses a weight or even the sample mass to apply force to the sample. Image analysis gives both the force and the stretching rates vs. time.	Melts	18
Converging die flow	Melt is fed through a conical or trumpet-shaped die that forces the melt to elongate. The stress is either deduced from the pressure drop, or by using birefringence.	Melts	27,28
Drum windup	A transient method whereby a polymer rod, fixed at one end, is attached by the other end to a cylinder, which is rotated. Two cylinders, one on each end, is an alternative design.	Melts	29
Pinch rolls	Pinch rolls, usually with roughened surfaces, are used on both ends to stretch the sample.	Melts	30

Name [a]	Description	Typical samples	Ref. [c]
Bubble inflation	A gas (usually [b]) is fed under a sheet that is confined at the edges to give either a hemispherical or domed bubble. The gas pressure and bubble shape lead to the stresses and rates.	Melts	31,32
Lubricated squeezing flow	A transient biaxial or planar extensional flow created by moving disks or plates that are lubricated to promote slip	Melts	34

[a] Some of these are merely descriptive and may not be universal.

[b] A fluid-driven commercial version was briefly available.

[c] Listed are examples; there are many other descriptions in the literature of similarly innovative devices.

[d] Perhaps the least expensive of the setups. See also Problem 2-8 for additional ideas.

2. Multiaxial Extension

Melts can and are stretched using planar and biaxial flows, as well as combinations of these. Again, the motivation is to probe the melt to reveal behavior that is, or is not, consistent with theoretical description. The practical application is that observations from such flows may be useful for fingerprinting the polymer and for predicting its processing behavior.

As with simple extensional flows, the techniques for achieving, controlling and observing planar and biaxial flows are numerous. Similarly, there are very few devices that are commercially available. The home-grown techniques for gathering information on transient biaxial flows include sheet inflation for both planar and biaxial extension (Figure 7-17), bubble inflation for biaxial extension, and lubricated squeezing flow (Figure 7-18). The latter is used mainly for biaxial deformation, but has been used for planar extension.

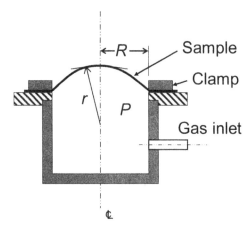

Figure 7-17. Sheet inflation as a method for stretching a sample biaxially. At higher deformations, the bubble tends to distort from its spherical shape, with only the very top remaining roughly spherical. A grid marked on the sheet, along with shape analysis from the side are essential for deducing the stress and deformation as a function of time.

The lubricated squeezing flow is perhaps of particular interest because it can be operated on any axial-motion rheometer with either position or load control. The principal concern is the lubricant. The nature and quantity of the lubricant often need to be determined by trial and error. Experiments by Bagley et al. suggested that the optimal combination for aqueous gels was a silicone-lubricated Teflon® surface.[33] For melts at high temperatures, a fluorocarbon treatment of the metal surface may be worth trying. The quality of the lubrication can be checked by varying the initial sample thickness, but applying the same deformation rate or stress. A more sophisticated approach is to pump in a thin layer of lubricant through porous surfaces, and control the flow according to the gap size, the pressure against the plate, and even radial position.[34]

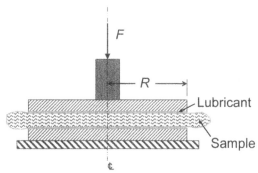

Figure 7-18. Lubricated squeeze flow to establish a biaxial extension. For planar extension, switch to rectangular coordinates, replacing the center line with a symmetry plane, and extend the plates in an out of the plane of the paper indefinitely (which usually translates to about 10 times the dimension of the plates in the flow direction).

Example 7-4: *Calculate the biaxial stress $\sigma_{rr} - \sigma_{zz}$ for the geometry shown in Figure 7-18. What is the pressure inside the sample?*

For the axial squeezing shown in Figure 7-18, the lubricant is supposed to reduce the shear stress τ_{rz} to zero. Thus, the equation of motion for the radial direction and slow flows reduces to

$$0 = \frac{d\sigma_{rr}}{dr} + \frac{\sigma_{rr} - \sigma_{\theta\theta}}{r} \tag{a}$$

The *z*-component does tell us that if τ_{rz} is indeed zero, then the total stress in the *z* direction (σ_{zz}) will be independent the value of *z*, i.e., σ_{zz} will be felt uniformly throughout the sample thickness, which is not surprising.

Given that we know the deformation must be biaxial, we know that the two extra stresses τ_{rr} and $\tau_{\theta\theta}$ must be equal, which means $\sigma_{rr} - \sigma_{\theta\theta}$ must be zero. Thus the stress state must be independent of radius, with σ_{zz} everywhere equal to $-F/\pi R^2$ (ssc), if we define *F* as positive for squeezing motion. As for the pressure, we have one total stress, $\sigma_{rr} = 0$ (or atmospheric pressure) at the edge. Furthermore, equation (a) and the fact that $\sigma_{\theta\theta} = \sigma_{rr}$ from symmetry tells us that both $\sigma_{\theta\theta}$ and σ_{rr} are zero. (Note that the corresponding extra stresses are most definitely not zero.) Our conclusion is the pressure, which is the negative (ssc) of the average applied stress, will be $F/3\pi R^2$.

Why isn't the pressure simply equal to $F/\pi R^2$? The reason is that the stretching of the sample is reducing the pressure; the relatively low pressure on the outside is in a sense pulling the sample out by its edges. Again we find that stresses are not the same as pressure.

3. Extensional rheometry summary

As extensional-flow experiments are not particularly easy to run, we might ask if there is any value in the information provided by the observations. Certainly, the very fact that well-thought-out descriptions of fluids have difficulty connecting extensional and shear material functions suggests that extension of melts and solutions provides information about the dynamics of the polymer molecules that cannot be gained by shear measurements alone. From a practical point of view, extensional experiments have proven useful for explaining behavior of polymer melts and solutions during processing involving extensional flows. Examples include spinning, blow molding, and thermoforming, which involve (roughly) simple (uniaxial), planar and biaxial extension, respectively.

D. SPECIALIZED GEOMETRIES

Described above are the most common techniques for the rheological examination of the polymer melts and solutions. The easy, automatic, least-troublesome methods prevail. Sadly, there are few resources available for development of new techniques, especially those suitable for difficult situations. Yes, there is periodic excitement over methods such as rheo-NMR, laser-Doppler velocimetry, nano-particle diffusion, etc., but exotic methods, if ever commercialized, usually disappear in a few years. As a result, one is likely to see dynamic linear viscoelastic response, combined with the Cox-Merz rule being used to characterize the non-Newtonian response of complex fluids.

Specialized geometries do have their place, as problems can't always be avoided and must be dealt with directly. We will deal with two examples below.

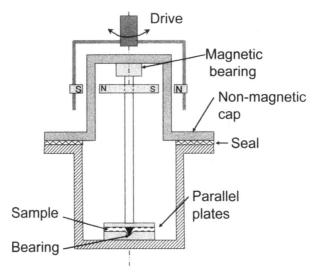

Figure 7-19. Schematic of gas-tight fixture for a rotational rheometer. Bearing systems, which are an extremely important aspect of the design, vary considerably. Some designs incorporate an optical encoder in the driven shaft, as the magnetic coupling is not perfect. Not shown are vacuum and gas lines.

1. Volatile loss (or gain)

Elementary thermodynamic theory points out clearly that good polymer solvents are likely to be low in molecular size and thus volatile. How does one characterize the flow properties of polymer solutions in these volatile solvents?

Certainly, the worst mistake is picking a geometry that would be highly sensitive to changes in properties due to loss of solvent at the sample-air interface. High on the list are the parallel-plate and cone-plate geometries, as a slight viscosity increase at the rim due to solvent loss will have a huge impact on the result. The Couette geometry is a much better choice.

Aside from the wise choice of a standard geometry, there are two approaches to keeping the solvent where it belongs. One is to physically block it from leaving. The other is to maintain a high solvent activity on the outside of the sample.

If one must absolutely maintain the composition of the sample invariant with time and temperature, the only procedure is to hermetically seal the sample. There are several ways to do this, both fancy and simple. The fancy, but expensive, method is to buy a fixture that is capable of sealing the sample in a vessel that resembles an autoclave. The vessel, once sealed, can be evacuated or pressurized with inert gas. A schematic of how these work is shown in Figure 7-19. The alternative design uses a torque tube to penetrate the sealed wall. The twist of the tube is measured.

The simpler and cheaper ways of sealing a sample are to use the falling- or rolling-ball geometry. In this way, the sample can be sealed into a tube under vacuum, so there is no question about changes in composition of the system. Sealed along with the sample is a precise sphere made from a dense and inert material. A good material choice is tungsten carbide, as it is dense, hard and relatively resistant to chemical attack. However, the much cheaper stainless-steel ball bearing will suffice. These two geometries are shown in Figure 7-20.

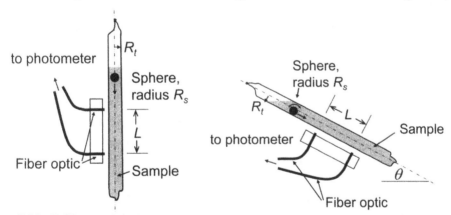

Figure 7-20. Falling- and rolling-ball geometries for measuring viscosity. The sphere is sized to place the flow regime well into the Newtonian region, thus yielding η_0. In the case of the falling-ball viscometer, the sphere can be replaced by an oblong object, which may allow non-Newtonian viscosity to be detected. The angle θ for the rolling sphere adds flexibility to control the time of descent. Velocity can be measured by passage time between the two fiber optics or magnetic pickups (for opaque fluids).

Other geometries have been proposed, but these two are the most used. The viscosities under ideal conditions are given approximately by the equations

$$V = \frac{d^2(\rho_B - \rho)g}{18\eta}\left(1 - 2.104\alpha + 2.09\alpha^3 - 0.95\alpha^5 + \cdots\right) \qquad (7\text{-}55)$$

for the falling ball,[35] and

$$V = \frac{d^2(\rho_B - \rho)g\sin\theta}{15\eta}(1-\alpha)^{5/2} \qquad (7\text{-}56)$$

for the rolling ball, where $\alpha = d/D$. D is the diameter of the tube, while d is the diameter of the ball. The densities ρ_B for the ball and ρ for the fluid should be evident; and, of course, g is the acceleration of gravity. Both of these equations assume Newtonian behavior and very low (< 0.1) Reynolds number, $\rho V d/\eta$. Also, the ball must reach its terminal velocity before it hits the measurement region, which in turn must be far enough above the bottom to avoid end effects. The falling ball must fall down the axis of the tube, which it may not want to do. The rolling-ball geometry avoids this problem.

Both of these geometries can behave badly with viscoelastic fluids when one attempts to use large, dense spheres to increase the velocity. As a result, one needs to exercise patience to find η_0 for polymeric fluids, especially melts, with these devices.

2. Prevention of slip

Slip at the polymer-metal interface seems to be a fact of life at high stresses. Slip is particularly prominent with filled polymers,[†††††] but can occur with neat melts as well. Solutions and suspensions often develop a layer of additive-free matrix near the wall, which produces an apparent slip due to its low viscosity.

While we have seen how to detect and correct for slip in capillary flow, it is often necessary to suppress slip to obtain the material properties of the sample at the higher stresses. Suppression of slip is particularly important for measurement of the yield stress. Rheologists have worked long and hard on this problem, and numerous solutions have been proposed. Some may work for one material, but fall short with another. Basically, they all consist of some texturing of the surface to increase the adhesion between the metal and polymer. The penalty is distortion of the geometry and disturbance of the flow pattern of the fluid.

[†††††] The term "filled" is commonly used to denote a formulation containing solid particles or chopped fibers. Polymers are filled principally to increase modulus and strength, but other properties may also be favorably modified. Processing generally gets more difficult.

A texturing pattern for flat surfaces has been proposed and tested in the parallel-plate geometry, with good results.[36]   The protrusions are less than a millimeter in size, which leads to an increase in the effective gap of only a few tenths of a millimeter for each textured surface.   For the double Couette geometry, slots in the rotor have been shown to work well with suspensions.[37]   The idea is that the material inside the slots will be forced to move with the rotor, and will connect well with the material next to the rotor.   Many other designs have been proposed for the roughened surface.

E. FLOW VISUALIZATION AND OTHER RHEO-OPTICAL METHODS

Rheometry methods abound, although the ones discussed above are used for easily 90% of all measurements appearing in the literature.   Some of the methods used for quality control are described in Chapter 11, but the results of these measurements (except for melt index) are not often published or specified.   Some methods, important at some time in history and for a limited range of materials, are no longer used due mainly to a lack of commercial instruments.   One source of such lore is the book by Van Wazer et al.[38]   We feel that most information about such tests should be on the Internet, but try, for example, a search on "Rossi-Peakes flow."

The one method that was used for many years and is still used with increasing sophistication is flow visualization.   In a nutshell, flow visualization is the process by which the velocity of the fluid is determined as a function of position and time.   Imagine velocity as a field of arrows (velocity vectors) pointing in the direction of flow and with a length that is proportional to the magnitude of the velocity.   For example, if could examine flow in a channel as a function of depth into the channel from the wall, then one could, in principle, get the velocity profile and determine the form of the viscosity function with a single observation.   Furthermore, we could examine the velocity profile along the length of the channel from entrance to exit, and tell where steady conditions had been achieved, and the presence of flow disturbances. Heady stuff.

1. Particle tracing

So, how is velocity inside a stream of fluid determined?   The earliest method exploited the presence of small particles in the fluid (there are many).   These particles were observed as they flowed past a fixed point. Photographs taken at known intervals would be analyzed to find the position of each particle with time, and therefore its velocity.   A more modern method is to illuminate the particles for a very precise time interval, and examine the trace of the reflected light on a photograph exposed over the same interval. This technique is called streak photography. The direction of the streak and its length give the velocity vectors at all locations, simultaneously.

A much more elaborate method of measuring velocity removes the need for photographs. This method, known as Laser-Doppler Velocimetry, relies on the precise wave properties of laser illumination. If the beam from a single laser is split and then rejoined, an interference pattern is established at the point where they join. The pattern has fringes spaced at precisely known intervals that depend on the wavelength of the light λ and the angle θ at which the two beams meet. Specifically, the spacing δ is

$$\delta = \frac{\lambda}{2\sin\left(\dfrac{\theta}{2}\right)} \tag{7-57}$$

Most simply, the fringes reflect light from a particle moving with the fluid, and the frequency f of the reflection pulses gives the velocity, V_P. The appropriate equation is

$$V_P = \frac{\lambda f}{2\sin\left(\dfrac{\theta}{2}\right)} \tag{7-58}$$

An alternative to the explanation just given is the one that gives the method its name. Imagine a single laser bean entering the fluid; the light reflected from this particle in a forward direction will be Doppler-shifted to a higher frequency. If we mix this signal with the original light, the beat frequency, which is the difference in the frequencies of the incident and reflected light, gives the particle velocity. This varies with the detector angle, so the direction of the particle (maximum Doppler shift) can be found. This is exactly the same principle that is used to catch speeders with either a radar or laser source.

Laser light can do other things such as heating the fluid or bleaching a dye dissolved in the fluid. With strong laser light, a pattern of low refractive index can be created, which moves with the fluid. A measuring pattern then will detect the burned-in pattern, and the beat frequency of the two will give the velocity. The photo bleaching of dye works in a similar fashion, but usually requires short-wavelength light, which may not penetrate the fluid well.

Flow "visualization" can also be done using magnetic resonance imaging (MRI). MRI is an established method for imaging hydrogen-rich materials. The classical technique is to use a magnetic field gradient to produce resonance at a known location in the material. This location can be moved by changing the field strength using auxiliary electromagnets called gradient coils.   The magnitude of the resonance signal at various locations provides the image.

But, how can one use this to measure velocity?   The secret is the fact that the perturbed spin state of the nucleus, brought about by application of a field pulse orthogonal to the main field, can persist for a certain amount of time. This time, referred to as reorientation time, is usually a very small fraction of a second.   If the nucleus is moving, its phase changes, relative to a stationary nucleus, and this phase change leads to a measurable difference.   Of course, Brownian motion acts similarly, so this must be taken into account in the analysis.

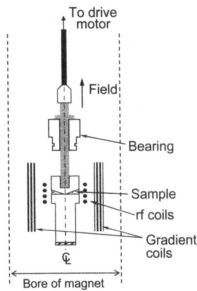

Figure 7-21. Schematic of rheo-NMR fixtures. Evident are the gradient coils for moving the location of the field required for resonance, and the radio-frequency (rf) coils for generating and measuring the "echo" signal. The fixture is designed for a vertical bore (common in laboratory instruments) with a vertical magnetic field. Torque is generally not measured, although in principle the bottom of the bore is accessible for such. [Adapted with permission from P. T. Callaghan, [40] *Encyclopedia of Nuclear Magnetic Resonance, Vol. 9: Advances in NMR* (D. M. Grant and R. K. Harris, eds.), Wiley-Blackwell, 2002.]

In principle, any fluid containing the right nucleus (mostly hydrogen) can be analyzed.   Opaque or colored samples are fine.   As signal strength depends upon the density of protons brought into the field by the sample, usually experiments are run using hydrogen-rich fluids, such as water solutions.   A severe limitation of the technique is the need for non-metallic containers for the fluid, as metals will interfere with the oscillating electro-magnetic fields. (This is used to advantage for detecting weapons at airports.)   Furthermore, all the equipment used to deform the fluid must be compatible with the high magnetic field and fit into a relatively small volume. The usual laboratory magnet has a "bore" about 80 mm in diameter, and the active volume can be buried tens of centimeters into the bore. The advent of "big bore" MRI equipment has removed some of the space constraints, but these magnets are superexpensive. Couette rheo-NMR fixtures fitting a 10-mm bore have been offered.[39] Figure 7-21 shows a schematic of one with cone-and-plate fixtures.

What magnitudes of spatial and velocity resolution are achievable with rheo-NMR? Spatial resolution of around 1 µm is now claimed, with velocity resolutions around 10 µm/s. But the real beauty of rheo-NMR is the ability to study the deformation of very complex fluids in great detail. For example, shear banding and other instabilities of suspensions in rotational flows are easily detected.[40]

2. Bulk rheo-optical techniques

Particle tracing with light requires fairly transparent fluids and a sparse concentration of particles. Transparent fluids are not needed for NMR velocity measurements, but melts at high temperatures are not easy to handle. Furthermore, particle tracing methods do not probe the local stress distribution. But other rheo-optical methods exist that can provide information about stress distributions and structural changes in the fluid. Birefringence and scattering are two of these methods that have received the most attention, and will be described briefly. The popularity of these methods has resulted in commercial offerings of optical fixtures for rotational rheometers and an optical stage for a microscope.

Birefringence is a property of materials that have directionally dependent structures, which results invariably in directionally dependent optical properties. Sometimes these differences are very small; in other cases they are large and easy to detect. Of importance to polymer rheologists is the directionally dependent structure that results when a polymer melt or solution is oriented. If the sample is transparent, this anisotropy can be detected using polarized light. The method is best understood by running the desk-top experiment described in Appendix 7-5. This experiment uses everyday equipment available to anyone, and can be described as the observation of a change of intensity of light transmitted through crossed polarizers. Not surprisingly, the first polarizer encountered by the light is called the *polarizer*, and its job is to produce linearly polarized light. The second polarizer is called the *analyzer*. The sample is held between the polarizer and the analyzer such that its principal-stress directions are at 45° to the direction of the incident polarized light. The optical methods and equipment used in most laboratories are somewhat more complex, but are aimed at the same goal: measuring the birefringence. The argument is that the principal stress and the birefringence are related, sometimes linearly, as given by the definition

$$n_i - n_j = C_{ij}(\sigma_i - \sigma_j) \tag{7-59}$$

where n_i are the refractive indices and σ_i are the principal stresses. The directions i and j are in the plane whose normal is in the viewing direction. Thus, if the stress-optical coefficient C_{ij} is known, then the entire stress tensor can be deduced by examination of the sample in three directions. Usually, this is not that easy, and the method is generally confined to mapping of a particular normal-stress difference with position in the sample, or with time at a particular position for a two-dimensional flow. An example of the latter is shown in Figure 7-22. The changing intensity was due to the build-up of stress on starting a flow in a slit die. The reason for the oscillation is revealed in Appendix 7-5 and Problem 7-22. Note that the values of C_{ij} may depend on stress, and may even change sign.

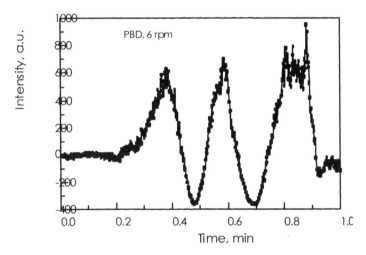

Figure 7-22. Time-resolved, transmitted light intensity changes during the start-up of the flow of polybutadiene in a slit die. Knowing the stress-optical coefficients for this polymer and the height of the channel, an average normal stress difference $(\sigma_{11} - \sigma_{33}) = N_1 + N_2$ can be found. Calculating the average stress at the peak intensity is the subject of Problem 7-22. The higher-frequency oscillations on the third peak can be interpreted as flow instability. (Reprinted from Y.-W. Inn et al.[41] with permission of WILEY-VCH Verlag.)

Another popular rheo-optical technique is low-angle light scattering (LALS) during deformation, especially shearing.[‡‡‡‡‡] Why do this? Unlike microscopy, LALS is not sensitive to the speed of the sample through the optical path, and gives instantaneous results. Using a focused laser, the scattering volume can be made extremely small, thus allowing phenomena to be studied that occur in a very small locality.

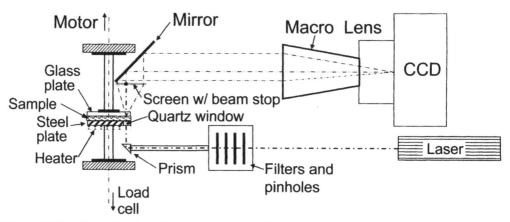

Figure 7-23. Example setup for measurement of low-angle light scattering from a deforming sample. The shear stress can be measured simultaneously (see Problem 7-23). [Adapted with permission from R.-J. Wu, M. T. Shaw and R. A. Weiss, *Rev. Sci. Instr.*, **66**, 2914-2921 (1995). Copyright 1995, American Institute of Physics.][42]

What are these phenomena? Light is scattered by refractive index changes occurring over length scales comparable to the wavelength of light. As with x-ray diffraction, the smaller the structure, the larger the scattering angle. Thus LALS is used to capture changes in structure that are quite large. (Knowing the scattering angle θ in radians, the size d can be approximated by the Bragg relationship $d = n\lambda/2 \sin\theta \sim \lambda/2\theta$ at low angle and, for first order, $n = 1$. Thus if there is light intensity found at a scattering angle of 5°, the source is likely to have refractive index modulation of around 4 µm.) By having a very strong light source and a very sensitive detector, very small changes in composition, stress and orientation can be found. In fact, too much scattering is highly undesirable. The scattering in pigmented resins, for example, approaches the diffusion limit, that is, where a photon enters the sample, bounces around for a while and then finally diffuses out in an arbitrary direction. Most detail is lost.

[‡‡‡‡‡] Also called LALLS for low-angle laser light scattering. As everyone uses a laser as the source, this seems a bit of an overkill. LALS is also known as SALS, for small-angle light scattering.

A limitation of scattering is that one must make some assumptions about the structure before drawing any conclusions. Unlike microscopy, the LALS pattern is not always unique to a single structure. Usually this is not a serious limitation, but it can be.

Figure 7-23 shows a typical scattering setup using the parallel-plate geometry. A result from this equipment is shown in Figure 7-24.

Want smaller size scales or different contrast? X-ray and neutron scattering can then be used. The cost and complexity of these techniques is beyond most labs; however, national facilities are available, along with a selection of rheometers.

Figure 7-24. SALS pattern for a near-critical 25/75 SAN/PMMA blend subjected to a shear rate of 0.02 s^{-1} and a shear strain of 7.9. The flow direction is vertical, while the light path is in the gradient (2) direction. The dark area on the left is due to the mirror (see Figure 7-23), which blocks some of the scattered light. A micrograph obtained on a quenched sample subjected to the same history had microstructure consistent with this pattern.   The dark circle in the center is due to the beam stop, which blocks unscattered light from the laser.[43] (Reproduced with permission from Z. Hong, M. T. Shaw and R. A Weiss, *Macromolecules*, **31**, 6211–6216. Copyright 1998, American Chemical Society.)

MICRO AND NANO RHEOLOGY

The search for methods to study the motions of individual polymer molecules has been a long one, but new methods have produced considerable progress. These vary from imaging of single, isolated polymer coils, to drag measurements using nanoparticles, to observations of Brownian motion. While these methods can determine local values of rheological properties, they are not designed to replace bulk measurements. The hope is, however, that the behavior of the bulk can be related to happenings at length scales comparable to that of polymer chains.

1. Uncoiling of individual polymer chains

While not strictly a measure of stress and strain, the observation of the behavior of single chains in a flow field has led to insight into the behavior of the bulk liquid. The trick here is to tag a single chain with fluorescing moieties that allow the shape of the molecule to be observed when placed in a flow field (usually shear or planar extension). The observation is the light given off by the fluorescing groups, not the chain itself. To obtain the needed spatial resolution, confocal microscopy is used. Confocal microscopy can be described crudely as scanning light microscopy. Light is directed a tiny volume with a know location, and the sample is moved in small increments to collect intensity or phase information from each volume element. The image is formed by adding all these elements together. The observation from these experiments is the very slow uncoiling of the molecule, rather than a simultaneous, affine change in coil shape.

2. Observation of Brownian motion

It has been known for years that submicroscopic particles can be observed using oblique illumination. If the illumination comes in the sample at a very large angle off the axis of the microscope, none of it reaches the objective, and the view is dark. This doesn't sound like the best situation for viewing anything. However, if there is something in the sample that scatters light, this light is gathered by the objective and can be observed. The particle can't be seen, but its path can. This was the method by which Brownian motion was first observed.

Einstein described the distance traveled by a particle in terms of its diffusivity D as

$$\langle x^2 \rangle = 2Dt \tag{7-60}$$

where t is time and $\langle x^2 \rangle$ is the average square of the distance traveled. As one might expect, the diffusivity is connected directly to the viscosity via the equation

$$D = kT/d \tag{7-61}$$

where d is the drag coefficient. For spheres at low Reynolds number, the drag coefficient is given by the Stokes-Einstein equation

$$d = 6\pi R \eta \tag{7-62}$$

where η is the viscosity and R is the radius of the sphere. (Note this is not the same as the dimensionless drag coefficient used by engineers.) By combining the above three equations, we have the following equation for viscosity:

$$\eta = \frac{kT}{3\pi R\left(\dfrac{\langle x^2 \rangle}{t}\right)} \tag{7-63}$$

What "viscosity" is this? Certainly if the particles are comparable in size to the polymer molecules, the local viscosity may not be at all similar to the bulk viscosity and can be expected to be sensitive to particle size. These differences are important, however, for examining assumptions involved in kinetic descriptions of rheological response. An example of the application of this technique is to measure the excursion of a polymer chain from its confining tube during reptation.[44]

Rotational diffusion of nano rods is another method. As both methods require lots of observations to establish precise averages, it pays to observe an assembly of particles simultaneously.

3. Motion of nanoparticles subject to a known force

A nanoparticle is batted around by collisions with polymer segments and solvent molecules, but the particle also can be forcefully moved by application of a field, usually a magnetic field. As the size of the particle decreases, the force exerted by a field drops off in proportion to the particle's volume, i.e., as R^3. The drag on the particle, however, drops off only in proportion to R. Thus moving nanoparticles with fields can be difficult (see Problem 7-20). However, very strong magnetic fields are readily accessible, and small magnetic particles (usually magnetite) are offered commercially. Although a uniform magnetic field will induce a magnetic dipole in a particle, the dipole will have no net force on it. To drag the particle, a field gradient is needed, which is exactly why a steel tool goes flying into an MRI magnet if brought too close.

The difficult aspects of this conceptually easy experiment are the observations of the particle and the calibration of the magnetic field gradient. Observing particle motion in high magnetic fields can be done using MRI; however, simpler setups using very small fluid volumes and electromagnets have been successful. An advantage of an electromagnet over a super-conducting magnet is that its field can be varied easily and turned off. Thus the viscoelastic properties as well as the steady-flow rheology can be studied.

As is amply illustrated in Problem 7-20, normal gravity is not the best body force for moving small particles through viscous fluids. However, ultracentrifuges can produce accelerations that are thousands of times stronger. As is well documented in the polymer literature, centrifuges can even separate polymer molecules from solvents. The basic equation for the velocity of an isolated particle is the same as for equation (7-55) for sedimentation: a balance of the buoyant, drag, and gravitational forces (see Problem 7-25). The result is

$$\eta = \frac{d_p^2(\rho_p - \rho_f)g_r}{18\,V} \tag{7-64}$$

where d_p and ρ_p are the diameter and density of the particle, respectively, ρ_f is the density of the fluid, g_r is the applied gravitational field in the centrifuge, and V is the particle velocity. Obviously, tracking one micro or nanoparticles would be difficult, so an ensemble of particles is viewed as they pass through two points of known spacing.

APPENDIX 7-1: NUMERICAL DERIVATIVES

There are many instances in the rheometry of non-Newtonian fluids where numerical derivatives of data are needed to get the final result. As real data (experimental observations) have inherent random error, finding derivatives is not a trivial issue. There are some reasonable methods for getting a result, but none of them are perfect, and some are best avoided. First, let's discuss the latter.

All functional descriptions of the entire data set should be avoided because the result will be forever biased by the choice of the function. Especially eschew the use of a single polynomial to describe the set. Polynomials have very bad behavior near the ends of the data set, as all polynomials go either to $+\infty$ or $-\infty$, as do their derivatives. In contrast, we expect the slope of the flow curve to approach 1.0 at low shear rates and decrease at higher shear rates. While the fit of a polynomial to the data may look fine, the derivatives can be very seriously off near the ends of the set, and reflect the nature of the function, not the data. A GNF model (Chapter 5) will behave more realistically, but again the result will be stamped forever with the form of the particular model used.

Finite-difference methods are another choice, and form the basis for some better methods. For real data, finite-difference methods can yield very noisy results, but there may be an opportunity to remove the noise later on. For the first cut, use divided differences to find the derivatives y_i', that is,

$$y_i' = \frac{y_{i+1} - y_{i-1}}{2\Delta x} \tag{a}$$

for data spaced evenly at intervals of $\Delta x = x_i - x_{i-1}$. This formula is just the average of slope from the ith point to the before and after; thus there is a bit of noise-reducing action. Unfortunately, this method drops two points, one from each end of the data set. Sure, the forward and backward differences can be used for the first and last point, respectively, but these two points will have more error than the others, and thus could distort the final result. Take some comfort in knowing that there really is no good way of handling the slope right up to the ends of the set. Additionally, the ends of the set usually exhibit all the systematic problems of an instrument working at the limits of its range, so it may actually be a good idea to get rid of them.

If the points are not equally spaced, then simply average the forward and backward differences, i.e.,

$$y_i' = \left(\frac{y_{i+1} - y_i}{x_{i+1} - x_i} + \frac{y_i - y_{i-1}}{x_i - x_{i-1}} \right) \Big/ 2 \tag{b}$$

These formulas are easy to set up in a spread sheet. Start with the second row in the data set, and enter an equation according to the above formula. If the spread sheet has a built-in differential operation, try it out too and compare with your results.

As one might expect, more sophisticated methods are available. For example, the authors of *Numerical Recipes*[45] and Hershey et al.[46] recommend a low-order polynomial, e.g., second order (quadratic) through an odd number of points to calculate the derivative at the center point. So, if five points are used, then two points at each end of the set are lost. This may too much of a sacrifice given a small data set.

On the expectation that the trend of y' with x will be of lower effective order than y vs. x, some cautious smoothing of the derivatives might be reasonable. Dozens of noise-reduction methods (filters) are available, but again the handling of the ends of the data set is often a problem. Certainly if one plans to model the set at the very end (say, describe a flow curve with a GNF), then it may be better to let the model do the smoothing. Granted, the parameters will have higher standard errors, but the result will at least be physically realistic.

APPENDIX 7-2: VELOCITY-PROFILE CORRECTION FOR NON-NEWTONIAN FLUIDS

The problem of finding a general formulation for the wall shear rate in capillary flow is approached below. This is a good introduction to two tricks for handling integral equations, i.e., equations involving an integral where the required information is annoyingly out of reach, buried in the integrand.

We start off with equation (7-23) for the total flow through the capillary. It's reproduced below:

$$Q = \int_0^R V_z(r)\,2\pi r\,dr \tag{7-23}$$

This equation is simply a mass balance, assuming the fluid is incompressible. Again, the goal is to extract a velocity gradient from the velocity in the integrand. We do not know what the velocity profile looks like, but by varying R we can, in a sense, explore the profile. To see this, imagine a large capillary and a smaller capillary with half the radius. If the small capillary is run with half the shear stress as the large one, the flow through the small one will be identical to the flow through the center part of the big one, up to $R/2$. The shear rate at the wall of the small capillary will be the same as that of the big one if the latter is run at the same stress. Thus by varying shear stress, we are in effect exploring the flow with radius.

As we want to get the velocity gradient from the velocity, an appropriate move is to integrate by parts. The formula for integration by parts is

$$\int_a^b u\,dv = uv\Big|_a^b - \int_a^b v\,du \tag{a}$$

Clearly, we want to pick the values for u and v carefully to avoid complications. As the velocity gradient is sought, picking $u = V_z(r)$ seems like a good choice. This leaves $2\pi r\,dr = dv$. Anything can be done with the 2 and the π, but if we examine the combination $2r\,dr$, it looks like the derivative of r^2. Summarizing:

$$u = V_z(r); \quad du = \frac{dV_z(r)}{dr}\,dr \tag{b}$$

$$dv = 2r\,dr; \quad v = r^2 \tag{c}$$

The π, of course, gets pulled out of the integral. Then

$$Q = \pi \left[V_z(r)r^2 \Big|_0^R - \int_0^R r^2 \left(\frac{dV_z(r)}{dr} \right) dr \right] \tag{d}$$

where the first term in the square brackets is the uv part. Note that we can toss this off as being zero using the assumption is that $V_z(R)$ is zero, i.e., no slip at the wall. Mooney is credited with calling instead for a slip velocity V_s at the wall that depends on shear stress only. This aspect is left for Appendix 7-3. So, for the moment we assume no slip and examine what we have done. Recognizing that the shear rate $\dot{\gamma}(r)$ magnitude is the negative of velocity gradient in this flow gives

$$Q = \pi \int_0^R \dot{\gamma}(r)r^2 dr \tag{e}$$

Some progress has been made in that the equation now contains the shear rate rather than the velocity. However, as with the velocity, the shear rate is in the integral. Remembering that we want to explore inside the capillary by varying the shear stress, it seems reasonable at this point to change the radius variable to shear stress. The transform relationships are simply

$$\tau = \frac{r}{R}\tau_W; \quad r = \frac{R}{\tau_W}\tau; \quad dr = \frac{R}{\tau_W}d\tau \tag{f}$$

while the limits are

$$\tau = 0 \ @ \ r = 0 \quad \text{and} \quad \tau = \tau_W \ @ \ r = R \tag{g}$$

Again, these are independent of the nature of the fluid.

Making these substitutions and moving constant terms outside the integral gives

$$Q = \frac{\pi R^3}{\tau_w^3} \int_0^{\tau_W} \tau^2 \dot{\gamma}(\tau) d\tau \tag{h}$$

To prepare for the next step, the wall shear stress is brought over to the left-hand side of the equation, along with the πR^3, yielding

$$\frac{\tau_W^3 Q}{\pi R^3} = \int_0^{\tau_W} \tau^2 \dot{\gamma}(\tau) d\tau \tag{i}$$

Now it is time for the second trick: differentiating the integral by τ_W. This is done using Leibniz's rule,[§§§§§] and the result is quite remarkable in that most of the complication is not with the integral, but with the left-hand side. The reason the right-hand side is simple is that τ_W appears *only* in the upper limit. The result, then, is

$$\frac{Q(3\tau_W^2)}{\pi R^3} + \frac{\tau_W^3}{\pi R^3}\frac{dQ}{d\tau_W} = \tau_W^2\dot{\gamma}(\tau_W) \tag{j}$$

By this operation, the stress-dependent shear rate at the wall has been extracted from the integral. Certainly a good trick to know about.

The final step is to solve for the shear rate and clean things up. Dividing through by τ_w^2 and putting the shear rate on the left-hand side gives

$$\dot{\gamma}(\tau_W) = \frac{3Q}{\pi R^3} + \frac{\tau_W}{\pi R^3}\frac{dQ}{d\tau_W} \tag{k}$$

At this point, it is customary to formulate the right-hand side as the Newtonian shear rate, multiplied by a correction factor. Thus, we pull out the factor $4Q/\pi R^3$ to get

$$\dot{\gamma}_W = \frac{4Q}{\pi r^3}\left[\frac{3}{4} + \frac{1}{4}\frac{\tau_W}{Q}\frac{dQ}{d\tau_w}\right] \tag{l}$$

For the second term in the brackets, it is traditional to convert the Q to $4Q/\pi R^3$, which doesn't change anything except the appearance. The result is equation (7-30), which is reproduced below:

$$\dot{\gamma}_W = \frac{4Q}{\pi R^3}\left[\frac{3}{4} + \frac{1}{4}\frac{d\ln\left(4Q/\pi R^3\right)}{d\ln\tau_W}\right] \tag{7-30}$$

[§§§§§] Leibniz's rule for differentiating an integral is a bit complicated looking, but very systematic. As it leads potentially to three terms including another integral, it is best applied to integral equations that do not have the variable of differentiation in the integrand. The formula is reproduced below:

$$\frac{d}{da}\int_p^q f(x,a)dx = \int_p^q \frac{\partial}{\partial a}[f(x,a)]dx + f(q,a)\frac{dq}{da} - f(p,a)\frac{dp}{da}$$

APPENDIX 7-3: INCORPORATION OF SLIP INTO THE VELOCITY-PROFILE CORRECTION—THE MOONEY CORRECTION

The Mooney correction[47] follows the same path as the Weissenberg-Rabinowitsch correction developed in Appendix 7-2. The departure point is equation (d) in Appendix 7-2; it is reproduced below:

$$Q = \pi \left[V_z(r)r^2 \Big|_0^R - \int_0^R r^2 \left(\frac{dV_z(r)}{dr} \right) dr \right] \tag{a}$$

Rather than tossing the uv term, we allow for a non-zero velocity at the wall, V_s, that depends on the wall shear stress, τ_W. The upper limit then gives

$$Q = \pi R^2 V_z(R) - \int_0^R r^2 \left(\frac{dV_z(r)}{dr} \right) dr \tag{b}$$

The first term can be thought of as a plug of fluid added to the base of the usual velocity profile. This is sketched in Figure 7-25.

From this point, the handling of the integral is exactly the same as the development in Appendix 7-2. Accordingly, the variable of integration is converted from r to τ using the substitutions shown in Appendix 7-2. These are reproduced below:

$$\tau = \frac{r}{R}\tau_W; \quad r = \frac{R}{\tau_W}\tau; \quad dr = \frac{R}{\tau_W}d\tau \tag{c}$$

Of course, $V_z(R)$ is the same as the slip velocity V_s. Making these substitutions gives

$$Q = \pi R^2 V_s + \frac{\pi R^3}{\tau_W^3} \int_0^{\tau_W} \tau^2 \dot{\gamma}(\tau) d\tau \tag{d}$$

which is similar to equation (h) in Appendix 7-2. Dividing by R^3 and multiplying by 4 gives the final result

$$\frac{4Q}{\pi R^3} = \frac{4V_s}{R} + \frac{4}{\tau_W^3} \int_0^{\tau_W} \tau^2 \dot{\gamma}(\tau) d\tau \tag{e}$$

Why is this the final result? Well, at constant τ_W the entire second term on the right-hand side is simply a constant. So, a plot of $4Q/\pi R^3$ vs. $1/R$ at constant τ_W—the *Mooney plot*—will result in a straight line with a slope of $4V_s$. The intercept will be the integral term, which is equivalent to the no-slip flow expressed in equation (i) of Appendix 7-2. The set of intercepts gathered at various values of τ_W can then be processed using the WRM equation in the usual fashion to find the true shear rate at the wall. Thus the Mooney plot eliminates the effect of slip by extrapolating to a capillary of infinite radius.

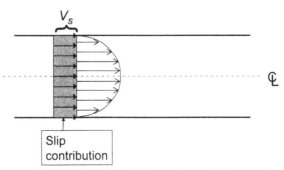

Figure 7-25. Illustration of the contribution of slip to the total flow rate in a capillary. The slip "plug" (shaded) adds to the pressure-driven flow (not shaded) to give the total flow.

The slip contribution can be viewed as a plug moving at velocity V_s. The flow contribution of the plug will increase as R^2. The pressure-driven flow contribution, though, will scale as R^3, as can be seen from equation (i) in Appendix 7-2. Thus the significance of slip decreases as the radius of the capillary increases and is eliminated entirely as $R \to \infty$.

APPENDIX 7-4: NORMAL STRESSES USING THE CONE-AND-PLATE GEOMETRY

The derivation of normal stresses present in a cone-and-plate rheometer is provided in most textbooks. However, the math can get somewhat ugly. As was discussed in Section B-1 of Chapter 7, the problem with the circular flow is that the tension generated in the flow direction constricts the material at lower radius, and this pressure keeps adding up at lower and lower radius. With the cone and plate, the compressive radial stress not only adds up with lower radius, but keeps acting on a smaller and smaller area. This can result in truly spectacular stresses acting against the plate surfaces.

At first we will attempt to follow what was so easily done in calculating the moment generated by the cone-and-plate fixture. In the same fashion, we use cylindrical coordinates and add up the z-direction normal stresses acting on the plate. Thus the normal force is

$$F = -\int_0^R \sigma_{zz}(r)\,2\pi r\,dr \qquad\qquad \text{(ssc) (a)}$$

from a force balance on the plate in the axial direction. [The minus sign is to give a positive value for F, because σ_{zz} will be negative (ssc), i.e., compressive.] In this development, we are assuming that atmospheric pressure is zero.

Of course, σ_{zz} has two contributions, i.e., $\sigma_{zz} = \tau_{zz} - p$ (ssc). The extra stress τ_{zz} is not a function of r, but p most definitely is. This leads to the equation

$$F = -\pi R^2 \tau_{zz} + \int_0^R p(r)\,2\pi r\,dr \qquad\qquad \text{(ssc) (b)}$$

The first term represents the contribution due to shearing, while the second contains the contribution due to the compression brought about by the tensile stress in the rotation (θ) direction.

To gain some information on $p(r)$, we need to use the equation of motion. The r component seems to be the most appropriate. In the absence of inertial and body forces, the result is

$$\frac{\partial p}{\partial r} = \frac{1}{r}\frac{\partial}{\partial r}\left(r\tau_{rr}\right) + \frac{1}{r}\frac{\partial \tau_{\theta r}}{\partial \theta} - \frac{\tau_{\theta\theta}}{r} + \frac{\partial \tau_{rz}}{\partial z} \qquad\qquad \text{(ssc) (c)}$$

In view of the symmetry and homogeneous flow, the derivatives of the extra stress with respect to r, z and θ do not contribute. The remaining terms, after expanding the derivative with r and combining terms, are simply

$$\frac{dp}{dr} = \frac{\tau_{rr} - \tau_{\theta\theta}}{r} \qquad\qquad \text{(ssc) (d)}$$

Note we have replaced the partial derivative with the total derivative because nothing changes in the z and θ directions.

Equation (d) is in the form of $dy/dx = a/x$, where a is a constant. This gives the slope of the pressure, i.e.,

$$\frac{dp}{d \ln r} = \tau_{rr} - \tau_{\theta\theta} \qquad\qquad \text{(ssc) (e)}$$

Because θ is the flow direction and r is the neutral direction, we can write this difference with combinations of $N_1 = \tau_{\theta\theta} - \tau_{zz}$ and $N_2 = \tau_{zz} - \tau_{rr}$. The winning combination is $-(N_1 + N_2)$. As $p = \tau_{zz} - \sigma_{zz}$, the slope of the total stress σ_{zz}

with radius will be the same as for p. Recall that flush transducers would measure $-\sigma_{zz}$, not p.

While this development looks very straightforward, it is wrong. There is no error in the algebra; instead, there is a basic error in the geometry. Our result would be okay if we had an arrangement such that the fixtures were parallel disks, but one disk was divided into rings that were driven in a tricky fashion such that the angular velocity $\Omega(r)$ of each ring was inversely proportional to its radial position, i.e., $\Omega(r) = a/r$, and not a constant, as with the normal parallel-plate geometry. This concept is depicted in Figure 7-26, where the constant tangential velocity $\Omega(r)r$ is determined by the drive pinion rotational rate only.

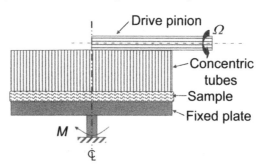

Figure 7-26. Design showing how (in principle) to make a set of parallel-plate fixtures with uniform shear rate. The concentric tubes are able to rotate around the vertical axis (dot-dash line), but are unable (somehow) to slide axially. Each tube has teeth cut into its top edge to engage the drive pinion such that their tangential velocities are all the same.

Our mistake can be understood by considering each layer of fluid to be in a state of tension due to the flow. Replace the material with wound up elastic thread of constant tension corresponding to the constant conditions in the cone-and-plate fixtures. This thread is to be wound around the axis in neat layers. Unlike the imaginary device depicted in Figure 7-26, there are every more threads in each layer as one moves outward in the cone-and-plate fixtures. This results in proportionally more constrictive force. We need to take this into account.

To do this, we switch to spherical coordinates.[******]   Then, the r component of the equation of motion is

$$\frac{\partial p}{\partial r} = \frac{\partial \tau_{rr}}{\partial r} + \frac{2\tau_{rr}}{r} - \frac{\tau_{\phi\phi} + \tau_{\theta\theta}}{r} \qquad \text{(ssc) (f)}$$

with the same assumptions.   Again, on recognizing that the flow is homogeneous and combining terms, we get

---

[******] It would be a good idea to review the subscript assignments for spherical coordinates.

$$\frac{dp}{dr} = -\frac{N_1 + 2N_2}{r} \tag{g}$$

Equation (g) must be integrated to give the radial dependence of the stress needed in the force balance

$$F = -\int_0^R \sigma_{\theta\theta}(r) 2\pi r dr \tag{ssc (h)}$$

where the minus sign is entered to give a positive value of F. (Recall that θ is the 2 direction.)   Our only problem with the integration of equation (g) is evaluation of the constant of integration C in

$$p(r) = -(N_1 + 2N_2)\ln r + C \tag{i}$$

At $r = R$, we are tempted to set $p = P_A$, the atmospheric pressure, but this would be a mistake because P_A acting on the curved outer surface of the sample produces a radial stress that is equal to $-\sigma_{rr}(R)$. Because of the homogeneity of the flow, the pressure gradient dp/dr is also the same as any of the total-stress gradients, say, $-d\sigma_{rr}/dr$. So it makes sense to integrate equation (g) over $\sigma_{rr}(r)$ instead of $p(r)$ to give

$$\sigma_{rr} = (N_1 + 2N_2)\ln\frac{r}{R} + \sigma_{rr}(R) \tag{j}$$

Rather than plugging in $-P_A$ for $\sigma_{rr}(R)$, we will first deal with the fact that equation (h) needs $\sigma_{\theta\theta}$, not σ_{rr}. The difference is $N_2 = \sigma_{\theta\theta} - \sigma_{rr}$. At this point, we return to our usual assumption of $P_A = 0$, and get the equation for the stress inside the gap, namely,

$$\sigma_{\theta\theta}(r) = (N_1 + 2N_2)\ln\frac{r}{R} - N_2 \tag{k}$$

Of course, if the atmospheric pressure were not zero, we would need to include it in the pressure, but it would not contribute to F because the atmosphere also presses on the outside surface of the plate.

At this point, we are ready to substitute equation (k) into the integral, equation (h), and perform the integration, remembering that N_1 and N_2 are not dependent on radius. The ready-to-integrate equation is

$$F = -2\pi(N_1 + 2N_2)\int_0^R r\ln\frac{r}{R}\,dr + 2\pi N_2\int_0^R r\,dr \tag{l}$$

The first integral requires a look at the tables,[††††††] where we find that

$$\int x\ln\left(\frac{x}{a}\right)dx = \frac{1}{4}x^2\left[2\ln\left(\frac{x}{a}\right)-1\right] \tag{m}$$

Equation (l) then shapes up to be

$$F = -2\pi(N_1+2N_2)\frac{1}{4}\left[r^2(2\ln\frac{r}{R}-1)\right]_0^R + \pi R^2 N_2 \tag{n}$$

It definitely looks like we will have a problem at the lower limit, but the r^2 overwhelms the log term, so the lower limit is zero. Thus we have the (almost) final result of

$$F = -2\pi(N_1 + 2N_2)\frac{1}{4}[-R^2] + \pi R^2 N_2 \tag{o}$$

after entering the limit at $r = R$. Simplifying this gives

$$F = \frac{\pi R^2}{2}N_1 \tag{p}$$

where N_2 has disappeared! Solving for N_1 gives equation (7-46).

APPENDIX 7-5: DESKTOP RHEO-OPTICAL EXPERIMENT

One can gain experience with birefringence by stretching the gummy stuff that is used to stick the "credit card" to paper in the many offerings for credit cards that arrive in the mail. Rescue this material and allow it to relax. This will be the polymer sample we will examine in an extensional flow. A helper is required to do the stretching, while you peer at the sample through a crossed pair of linear polarizers.

[††††††] Or, a quick trip to the online integrator maintained by Wolfram, Inc., the publishers of Mathematica®. The current website is: http://integrals.wolfram.com/index.jsp.

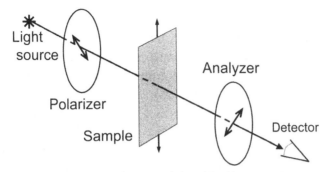

Figure 7-27. Simple optical setup for examining birefringence in a specimen subject to extensional deformation in the vertical direction. The crossed linear polarizers give a dark field.

Where does one get polarizers?   Some types of glasses for looking at 3D movies have linear polarizers as "lenses."   Polaroid sunglasses also have polarizers.   Good (expensive) Polaroid® sunglasses have excellent polarizers, but you may need to find two pairs, unless you are able to remove the lenses from the frames.   Polarizers are "crossed" when they cut the transmitted light to a minimum on rotation of one relative to another.   Examination of Figure 7-27 will help you to set up the optical "train" correctly.   As the helper stretches the gum, observe that the transmitted light increases and then decreases.   It increases because the polymer molecules in the gum are being oriented in response to the imposed stress.   Then why does it decrease?   To understand this, we need to understand what is going on.

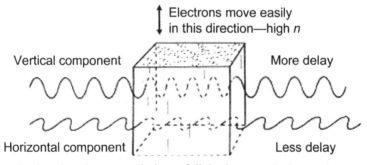

Figure 7-28.   Explanation for transmission of light by stretched sample (center) through crossed polarizers at −45° and +45° to stretch direction, as shown in Figure 7-27.

Light, being electromagnetic radiation, has an associated oscillating electric field.   A source of light such as an incandescent bulb emits radiation that has electric fields oscillating in planes tilted in all directions.   Such light is not polarized.   Polarizers are designed to pass light with electric vectors oscillating in the same plane and block light with an electric vectors oscillating in a plane perpendicular to the first.   The polarization direction is designated in Figure 7-

27 with a double-ended arrow. But which direction is that? The polarizer looks perfectly uniform, and may not have any markings.

One can find the polarization direction by examining light that is polarized in a known direction. If this is beginning to sound like a chicken-and-egg problem, rest assured there is a readily available source of polarized radiation with known polarization direction. This is light that is reflected off of a shiny dark surface, such as the hood of a (clean) car with dark paint. (That's why polarizing sunglasses help to cut the "glare.") As you face the source of the light, the polarization direction will be in the plane of the black surface, and perpendicular to the plane described by the path of the light as it moves from the source, to the surface and then to your eye. Knowing this, if you have handy a pair of Polaroid sunglasses, you have a reference to polarization direction—it's up and down. The lens will pass light that is polarized vertically, but will block light polarized horizontally (i.e., the glare).

Some computer displays provide a handy source of polarized light, and the light can be manipulated to provide almost a single wavelength, which will help with the experiment. If this is the case, select a drawing program and draw a large square on the screen. Adjust the color in the square to an intense red, and check the polarization again with one lens of your polarizing sunglasses. This optic is referred to as the analyzer. If your display is not polarized, revert to the previously described apparatus.

Now comes the critical part. Hold the analyzer such that the light is as completely blocked off as possible. This conditioned is called "crossed polarizers;" the polarizer and analyzer optics are producing a dark field. Ask the helper to hold the gum such that the stretch direction will be at 45° to the polarization direction of the source (and therefore the preferred direction of the analyzer). If the gum is perfectly relaxed, it will be dark under these conditions. Generally, it will not be perfectly dark due to residual orientation, contamination, roughness on the surface, etc., but these will be minor. While observing the transmitted light, stretch the sample. You should see the sample brighten; and, if all goes well, you may see it darken at higher strains as the transmitted components come back into phase.

PROBLEMS

7-1. A new flow-visualization technique for pressure-driven flows is being explored. The sample is prepared with a uniform concentration of microscopic particles, and the particles are assumed to stay uniformly distributed during testing. The concentration should be low enough so the particles do not interfere with each other. On flowing through a slit geometry, the particles at various depths into the flow stream are examined with a microscope that can focus at a given depth into the fluid. This is often called "optical slicing." Their velocities are not measured; instead, the particles passing a reference line are simply counted for a fixed time period. Of course, some judgment is required to decide if a particle is within the optical slice.

(a) If the particle concentration is N per cubic millimeter, derive the relationship between velocity V in mm/s and particle counting rate n in particles per second per millimeter of width if the "optical slice" thickness is δ_s.

(b) Calculate and plot the distribution of normalized counting rates from the wall to the midplane of a slit-flow channel with a 1-mm gap, assuming a Newtonian fluid and a location of optical slice that is remote from the edges of the slit. (Hint: normalize the counting rates for the total particle flow rate.)

(c) Data for a polymer melt are listed in the table below for tube flow where the optical slice is coincident with the axis of the tube. Plot this data and compare with the results of part b. Is the melt shear-thinning?

(d) Calculate the parameters for a power-law model that is the best description for the data of part c. Report the parameter values, along with their 95% confidence intervals.

(e) Suggest some problems with this method in comparison to particle image velocimetry via streak photography (look up this term).

Bin #	Bin location [a]	Counting rate, particles/min
1	0.05	77
2	0.12	77
3	0.22	71
4	0.33	65
5	0.39	57
6	0.45	52
7	0.56	42
8	0.67	34
9	0.78	22
10	0.91	10

[a] Location refers to the center of the bin and is given in terms of the tube radius R; thus, Bin 5 is centered at 0.39 R.

7-2. (Open end) The so-called barrel correction in capillary rheometry combines a slight pressure drop through the reservoir (important when the force on the ram is used to estimate the pressure) and a contribution due to the contraction ratio. A mechanical-energy balance suggests a pressure drop for Newtonian fluids that incorporates the kinetic energy of the fluid before and after the entrance. Explore what this amounts to for a polymeric fluid with a viscosity of 10 kPa s going through a 0.03-in. diameter capillary that has a length of 1 in. The barrel diameter is 0.375 in. Carefully list any assumptions.

7-3. Some experimentalists have attempted to find the entrance pressure drop directly by using a capillary with a length of zero. The schematic below shows how these are designed. Explain why the pressure drop through this geometry might be less than the extrapolated value based on longer capillaries at high shear rates. (Hint: think in terms of stored energy.) Find the points in the Bagley plot in Problem 7-4 that would tend to support this view.

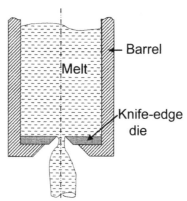

Figure 7-29. Schematic of a knife-edge or orifice die, i.e., a capillary with an $L/D = 0$.

7-4. (Computer) The graph below is from the 1957 paper of Bagley[1] showing the original "Bagley plot" for a 2.9-MI LDPE. Use the information in this graph, along with the Rabinowitsch correction, to determine the viscosity corresponding to the nominal shear rate ($4Q/\pi R^3$) of 90 s^{-1}. Report your answer in units of Pa s, along with the corrected shear rate.

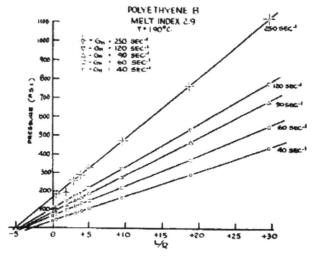

Figure 7-30. Copy of original Bagley plot. [Reprinted with permission from E. B. Bagley, *J. Appl. Phys.*, **28**, 624-627 (1957). Copyright 1957, American Institute of Physics.][1]

7-5. The velocity profile in a capillary for a Newtonian fluid is given by equation (7-11). Its form suggests that for a power-law fluid, the profile might be given by the equation

$$\frac{V_z(r)}{V_0} = 1 - \left(\frac{r}{R}\right)^k \qquad (7\text{-}65)$$

where k has some relationship to n of the power-law equation $\eta = m\dot{\gamma}^{n-1}$. Find out if this equation is correct or not. If it is correct, relate k to n.

7-6. The capillary data[48] entered below are for a 1.5% solution of carboxymethyl cellulose (CMC), a water-soluble polymer often used to thicken foods such as milkshakes.

(a) Run a simple check for the presence of slip using selected capillaries.

(b) Process these data using the Bagley end correction, the Mooney slip correction (if needed) and, finally, the Weissenberg-Rabinowitsch-Mooney (WRM) equation. Plot viscosity vs. shear rate using both the uncorrected $(4Q/\pi R^3)$ and corrected shear rates. Comment on these. Include in your plot the totally uncorrected data.

(c) Plot your results as viscosity vs. stress and compare with the effect of the WRM correction with a simple shift along the shear-stress axis of −0.097 decades, corresponding to an assignment of the viscosity to a shear stress of 4/5 of the actual stress.

D, cm	L, cm	ΔP, psi	Q, cm³/s
0.271	94.4	20.0	0.591
		39.3	2.95
		59.5	8.32
		85.8	21.0
0.271	67.4	15.0	0.733
		20.0	1.26
		39.5	7.06
		58.0	17.6
		79.5	35.1
0.182	63.4	20.5	0.192
		28.0	0.477
		37.5	0.801
		59.0	2.58
		79.5	5.56
		100.3	9.34
0.182	45.2	9.0	0.067
		20.5	0.35
		31.0	0.972
		45.0	2.72
		59.5	5.75
		75.0	9.84
0.0856	29.7	41.0	0.0906
		81.0	0.581
		100.0	1.0
		126.5	1.74
		154.0	2.66
0.0856	20.95	31.3	0.108
		55.0	0.506
		70.0	0.947
		91.5	1.81
		113.5	3.2
		134.0	3.85

7-7. For tube flow, is the power-law index n the same for a power-law fluid if $4Q/\pi R^3$ is used, as opposed to $\dot{\gamma}_W$? Recall that for a power-law fluid, $\tau_{rz} = m\dot{\gamma}^{n-1}\dot{\gamma}_{rz}$, where $\dot{\gamma}$ is the magnitude of the shear rate. Show your reasoning.

7-8. If only one observation is available, the WRM correction cannot be run. However, approximations have been proposed. One[49] is to use the wall shear stress and $4Q/\pi R^3$ to calculate the viscosity, but assign this value to 4/5 of the wall shear stress, not to the actual shear stress. Test this approximation for a power-law fluid with an exponent $n = 0.5$.

7-9. Derive equation (7-38) in more detail than what was provided in the text.

7-10. The velocities associated with the data in Problem 7-6 are quite high.

(a) Using the mechanical-energy balance, decide if the kinetic-energy contribution to the pressure drop is important, or not.

(b) What influence would kinetic energy have on the Bagley plot and the corrected shear stress?

7-11. In the Couette fixture, the bob design featuring a cone on the bottom is very popular because if makes maximum use of the sample. What angle should the cone have so that the entire sample is subjected to the same shear rate?

7-12. One method for eliminating concerns about the contributions of top and bottom of the bob of the Couette geometry is to fill the annular gap only partially, but to different heights. If the height of the fluid along the side of the bob can be measured, what procedure should be followed to find the true shear stress in the fluid? Assume a very small gap.

7-13. Using the definitions of N_1 and N_2 given in equation (6-1) and equation (6-10), respectively, show that equation (7-50) does indeed follow from equation (7-49).

7-14. Test out the single-point method suggested for the parallel-plate fixture using a power-law fluid. Report the maximum error in viscosity as a function of the power-law exponent n.

7-15. The cone of a 50-mm cone-and-plate fixture set of one rheometer was measured with a digital caliper and found to be 49.99 ± 0.01 mm in diameter, where the ± 0.01 is the unbiased estimate of the standard error of the mean of four measurements taken around the circumference. The plate half of the set was a bit dirty in spots from hard use, but was measured at 50.01 ± 0.02 mm again in the same fashion and with the same caliper. The cone angle was measured by the manufacturer as 0.0401 rad. Are these fixtures consistent with the aim of establishing a flow with spherical symmetry?

7-16. Verify that the combination $N_1 + 2N_2$ shown in equation (g) of Appendix 7-4 is consistent with the combination of stresses shown in equation (f).

7-18. Verify that the use of spherical coordinates will give the same value for torque as that shown in equation (7-38) for the cone-and-plate rheometer.

7-19. Example 7-4 shows how to run a biaxial extension experiment at constant applied stress using the lubricated squeezing-flow geometry. If the machine you have available is position controlled, how must the crosshead position be programmed to give constant stretching rate, dV_r/dr?

7-20. From previous measurements, we know that silicone SE-30 resin has a zero-shear-rate viscosity of about 20 kPa s at 30 °C; however, we are anxious to get a more precise value using a rolling-ball viscometer with a tube diameter of 8 mm. The most appropriate ball available is 3 mm in diameter and made from tungsten carbide. What will be the transit time between two marks 1 cm apart if the tilt angle is held at 30°?

7-21. The discussion in Chapter 7 suggests that the speed resolution of rheo-NMR is around 10 μm/s; that is, this speed can be distinguished from a speed of zero, but lower speeds cannot, at least reliably. Compare this value with the resolution of a laser-Doppler setup with He-Ne lasers crossed at 5° (included angle). The lasers can be turned on for only 1 s before significantly heating the fluid in the path of a particle.

7-22. The data in Figure 7-22 were gathered using a slit die with a 2-mm gap. The illumination was a He-Ne laser. The polarizer and analyzer were held at −45° and +45 with respect to the flow direction. In this geometry, maximum brightness is obtained when the "path length" for component of polarized light in the flow direction (x_1) is exactly a half wavelength longer than for light polarized in the transverse direction (x_3), i.e., the phase difference is π. Path length is put in quotes because, of course, the actual physical distance, L, is the same, but the light goes slower if it is polarized in the direction of the higher refractive index; thus, its phase ϕ has shifted by π for maximum brightness. Of course, both components are in phase before entering the path, and their total phase on exiting will be ωt, where ω is the frequency of vibration (rad/s) and t is the time of transit.

(a) With this information at hand, calculate the birefringence value at the first maximum.

(b) Assuming the stress-optical coefficient C_{13} is that for PBD with 47% trans 1,4; 43% cis 1,4 and 10% 1,2 configuration (3.6×10^{-9} Pa^{-1} at RT),[50] calculate the value of $N_1 + N_2$.

7-23. Figure 7-23 shows a setup for light scattering from a sample confined between parallel plates. The caption suggests, somewhat incorrectly, that the shear stress can be measured simultaneously, presumably in the scattering volume. Considering the single-point method described in Chapter 7 for this geometry, what would be a good location for the light path, and what would the shear stress be at this point, knowing the torque M, the gap H, the plate radius R and rotational speed Ω?

7-24. The intrinsic birefringence of a polymer is considered to be that measured if all the chains were stretched straight in a single direction. Of course, this is virtually impossible to do, but can be approached in a uniaxial extension experiment. High refractive index is associated with ease of electron movement, which is often, but not always, in the direction of the main chain. According to Ferry,[51] a PBD containing a significant amount of 1,2 isomer has a negative birefringence. What might be the reason for this, and what other polymers might exhibit similar behavior?

7-25. Verify equation (7-64) for the sedimentation of a spherical particle starting with the relationships for buoyancy and drag force on a moving sphere.

7-26. (Challenging) Find a resource that describes the force exerted by a magnetic field on a paramagnetic particle.

(a) As a first step, examine the units of magnetic field strength vs. magnetic flux density. How are they related? Find the relationship between the units Gauss and Tesla. Which is greater, and by how much?

(b) (Open end) A Helmholtz coil is a convenient way of setting up a known field gradient. The expression for the gradient of a single coil is

$$\frac{dB}{dz} = \frac{3\mu_0 N i R^2 z}{\left(R^2 + z^2\right)^{5/2}}$$

where B is the magnetic flux density, z is the distance in the axial direction up or down through the coil of N turns carrying a current i. The magnetic susceptibility of space is μ_0. The SI value of μ_0 is $4\pi \times 10^{-7}$, with units of kg m/s^{2}A^2 [=] N/A^2. Rather than fooling with a Helmholtz coil, though, you propose to use your local NMR superconducting solenoid magnet. Find the expression for the on-axis field produced by a solenoid, and program this onto a spread sheet. Using the geometry of the NMR magnet at your institution, calculate the location and strength of the maximum field gradient that is accessible for your experiment. (If you do not have a magnet nearby, use $R_i = 88$ mm, $R_o = 237$ mm, $L = 429$ mm and $z_1 = 507$ mm for inside and outside radii of the solenoid, respectively; the length of the solenoid; and the distance from the center of the solenoid to the outside at the bottom of the magnet. The strength of the field at the center of the solenoid is 7.05 T.)

(c) The force produced by a field gradient dB/dz on a particle is

$$F = \mu \frac{dB}{dz}$$

where μ is the magnetic dipole induced by the field. Find the value of μ in the position you have chosen for a magnetite particle 100 nm in diameter.

(d) Using the particle described in part (c), calculate and compare the distance moved by the particle in your field from part (b) vs. the average distance moved by Brownian motion.

7-27. Using an old-fashioned cork screw to remove the cork from a wine bottle can involve a force of roughly 20 lb$_f$, or even more if the bottle is old. To reduce this effort (and cork breakage), a device was introduced that blew the cork out by pressurizing the space in the bottle with CO_2 delivered via a needle.

(a)   Calculate the pressure required to remove the cork assuming a cork diameter of 2.0 cm and a length of 4.3 cm.

(b)   Find the shear stress at the cork-glass interface.

(c)   Plot the shear-stress profile inside the cork.

(d)   Estimate the peak tensile stress in the glass at the peak pressure. Use the thin-shell approximation and check using the appropriate equation of motion. A typical bottle for still wine is 3 in. in diameter with 2-mm-thick glass.

(e)   What is the influence on the stress in the glass due to the wine in the bottle.

7-28. (Computer) The parallel-plate geometry is a favorite for quick measurement of viscosity. The unwary user may, however, report the apparent (Newtonian) viscosity, as listed by some software packages, rather than the actual viscosity. Assuming the fluid obeys the two-parameter Ferry equation, what will be the error if the shear rate is equal to the characteristic shear rate, i.e., just entering the non-Newtonian region?

REFERENCES

1. E. B. Bagley, "End corrections in the capillary flow of polyethylene," *J. Appl. Phys.*, **28**, 624–627 (1957).

2. J. M. Dealy, "Weissenberg and Deborah numbers—Their definition and use," *Rheol. Bull.*, **79**(2), 14–18 (2010).

3. F. Koran and J. M. Dealy, "A high pressure sliding plate rheometer for polymer melts," *J. Rheol.*, **43**, 1279–1290 (1999).

4. P. T. Mather, A. Romo-Uribe, C. D. Han and S. S. Kim, "Rheo-optical evidence of a flow-induced isotropic-nematic transition in a thermotropic liquid-crystalline polymer," *Macromolecules*, **30**, 7977–7989 (1997).

5. Y. An and M. T. Shaw, "Actuating properties of soft gels with ordered iron particles: Basis for a shear actuator," *Smart Mater. Struct.*, **12**, 157–163 (2003).

6. M. Mackay and G. K. Dick, "A comparison of the dimensions of a truncated cone measured with various techniques: The cone measurement project," *J. Rheol.*, **39**, 673–677 (1995).

7. R. I. Tanner, "A theory of die-swell," *J. Polym. Sci. Part A: Polym. Chem.*, **8**, 2067–2078 (1970).

8. N. Y. Tuna and B. A. Finlayson, "Exit pressure experiments for low density polyethylene melts," *J. Rheol.* **32**, 285–308 (1988).

9. L. M. Quinzani, R. C. Armstrong, and R. A. Brown, "Use of coupled birefringence and LDV studies of flow through a planar contraction to test constitutive equations for concentrated polymer solutions," *J. Rheol.*, **39**, 1201–1228 (1995).

10. H. J. Park, D. G. Kiriakidis, E. Mitsoulis, and K. J. Lee, "Birefringence studies in die flows of an HDPE melt," *J. Rheol.*, **36**, 1563–1583 (1992).

11. T. Hasegawa, H. Asama and T. Narumi, *Nihon Reoroji Gakkaishi (J. Rheol. Japan)*, **31**, 243–252 (2003).

12. J. C. Savins, "A pitot tube method for measuring the first normal stress difference and its influence on laminar velocity profile determinations," *AIChE J.*, **11**, 673–677 (1965).

13. S. Okubo and Y. Hori, "Experimental determination of secondary normal stress difference in annular flow of polymer melts," *J. Rheol.*, **24**, 275–286 (1980).

14. D. G. Baird, "A possible method for determining normal stress differences from hole pressure error data," *Trans. Soc. Rheol.*, **192**, 147–151 (1975).

15. D. Hadjistamov, "Determination of the onset of shear thinning of polydimethyl siloxane," *J. Appl. Polym. Sci.*, **108**, 2356–2364 (2008).

16. S.-G. Baek and J. J. Magda, "Monolithic rheometer plate fabricated using silicon micromachining technology and containing miniature pressure sensors for N_1 and N_2 measurements," *J. Rheol.*, **47**, 1249–1260 (2003).

17. F. T. Trouton, "On the coefficient of viscous traction and its relation to that of viscosity," *Proc. Roy. Soc.*, **A77**, 426–440 (1906).

18. I.-J. Chen, G. E. Hagler, L. E. Abbott, D. C. Bogue and J. L. White, "Interpretation of tensile and melt spinning experiments of low density and high density polyethylene," *Trans. Soc. Rheol.*, **16**, 473–494 (1972).

19. M. T. Shaw and F. B. Lin, "Material functions in extension using arbitrary deformation programs," in *Current Topics in Polymer Science Vol. II* (S. Inoue, L. A. Utracki and R. M. Ottenbrite, eds.), Hanser, Munich, 1987, pp. 137–148.

20. M. L. Sentmanat, "Miniature universal testing platform: from extensional melt rheology to solid-state deformation behavior," *Rheol. Acta*, **43**, 657–669 (2004).

21. P. H. Paul, M. G. Garguilo and O. J. Rakestraw, "Imaging of pressure- and electrokinetically driven flows through open capillaries," *Anal. Chem.*, **70**, 2459–2467 (1998).

22. C. D. McGrady and D. G. Baird, "Note: Method for overcoming ductile failure in Münstedt-type extensional rheometers," *J. Rheol.*, **53**, 539–445 (2009).

23. K. K. Chao and M. C. Williams, "The ductless siphon: A useful test for evaluating dilute polymer solution elongational behavior. Consistency with molecular theory and parameters," *J. Rheol.*, **27**, 451–474 (1983).

24. T. Sridhar and R. K. Gupta, "Material properties of viscoelastic liquids in uniaxial extension," *J. Rheol.*, **35**, 363–377 (1991).

25. W. R. Burghardt and J-M. Li, "Uniaxial extensional characterization of a shear thinning fluid using axisymmetric flow birefringence," *J. Rheol.*, **43**, 147–165 (1999).

26. H. Münstedt, "New universal extensional rheometer for polymer melts. measurements on a polystyrene sample," *J. Rheol.*, **23**, 421–436 (1979).

27. J. R. Collier, O. Romanoschi and S. Petrovan, "Elongational rheology of polymer melts and solutions," *J. Appl. Polym. Sci.*, **69**, 2357–2367 (1998).

28. D. G. Baird, T. W. Chan, C. McGrady and S. M. Mazahir, "Evaluation of the use a semi-hyperbolic die for measuring elongational viscosity of polymer melts," *Appl. Rheol.*, **20**, 34900-1 – 34900-12 (2010).

29. M. Sentmanat, B. N. Wang and G. H. McKinley, "Measuring the transient extensional rheology of polyethylene melts using the SER universal testing platform," *J. Rheol.*, **49**, 585–606 (2005).

30. P. Hachmann and J. Meissner, "Rheometer for equibiaxial and planar elongations of polymer melts," *J. Rheol.*, **47**, 989–1010 (2003).

31. D. D. Joye, G. W. Poehlein and C. D. Denson, "A bubble inflation technique for the measurement of viscoelastic properties in equal biaxial extensional flow," *J. Rheol.*, **16**, 421–445 (1972).

32. P. A. O'Connell and G. B. McKenna, "Novel nanobubble inflation method for determining the viscoelastic properties of ultrathin polymer films," *Rev. Sci. Instrum.*, **78**, 013901–013911 (2007).

33. E. B. Bagley and D. D. Christianson. "Uniaxial compression of viscoelastic rings-effect of friction at the platen/sample interface for gels and doughs," *J. Rheol.*, **32**, 555–573 (1988).

34. D. C. Venerus, T.-Y. Shiu, T. Kashyap and J. Hosttetler, "Continuous lubricated squeezing flow: A novel technique for equibiaxial elongational viscosity measurements on polymer melts," *J. Rheol.*, **54**, 1083–1095 (2010).

35. R. L. Powell, L. A. Mondy, G. G. Stoker, W. J. Milliken and A. L. Graham, "Development of a falling ball rheometer with applications to opaque systems: measurements of the rheology of suspensions of rods," *J. Rheol.*, **33**, 1173–1188 (1989).

36. C. S. Nickerson and J. A. Kornfield, "A 'cleat' geometry for suppressing wall slip," *J. Rheol.*, **49**, 865–874 (2005).

37. L. Zhu, N. Sun, K. Papadopoulos and D. De Kee, "A slotted plate device for measuring yield stress," *J. Rheol.*, **45**, 1105–1122 (2001).

38. J. R. Van Wazer, J. W. Lyons, K. Y. Kim and R. E. Colwell, *Viscosity and Flow Measurement*, Wiley, New York, 1963.

39. See, for example, the website: www.rheo-NMR.com.

40. P. T. Callaghan, "Rheo-NMR: A new window on the rheology of complex fluids," in *Encyclopedia of Nuclear Magnetic Resonance, Vol. 9: Advances in NMR* (D. M. Grant and R. K. Harris, eds.), Wiley, New York, 2002.

41. Y.-W. Inn, L. Wang and M. T. Shaw, "Efforts to find stick-slip flow in the land of a die under sharkskin melt fracture conditions: Polybutadiene," *Macromol. Symp.*, **158**, 65–75 (2000).

42. R.-J. Wu, M. T. Shaw and R. A. Weiss, "A rheo-light-scattering instrument for the study of phase behavior of polymer blends under simple-shear flow," *Rev. Sci. Instrum.*, **66**, 2914–2921 (1995).

43. Z. Hong, M. T. Shaw and R. A. Weiss, "Effect of shear flow on the morphology and phase behavior of a near-critical SAN/PMMA blend," *Macromolecules*, **31**, 6211–6216 (1998).

44. B. Wang, J. Guan, S. M. Anthony, S. C. Bae, K. S. Schweizer and S. Granick, "Confining potential when a biopolymer filament reptates," *Phys. Rev. Lett.*, **104**, 118301 (2010). (4 pages)

45. W. H. Press, B. P. Flannery, S. A. Teukolsky and W. T. Vetterling, *Numerical Recipes in FORTRAN: The Art of Scientific Computing*, 2nd edition, Cambridge University Press, Cambridge, 1992.

46. H. C. Hershey, J. L. Zakin and R. Simha, "Numerical Differentiation of Equally Spaced and Not Equally Spaced Experimental Data," *I&EC Fund.*, **6**, 413–421 (1967).

47. M. Mooney, "Explicit Formulas for Slip and Fluidity," *J. Rheol.*, **2**, 210–222 (1931).

48. S. Middleman, *The Flow of High Polymers*, Wiley-Interscience, New York, 1968.

49. M. T. Shaw and Z. Liu, "Single-point determination of nonlinear rheological data from parallel-plate torsional flow," *Appl. Rheol.*, **16**, 70–79 (2006).

50. J. E. Mark and M. A. Llorente, "Photoelastic studies of some polybutadiene networks," *Polym. J.*, **13**, 543–553 (1981).

51. J. W. M. Noordermeer and J. D. Ferry, "Nonlinear relaxation of stress and birefringence in simple extension of 1,2-polybutadiene," *J. Polym. Sci, Polym. Phys. Ed.*, **14**, 509–520 (1985).

8

Strain, Small and Large

This is the most difficult chapter in the book. It requires grasping a certain concept that is very foreign to many people. This concept is that of a constantly changing reference for measuring the strain in a material. Without this concept, we cannot deal with large strains applied to fluids.

How large is large when it comes to strains? For with very elastic solids, we can start to see the effects we are looking for at low strains. But typical polymer melts and solutions—non-Newtonian, viscoelastic fluids—are used at large strains. Indeed, for fluids, the steady state means that the strains, relative to the starting point, are indefinitely large. So for fluids at large strains, the present position will be assigned a strain of zero. Weird, yes, but the fact of the matter is, there is no other choice. Think about it; if the fluid has been flowing for a very long time, the beginning state has been lost and forgotten in the far-distant past.

A. DISPLACEMENT

Displacement is movement away from the reference position. If we move the entire sample, this is referred to as *bulk displacement*. As with rotation, bulk displacement fails to produce any deformation, assuming inertial effects are negligible. The common method to apply "inertial effects" is to shake the material. Try shaking a jar of silicone gum to get a feeling for the very large accelerations that would be needed to generate deformation.

However, displacement is an important aspect of large strain, so we need to be more precise. Thus, once again, we start with rectangular coordinates, keeping in mind that the origin and tilt of the coordinates are arbitrary, but are chosen to make life easier. Again, the displacements will be in the x direction, which we also call the 1 direction. The distance along in the 1 direction is x_1; 2 direction, x_2; and 3 direction, x_3.

Some other notational considerations: we use primed variable symbols to declare a past location, and unprimed symbols for the present position. We don't worry about future positions, as we are interested in strains and stresses now, and we implicitly assume that future strains cannot influence the present state.

With these things decided, we can write down some equations for displacements for simple shear strain of γ. The strain will be a function of time, and may have been applied for a very long time in a continuous fashion. We really don't care when it started. Seems strange, but watch. First we need a viable definition of strain. As with an odometer reading, we need a starting point and a finishing point, so we set the trip odometer to zero at the start, and read the mileage at the end. But suppose the odometer is broken, although the speedometer is still functioning. What can be done is take speed readings every second, and sum them to get the mileage. The distance traveled D will then be

$$D(t,t') = \int_t^{t'} S(s)ds \qquad (8\text{-}1)$$

where S is the speed, t is the arrival time, and t' represents any time during the trip. The variable s is a dummy integration variable taking on values from t' to t, and ultimately does not enter into the answer. Note that if $t' = t$, the distance traveled is zero, as it should be. Also note that because $t \geq t'$, which we have declared, D will have negative values. So, if one says "10 miles back (negative direction), we had a flat tire," no one has a problem.

If the speed is constant, then the distance is given by the simple expression

$$D(t, t') = S \times (t'-t) = -S \times (t-t') \qquad (8\text{-}2)$$

The multiply sign has been inserted to emphasize that the speed is multiplied by the time traveled, as opposed to the speed being a function of the time difference, which it may well be. From now on, the multiply sign will not be provided; use the units to distinguish the two possibilities.

As applied to displacement, the result is the same. As applied to strain, we have the very important definition

$$\gamma(t,t') = \int_t^{t'} \dot{\gamma}(s)ds \qquad (8\text{-}3)$$

where the symbol $\dot{\gamma}$ is our familiar shear rate.

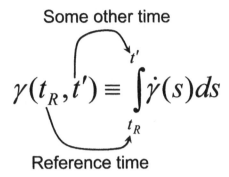

Some other time

$$\gamma(t_R, t') \equiv \int_{t_R}^{t'} \dot{\gamma}(s)\,ds$$

Reference time

Figure 8-1. Shear strain definition. If $t_R = 0$, the term "lab strain" is used. For liquids, $t_R = t$, the present time; in this case the strains are negative for positive shear rate.

Example 8-1: *Consider a typical shear startup experiment in which a sample is at rest for a long time, and then is sheared at a constant rate of 1 s^{-1} for ten seconds before stopping the shear. Draw strain vs. time using 11 s after the startup as the present time.*

Our instinct is to repeat what we have learned in linear viscoelasticity. Suppress this urge, and use equation (8-3), which we can integrate piecewise, or by inspection. At our reference time, which becomes the new zero time, the strain is 0, and was so for 1 s back. Traveling further back to the segment between -1 and -11 s, the strain decreases from zero to -10 at -11 s, where it remains for all time before that. The result is summarized in Figure 8-2.

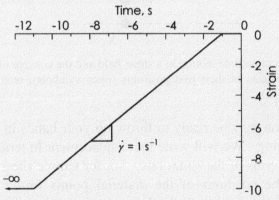

Figure 8-2. Strain history for a sample sheared for 10 s at a shear rate of 1 s^{-1}.

There are a couple of things to note about this result. One is that for positive deformation rates, the past strains, relative to the reference state at the present time, are negative. Especially note that the strain is not zero before the experiment starts. This sounds mad, but we shall see how this all works out. Note also that the shear rate is positive and looks the same as always. No change there; it retains its sense of being the derivative of the strain with time.

If you are having trouble at this point, stop. Go over Example 8-1 again. Do not go on until this exercise is completely natural.

So, with the concept of the present position as the reference in hand, we are prepared to talk about displacement. In the shearing example, the material particles are displaced from left to right. However, with the present as the reference state, we say the displacements are to the left, and negative. If we choose to use the symbol u for displacement, then we have the statement

$$u_1 = x_1' - x_1 \tag{8-4}$$

where we have used to subscript 1 to designate the x direction. The location on the x axis at the present time is x_1 (not primed), while the prime denotes x positions at past times. For flow in the positive x_1 direction, we can see that u_1 values (displacements at times past) are negative, as depicted in Figure 8-3.

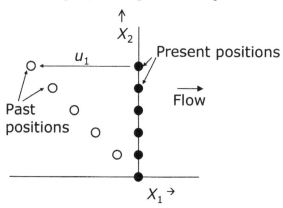

Figure 8-3. Depiction of particle motion in a shear field and the concept of a negative value of displacement u of particles in their past positions (open symbols), relative to their present positions (filled symbols).

At this juncture, you may be ready to throw up your hands in despair, but first let's check something. We will write the displacement in terms of the strain γ, and eventually in terms of the strain rate. So, for simple shear with the bottom plate stationary, the positions of the material points further away from the bottom will have been further to the left (negative x_1) at times past. This is illustrated in Figure 8-3. Can we write

$$u_1 = \gamma x_2 \tag{8-5}$$

without getting into difficulty? Looks like u_1 will increase with x_2, which is fine, but what about the sign? Recalling the γ is negative, we can see that this equation correctly predicts that the material points originated from a greater distance to the left. Thus, at least everything seems to be consistent.

It's high time to ask what all this buys us. The important aspect is that using the present as the reference and calculating displacements backwards, allows us to handle large strains; and, in particular, strains that have been going on forever. In linear viscoelasticity, this concept is snuck in when discussing continuous oscillation in dynamic experiments. In that analysis, the point where we want the stress is assigned $t' = 0$, whereas the experiment starts at $t' = -\infty$. Otherwise, we would really have some difficulty getting the answer.

B. INFINITESIMAL STRAIN

Speaking of linear viscoelasticity, it's also time to do a reality check, because our developing concept of *finite strain* must be consistent with results using *infinitesimal strain*. We will use the stress-growth experiment in shear as our test, where the strain increases linearly after the start of the experiment at zero time.

The familiar expression from linear viscoelasticity for stress growth is based on the Boltzmann equation

$$\boldsymbol{\tau}(t) = \int_{-\infty}^{t} G(t - t')\dot{\boldsymbol{\gamma}}\, dt' \qquad \text{(ssc)}^* \text{ (8-6a)}$$

or, for simple shear,

$$\tau(t) = \int_{-\infty}^{t} G(t - t')\dot{\gamma}\, dt' \qquad \text{(ssc) (8-6b)}$$

where $G(t - t')$ is the shear stress-relaxation modulus, τ is the shear stress and $\dot{\gamma}$ is the shear rate. (Note that we have dropped the subscripts from τ_{21} to simplify the presentation.) For (fsc), insert a negative sign before the integral. By way of review, we divide this integral into two parts: one for $-\infty < t' < 0$ and one for $0 \leq t' \leq t$, where the shearing starts at $t' = 0$. The first one is discarded because the rate of strain is zero, while the second is integrated to yield the familiar exponential growth. No problem.

---

* See Appendix 3-3 for definitions of (ssc) and (fsc).

Instead, suppose we wish to do this same problem in terms of shear strain, so we can compare it with the finite-strain concept? This is not particularly easy, but let's give it a try. The first step is to recall the formula for integration by parts:

$$\int u\,dv = uv - \int v\,du \qquad (8\text{-}7)$$

To keep things simple, we want to pick the u and v very wisely. First we recognize that the shear rate can be described by

$$\dot{\gamma}(t') = \frac{\partial \gamma(t')}{\partial t'} \qquad (8\text{-}8)$$

no matter what we chose for the reference position (review Figures 8-1 and 8-2). So the substitutions that make the most sense are

$$v = \gamma(t') \quad dv = \frac{\partial \gamma(t')}{\partial t'}\,dt' \qquad (8\text{-}9\text{a})$$

$$u = G(t - t') \quad du = \frac{\partial G(t - t')}{\partial t'}\,dt' \qquad (8\text{-}9\text{b})$$

Substituting these into equation (8-7), as applied to equation (8-6b), gives the result

$$\tau(t) = \gamma(t')G(t - t')\Big|_{-\infty}^{t} - \int_{-\infty}^{t} \gamma(t') \frac{\partial G(t - t')}{\partial t'}\,dt' \qquad \text{(ssc)} \ (8\text{-}10)$$

It would be very nice if the uv term disappeared, but note that the upper limit gives the result $\gamma(t)G(0)$, which most certainly is not zero if we are using the start of the deformation as the reference position. The lower limit, $\gamma(-\infty)G(\infty)$, will go to zero for two reasons: the strain at $-\infty$ is assumed to be zero; and, if we assume we have a fluid, $G(t)$ will eventually go to zero, and usually exponentially.

With this development in mind, an examination of the situation using the present position for the reference is in order. The starting point is exactly the same, because the shear rate is independent of the reference. However, the choice for the variable v in equation (8-7) requires a slight but important change to

$$v = \gamma(t,t') \quad dv = \frac{\partial \gamma(t,t')}{\partial t'} dt' \tag{8-11}$$

The variable u stays the same. Thus, integration by parts gives

$$\tau(t) = \gamma(t,t')G(t-t')\Big|_{-\infty}^{t} - \int_{-\infty}^{t} \gamma(t,t') \frac{\partial G(t-t')}{\partial t'} dt' \quad \text{(ssc) (8-12)}$$

Now, once again, we examine the uv term. The upper limit gives $\gamma(t,t)G(0)$, which is most definitely zero because $\gamma(t,t)$ is zero by definition and $G(0)$ cannot be indefinitely large for real materials. So, how about the lower limit, $\gamma(-\infty)G(\infty)$, which looks like it might be a problem due to the strain at $-\infty$ increasing in magnitude as the shearing continues. However, the exponentially decreasing modulus will always win in the end (see Problem 8-2).

After all this, we can safely state that for fluids, the conversion of the usual Boltzmann equation to strain is given by the expression

$$\tau(t) = - \int_{-\infty}^{t} \gamma(t,t') \frac{\partial G(t-t')}{\partial t'} dt' \quad \text{(ssc) (8-13)}$$

One small change is often made, because the derivative of G is negative. We can see this by using the Maxwell model expression, which is

$$G(x) = G_0 e^{-x/\tau}; \quad \frac{\partial G(x)}{\partial x} = -\frac{G_0}{\tau} e^{-x/\tau} \tag{8-14}$$

But our concern is the derivative of $G(t-t')$. This would also be negative if we follow the formula with $x = t - t'$. But if we use

$$\frac{\partial G(t-t')}{\partial t'} = \frac{G_0}{\tau} e^{-(t-t')/\tau} \tag{8-15}$$

the result is positive!

Why is this important? The reason is that the derivative, as shown in equation (8-15), is a material function called the *memory function*, and is usually given the symbol M.[†] As a material function, it is most appropriate to have it be a positive quantity. Thus

---

[†] Be aware that other definitions of the memory function $M(t)$ have been used; there appears to be no standardization. There was also an attempt, admittedly sensible, to change its name to

$$M(t,t') \equiv \frac{\partial G(t-t')}{\partial t'} \tag{8-16}$$

With this definition, equation (8-13) becomes

$$\tau(t) = -\int_{-\infty}^{t} M(t-t')\gamma(t,t')dt' \qquad \text{(ssc)} \tag{8-17}$$

where each term is now very precisely defined.

We might generalize equation (8-17) to three dimensions, as shown below:

$$\boldsymbol{\tau}(t) = -\int_{-\infty}^{t} M(t-t')\boldsymbol{\gamma}(t,t')dt' \qquad \text{(ssc)} \tag{8-18}$$

However, the three-dimensional array $\boldsymbol{\gamma}$ has not been defined. Therein lies the remainder of this chapter. But for small strains, we can assert that the answer is similar to the simple formula for the components of the rate-of-deformation tensor, $\dot{\boldsymbol{\gamma}}$. Following this we have that

$$\gamma_{ij} = \frac{\partial u_i}{\partial x_j} + \frac{\partial u_j}{\partial x_i} \tag{8-19}$$

where the u_i are the displacements.

Let's see if this works for simple shear where the shear rate is $\dot{\gamma}_0$ and the shearing has been going on for a long time. Revisiting equation (8-5) we have

$$u_1 = \gamma(t,t')x_2 \tag{8-5}$$

and displacements in all other directions are zero. Recall that at $t' = t$ the displacement u_1 in the 1 direction is zero and at $t' < t$ the displacement is negative. Following the prescription in equation (8-19) and applying the definition of shear strain given in equation (8-3) gives the simple result that

$$\gamma_{21} = \gamma_{12} = \int_{t}^{t'} \dot{\gamma}(s)ds = \dot{\gamma}_0(t'-t) \tag{8-20}$$

---

something like "forgetting function," because M leads to "forgetting" of stresses created by past strains. Attempts to change established names and conventions are usually futile.

as expected. But let's try to calculate the shear stress, and thus the viscosity. To keep things simple, we chose the Maxwell-model relaxation function,[‡] giving the expression

$$\tau(t) = -\int_{-\infty}^{t}\frac{G_0}{\tau}e^{-(t-t')/\tau}\dot{\gamma}_0(t'-t)dt' \qquad \text{(ssc) (8-21)}$$

Removing constants from the integral and getting rid of the minus sign gives

$$\tau(t) = \frac{G_0\dot{\gamma}_0}{\tau}e^{-t/\tau}\int_{-\infty}^{t}(t-t')e^{+t'/\tau}dt' \qquad \text{(ssc) (8-22)}$$

There are two integrals in this equation. As t is a constant with respect to the integration, the first term is easy; it's just $t\tau\exp(t/\tau)$,[§] as the lower limit is zero. The second integral may require a quick look at the tables or an online integrator.[**] The form is

$$\int xe^{x/a}dx = ae^{x/a}(x-a) \qquad (8\text{-}23)$$

The end result, after putting in the limits and recognizing the usual simplifications, is the totally expected fact that the viscosity of the Maxwell model is simply $G_0\tau$. This exercise seems like a lot of work for a simple result, but it does show that the use of the present as a reference works just fine.

What about normal stresses in simple shear for the model described by equation (8-18)? Let's simply apply the formula to calculate $N_1 = \tau_{11} - \tau_{22}$. For τ_{11}, we have the displacement gradients of u_1 to worry about. Specifically, according to equation (8-19), we need the gradient of u_1 in the 1 direction, but u_1 is only a function of x_2. For τ_{22}, it's the displacements in the 2 direction that count, but there are no displacements in that direction. Thus, equation (8-18) does not predict normal stresses in simple shear. This model, in spite of the use of the present configuration as the reference state, is still only another form of

---

[‡] There is some concern, of course, that a single-exponential relaxation may be a special case. However, to a large extent, any physical relaxation behavior can be expressed as sum of many exponential terms. Each of these terms can be integrated in the same fashion as the example, so the resulting terms will have the same form as above.

[§] Note the use of τ for both shear stress and characteristic time. The former is kept on the LHS.

[**] An example is the presently free integrator at http://integrals.wolfram.com/index.jsp. The status of this site, and all others, is subject to change.

the Boltzmann equation and predicts only linear responses. Something more severe needs to be done to yield nonlinear behavior.

Example 8-2: *(a) Find the transient "stress-growth" response of a Maxwell-model fluid in simple shear using the strain measure given by equation (8-3). (b)Compare this result with that found using the start-up point as the reference.*

(a) We start with equation (8-17), but immediately divide the right-hand side into the sum of two integrals, one from $-\infty$ to t_0, where t_0 is the point at which the shearing starts, and the other from $t' = t_0$ to $t' = t$. This program is similar to that depicted in Figure 8-2, except that the shearing does not stop after the startup.

So, the integrals to be evaluated are then

$$-\tau(t) = \int_{-\infty}^{t_0} M(t-t')\dot{\gamma}_0(t_0-t)dt' + \int_{t_0}^{t} M(t-t')\dot{\gamma}_0(t'-t)dt' \quad \text{(ssc)} \quad (8\text{-}24)$$

where the minus sign has been switched to the left-hand side, and the negative terms for the strains (see Figure 8-2)) have been left intact.[††] It is tempting to follow the familiar procedure for the Boltzmann equation based on the rate-of-deformation tensor, and set the first integral to zero because the shear rate there is zero. However, and this is important, the strain in that region is *not* zero (in spite of the fact that sample hasn't been strained!). The positive term $\dot{\gamma}_0$ is the (positive) shear rate *after* the startup, not before. The strain, with the present position as the reference, is $\dot{\gamma}_0(t_0-t)$ which is negative. Fortunately, for the purposes of the integration, it is a constant. Thus the first integral becomes, simply,

$$\int_{-\infty}^{t_0} M(t-t')\dot{\gamma}_0(t_0-t)dt' = \frac{-\dot{\gamma}_0(t-t_0)G_0 e^{-t/\tau}}{\tau}\int_{-\infty}^{t_0} e^{t'/\tau}dt' = -\dot{\gamma}_0(t-t_0)G_0 e^{-(t-t_0)/\tau}$$

$$\text{(ssc)} \quad (8\text{-}25)$$

Note that while this term exists, it is knocked down exponentially with time after the shearing starts. (The term from the lower limit of the integration disappears because it has a factor of $e^{-\infty}$.)

The second integral is the more tedious of the two in that it comprises two integrals similar to equation (8-21) except that the lower limits of the two are

---

[††] Half of the battle with solving problems of this type is keeping track of the signs. If working by hand, it is wise to check the sign of each term after each step to make sure a sign hasn't been dropped.

not zero. Rather than going through the evaluation of both, we will attempt the common substitution: $s = t - t'$; $ds = -dt'$. Then the integral becomes

$$\int_{t_0}^{t} M(t-t')\dot{\gamma}_0(t'-t)dt' = \dot{\gamma}_0 \int_{t-t_0}^{0} sM(s)ds = \frac{\dot{\gamma}_0 G_0}{\tau} \int_{t-t_0}^{0} se^{-s/\tau}ds \qquad (8\text{-}26)$$

where the two negative signs have cancelled out. Judging from equation (8-23), this integral will produce two terms in spite of the simplifying substitution. The result of the integration of the above, and its simplification are shown below:

$$\frac{\dot{\gamma}_0 G_0}{\tau} \int_{t-t_0}^{0} se^{-s/\tau}ds = -\dot{\gamma}_0 G_0 \left[e^{-s/\tau}(\tau+s)\right]_{t-t_0}^{0} = -\dot{\gamma}_0 G_0\left[\tau - e^{-(t-t_0)/\tau}(\tau+t-t_0)\right] \quad (8\text{-}27)$$

The limits have been inserted in the last expression. Fortunately for our sanity, the $t - t_0$ times the exponential and the front factor cancel with the result shown in equation (8-25). The remaining terms result in the absurdly simple and expected result shown below:

$$\tau(t) = \dot{\gamma}_0\tau G_0\left(1 - e^{-(t-t_0)/\tau}\right) = \dot{\gamma}_0\tau G_0\left(1 - e^{-t_L/\tau}\right) \qquad \text{(ssc)} \quad (8\text{-}28)$$

Here the symbol t_L has been used to denote laboratory time, that is, the time after the start of the shear.

(b) The solution to problem using the lab starting state as the reference state should be familiar already. However, we repeat it for comparison. Starting with the Boltzmann expression, equation (8-6b), after recalling that we cannot use equation (8-17), we break up the equation into the domain before the imposition of the constant strain rate, and after. Thus

$$\tau(t) = \int_{-\infty}^{0} G(t-t')\dot{\gamma}(t')dt' + \int_{0}^{t} G(t-t')\dot{\gamma}(t')dt' \qquad \text{(ssc)} \quad (8\text{-}29)$$

For the time interval from $-\infty < t' \le 0$, the shear rate is zero, so the entire integral disappears. For the second interval, the shear rate is constant at $\dot{\gamma}_0$, so the rate can be removed from the integral, as can the constant terms of the relaxation modulus. Thus

$$\tau(t) = \dot{\gamma}_0 G_0 e^{-t/\tau} \int_{0}^{t} e^{t'/\tau}dt' \qquad \text{(ssc)} \quad (8\text{-}30)$$

which can be easily integrated to give the final result

$$\tau(t) = \dot{\gamma}_0 G_0 e^{-t/\tau} \left[\tau e^{t'/\tau} \right]_0^t = \dot{\gamma}_0 \tau G_0 \left(1 - e^{-t/\tau} \right) \qquad \text{(ssc)} \quad \text{(8-31)}$$

This is the same as that shown in equation (8-28), as it must be.

While both parts of Example 8-2 gratifyingly result in the familiar and correct answer, it is disappointing that the use of the present time as the reference has resulted in more complicated algebra with no apparent gain over the usual approach from linear viscoelasticity starting with the Boltzmann equation. And, indeed, we shouldn't expect any change, because we start with the same expression, integrate by parts and introduce a linear shift in the time scale. What has happened, though, is we are in a position to introduce a nonlinear measure of strain that will replace that defined in equation (8-19). As we are dealing with large strains, even infinite strains, we need to use the present configuration as the reference, along with all the complications it entails.

C. HOOKEAN SOLIDS

Before exploring the realm of nonlinearity, we divert briefly to take advantage of the three-dimensional infinitesimal strain tensor that was introduced with equation (8-19). The reason for this is that it points out clearly the analogy between Newtonian fluids and perfectly elastic solids. Solids the same as fluids? This certainly makes little sense at first glance. However, the definition of a Hookean solid is

$$\boldsymbol{\tau} = G\boldsymbol{\gamma} \ \text{ or } \ \tau_{ij} = G\gamma_{ij} \qquad \text{(ssc)} \quad \text{(8-32a)}$$

$$\boldsymbol{\tau} = -G\boldsymbol{\gamma} \ \text{ or } \ \tau_{ij} = -G\gamma_{ij} \qquad \text{(fsc)} \quad \text{(8-32b)}$$

where G is the shear modulus and γ is the infinitesimal strain tensor. Compare this with the definition of a Newtonian fluid found in equation (4-2a). Thus, the stress field around a projection into a flowing stream of a Newtonian fluid could be examined by filling the channel with a gel, applying pressure and examining leisurely the birefringence patterns (see Chapter 7, Section E). No pumps, no leaks, no mess.

Example 8-3: *Find Young's modulus for a Hookean solid. Assume constant volume.*

Young's modulus is the slope of the tensile stress vs. tensile strain curve at zero strain. In (ssc) language, the tensile stress σ_T is $\sigma_T = \sigma_{11} - \sigma_{22} = \tau_{11} - \tau_{22}$. The subscripts used imply that the sample is stretched in the 1 direction and allowed to shrink freely and equally in the 2 and 3 directions. At low strains, the tensile strain ε is defined $\varepsilon = \Delta L/L_0$, where L_0 is the initial length of the test unit, and ΔL is the change in length. This is identical to displacement gradient $\partial u_1/\partial x_1$, i.e., the displacement per unit length.

With these definitions in mind, we calculate the stresses using equation (8-32). Thus

$$\tau_{11} = G\left(\frac{\partial u_1}{\partial x_1} + \frac{\partial u_1}{\partial x_1}\right) = 2G\varepsilon \qquad \text{(ssc) (8-33)}$$

Similarly, but recognizing the transverse displacement gradient is ½ of the longitudinal gradient, we have

$$\tau_{22} = G\left(\frac{\partial u_2}{\partial x_2} + \frac{\partial u_2}{\partial x_2}\right) = -G\varepsilon \qquad \text{(ssc) (8-34)}$$

The combination to get the tensile stress gives

$$\sigma_T = \tau_{11} - \tau_{22} = 2G\varepsilon - (-G\varepsilon) = 3G\varepsilon \qquad \text{(ssc) (8-35)}$$

The Young's modulus is thus $E = 3G$.

An important practical question is the existence of Hookean solids. Certainly, we can invent such, but do real materials of any sort have the essential properties of incompressibility[‡‡] and strictly linear behavior that is independent of time. The answer is no, as every real material will have losses even at very small strains and is somewhat compressible. But there are some grades of high-quality[§§] steel that, with proper treatment, can be essentially Hookean, at least at low strains in shear. As for polymers, well, for at short times, elastic gels are fairly Hookean.

[‡‡] See Problem 8-3 for a look at what happens if the volume is not quite constant.

[§§] A tuning fork may have a quality factor (Q factor) of 1000 or more. The Q factor is roughly the $2\pi/\tan \delta$. When a tuning fork is struck, the primary source of damping is air resistance.

D. FINITE STRAIN

To describe large strain (finite strain vs. infinitesimal strain) in a consistent fashion, we need to think about deformation in a very fundamental fashion. Shearing, extensional, biaxial, and planar deformations have all been defined and analyzed. But what is common to all of these? Can we look inside the polymer melt or solution and see the same thing going on regardless of the deformation mode?

The answer is conceptually something "polymer people" are already familiar with from the derivation of the kinetic theory of elasticity.[1] Recall that the important aspect of this theory was the change of length of network chains in every direction. It was assumed that the change of length developed a force in the chain that was related thermodynamically to the change in the configurations available to that chain. The resulting expression for the assembly of ideal chains is

$$\sigma_T = NRT \frac{\overline{r_0^2}}{r_f^2}\left(\lambda^2 - \frac{1}{\lambda}\right) \qquad \text{(ssc) (8-36)}$$

where N is the molar concentration of network chains, r_0 is the end-to-end distance of a network chain, and r_f is the end-to-end distance of the network chains if they were cut free from the network.

But what about polymer melts and solutions? Can we use the distance between two material points as a measure of deformation? It turns out, which is something we need to take on faith, that the square of the distance—a measure of magnitude regardless of position or direction—is the measure of strain we are looking for. Think about this in terms of simple shear, extension, etc. Is it possible to apply these deformations without changing the distances between material points? Clearly not. Can distance change without deforming the sample. No. Position can, angle can, but not distance.

With this discussion, we search now for a consistent method of describing strain using the square of distance and the present state as the reference to accommodate fluids. (With elastomers, the unstrained state was used.) The basic thought is to end up with something akin to

$$\tau_{ij} = G\left(\frac{l'}{l}\right)^2 \qquad (8-37)$$

where the unprimed is the reference length and the primed is a past length. While this would work for extension of elastomers, it certainly needs to be revamped and written in terms of displacements and displacement gradients.

Let's jump directly to the result of the development kinetic theory of rubber elasticity. In the simplest form, the tensile stress is of the form

$$\sigma_T = G\left(\lambda^2 - 1/\lambda\right) \qquad \text{(ssc) (8-38)}$$

where λ is the stretch ratio L/L_0. As it looks very much like this is a combination of two stresses, e.g., $\tau_{11} - \tau_{22}$, we could surmise that the extra stress τ_{11} is given by an expression similar to

$$\tau_{11} = G\left[\left(\frac{L}{L_0}\right)^2 - 1\right] \qquad \text{(ssc) (8-39)}$$

But what about the transverse stress, τ_{22}? Preservation of volume dictates that the ratio of the stretched width to the original width must go as $1/\lambda^{1/2}$. Unlike the development used for Poisson ratio (see Problem 8-3), this is true at large strains also! (See Problem 8-4.) Thus, the expression in the square brackets of equation (8-39) seems like a reasonable start on a measure for large strains. Without any more hesitation, we will see how this measure can be applied to polymer melts and solutions.

Our first job is to generalize the expression in equation (8-39) to three dimensions in the usual coordinate systems. Unfortunately, this is a bit complicated; instead we jump directly to an example of what we should expect to see for tensile strain. The expression for rectangular (x, y, z) coordinates is

$$C_{11}^{-1} - 1 = \left(\frac{\partial x}{\partial x'}\right)^2 + \left(\frac{\partial x}{\partial y'}\right)^2 + \left(\frac{\partial x}{\partial z'}\right)^2 - 1 \qquad \text{(8-40a)}$$

First of all, the left-hand side needs an explanation. This nomenclature results a bit from history and a bit from convenience. The C^{-1} is the traditional part; this is the usual symbol for the *Finger strain tensor*, although not "official" nomenclature. The -1 on the LHS is the result of a practical desire to have a strain measure that is zero at the reference state. It only affects the diagonal elements of the Finger tensor. The general expression for rectangular coordinates is

$$C_{ij}^{-1} - \delta_{ij} = \frac{\partial x_i}{\partial x_1'}\frac{\partial x_j}{\partial x_1'} + \frac{\partial x_i}{\partial x_2'}\frac{\partial x_j}{\partial x_2'} + \frac{\partial x_i}{\partial x_3'}\frac{\partial x_j}{\partial x_3'} - \delta_{ij} \qquad (8\text{-}40b)$$

where we have switched from x, y, z to x_1, x_2, x_3 notation. The symbol δ_{ij} is taken to be 1 if $i = j$, and 0 otherwise. ***

Before starting an example, the primed and unprimed symbols need to be explained. The primed symbols refer to past positions, as with the infinitesimal strain tensor, and the unprimed refers to the reference positions (the present positions for fluids). The bigger mystery, though, is the meaning of the derivatives. Perhaps an example will make these clearer.

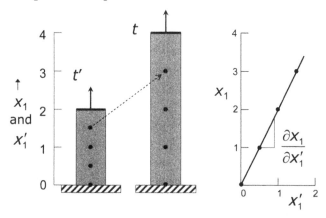

Figure 8-4. Drawing of an elongating sample, marked along its center line. After a 2× deformation, configurations at t' and t are shown, along with a plot to give the derivative.

Consider a strip of material that is being stretched. Marked along the center line are some dots. At the present time they are evenly spaced; and, if the deformation is homogeneous, they were always evenly spaced (Figure 8-4). The data from these two configurations is compiled in Table 8-1, and plotted on the right-hand side of Figure 8-4.

Table 8-1. Positions of markers shown in Figure 8-4.

x_1'	x_1
0	0
0.5	1
1.0	2
1.5	3

*** This strain tensor is the negative of that defined in Bird et al.[2] The symbol used in this reference is: $\gamma_{[0]}$. The reason for the sign difference has to do with (ssc) vs. (fsc).

To get $\partial x_1 / \partial x_1'$, we can plot x_1 vs. x_1' and find the slope, which is 2 in this example. More systematically, we can write the displacement functions, as was done with the infinitesimal strain tensor previously, and take the derivatives. For example, if the sample is being stretched at a constant extension rate $\dot{\varepsilon}$, the expression for the positions of past points with respect to the present can be written as

$$x_1' = \frac{x_1}{\exp[-\varepsilon(t,t')]} = \frac{x_1}{\exp[-\dot{\varepsilon}(t'-t)]} = \frac{x_1}{\exp[\dot{\varepsilon}(t-t')]} \tag{8-41}$$

where

$$\varepsilon(t,t') = \int_t^{t'} \dot{\varepsilon}(s)ds \tag{8-42}$$

in analogy to equation (8-3) for shear. Solving for x_1 gives

$$x_1 = x_1' e^{\dot{\varepsilon}(t-t')} \tag{8-43}$$

which can be differentiated easily to give

$$\frac{\partial x_1}{\partial x_1'} = e^{\dot{\varepsilon}(t-t')} \tag{8-44}$$

This is all that is needed to find the strain, because the other two terms are zero. Thus

$$C_{11}^{-1} - 1 = \left(\frac{\partial x_1}{\partial x_1'}\right)^2 = e^{2(t-t')} \tag{8-45}$$

Example 8-4: *Find all non-zero components of the Finger tensor for steady simple shearing motion that has been going on for some time.*

The first step is to write the displacement functions in terms of the shear rate $\dot{\gamma}_0$, which is positive. We have

$$x_1' = x_1 + x_2' \gamma(t,t') = x_1 - x_2' \dot{\gamma}_0(t-t') \tag{a}$$

$$x_2' = x_2 \tag{b}$$

$$x_3' = x_3 \tag{c}$$

for the three coordinates. For simple shear, the position associated with the time-dependent strain can be primed or not, because $x_2' = x_2$. Note the change of signs in equation (a) on performing the integration according to equation (8-3) for constant shear rate.

On examining equation (8-40), we can see there are several terms that will survive the differentiation part of this expression. For example for the 1,1 component ($i = 1$ and $j = 1$) carries with it the first term, giving 1, and the second term, which gives $[\dot{\gamma}_0(t - t')]^2$. The former is taken out by the $\delta_{11} = 1$. The third differentiation term is zero. For the 2,2 component, the second differentiation term gives 1, which is cancelled by the $\delta_{22} = 1$.

The off-diagonal components are less numerous, with only the 1,2 and 2,1 surviving. For the 1,2 and 2,1 components, which are the same, the second differentiation results in $\dot{\gamma}_0(t - t')$. Note that this is positive.

The end result of this in matrix form is

$$\mathbf{C}^{-1} - \delta = \begin{pmatrix} [\dot{\gamma}_0(t-t')]^2 & \dot{\gamma}_0(t-t') & 0 \\ \dot{\gamma}_0(t-t') & 0 & 0 \\ 0 & 0 & 0 \end{pmatrix} \tag{d}$$

Note that we are very careful to avoid calling these terms "strain," even though they look very much like strain. However, unlike strain, they are neither referenced to initial configuration (which is long forgotten) or to the present. Furthermore, if we plot these terms versus t', the slope will be negative (they get smaller as t' approaches t); but, instead, the shear rate is positive. Thus, the expression "relative strain."

Given this strain measure, what can be done with it? It doesn't take much imagination to substitute it into the Boltzmann expression and predict stresses. So, that's exactly what we will do.

E. THE LODGE ELASTIC FLUID AND VARIANTS

The molecular development of the Lodge Elastic Fluid (LEF) started with the concept of a network, but with "crosslinks" that have a fleeting existence. As we have seen, the Finger relative strain tensor gives tensile behavior that mimics that of an ideal elastomer. In other words, this measure of strain predicts the entire stress state for the ideal elastomer subjected to any deformation. The possibility that it could also be applied to polymer solutions and melts is really quite exciting.

1. The base-case LEF: formulation and properties in shear

Equation (8-46) describes the simplest form of the LEF, where the memory function M depends only on time:

$$\tau(t) = \int_{-\infty}^{t} M(t-t')\left[\mathbf{C}^{-1}(t,t')-\boldsymbol{\delta}\right]dt' \qquad \text{(ssc)} \quad \text{(8-46)}$$

The square brackets contain the form of the Finger tensor described in Section D above.

Example 8-5: *Using the LEF expression, predict the viscosity of a fluid subject to a steady shear rate $\dot{\gamma}_0$ and with a Maxwell relaxation function.*

The single Maxwell element relaxation function is $G(t; G_0, \tau) = G_0 \exp(-t/\tau)$, where the key variable for this problem is t. The parameters, the zero-time modulus G_0 and the relaxation time τ are assumed to be constant. The corresponding memory function is the negative derivative of the modulus derivative, as described earlier. For the Maxwell relaxation, the memory function is $M(t; G_0, \tau) = (G_0/\tau) \exp(-t/\tau)$.

First, let's look at the shear stress and, consequently the viscosity. According to equation (8-46), along with the Maxwell memory function and the 2,1 component from equation (d) in Example 8-4, we can write the expression

$$\tau_{21}(t) = \int_{-\infty}^{t} (G_0/\tau) e^{-(t-t')/\tau}\left[\dot{\gamma}_0(t-t')\right]dt' \qquad \text{(ssc)} \quad \text{(a)}$$

For (fsc), a negative sign will be needed in front of the integral.

At this point it is convenient to make a substitution for $(t - t')$. This doesn't need to be done; it is merely convenient. When doing this, be sure to keep track of signs and remember to replace the limits. The substitutions is listed below:

$$s = t - t'$$
$$ds = -dt' \quad \text{or} \quad dt' = -ds$$
$$@ \ t' = -\infty, \ s = +\infty$$
$$@ \ t' = t, s = 0$$

Taking the constant terms outside the integral leaves a simple integration of the exponential and making the above substitutions we have

$$\tau_{21}(t) = -\frac{G_0 \dot{\gamma}_0}{\tau} \int_{\infty}^{0} e^{-s/\tau} s \, ds \qquad \text{(ssc)} \quad \text{(b)}$$

At this point, one might be concerned that the argument t has been lost on the RHS. Certainly this is a legitimate concern; however, this deformation has been going on for an infinite amount of time, and thus the stress will not be a function of t. (See Problem 8-17.)

Referring to the integral tables, or an appropriate website, for help with the integration, yields the expression in the square brackets below:

$$\tau_{21} = -\frac{G_0 \dot{\gamma}_0}{\tau}\left[-\tau e^{-s/\tau}(s+\tau)\right]_{\infty}^{0} \qquad \text{(ssc) (c)}$$

Canceling the negative sign and the τ's, and plugging in the limits gives the result

$$\tau_{21} = G_0 \dot{\gamma}_0 \tau \qquad \text{(ssc) (d)}$$

Note that the product $(\infty + \tau)\exp(-\infty/\tau)$ resulting from the lower limit is very much zero because the exponent overwhelms the linear factor. This is, in effect, the result of having a fluid.

Our result shows, importantly, that the viscosity of a LEF is independent of shear rate, which is the same behavior as that with the infinitesimal strain tensor. This is a bit disappointing, but can be repaired, as we shall see.

Following on Example 8-5, let's explore if the LEF vs. infinitesimal strain tensor gives something more interesting for normal stress differences. And indeed it does. Examining equation (d) of Example 8-5, we can see that unlike the infinitesimal expression, the Finger strain tensor for shear has terms on the diagonal. Following the procedure shown in Example 8-5, we can calculate the first and second normal-stress differences.

Example 8-6: *Calculate the first and second normal-stress differences in simple shear for the LEF with a single-element Maxwell stress-relaxation modulus.*

For the extra stress τ_{11}, we have the expression

$$\tau_{11}(t) = \int_{-\infty}^{t}(G_0/\tau)e^{-(t-t')/\tau}\left[C_{11}^{-1}(t,t')-1\right]dt' \qquad \text{(ssc) (a)}$$

Substituting for the strain using the expression in equation (d) of Example 8-4 gives

$$\tau_{11}(t) = \frac{G_0 \dot{\gamma}_0^2}{\tau} \int_{-\infty}^{t} e^{-(t-t')/\tau}(t-t')^2 \, dt' \qquad \text{(ssc) (b)}$$

Again the substitution of s for $(t - t')$ looks attractive, giving

$$\tau_{11}(t) = -\frac{G_0 \dot{\gamma}_0^2}{\tau} \int_{\infty}^{0} s^2 e^{-s/\tau} \, ds \qquad \text{(ssc) (c)}$$

This integral definitely requires the tables, which give the result

$$\tau_{11} = -\frac{G_0 \dot{\gamma}_0^2}{\tau} \left[-\tau e^{-s/\tau}(2\tau^2 + 2\tau s + s^2) \right]_{\infty}^{0} \qquad \text{(ssc) (d)}$$

Canceling the minus signs and the τ's, followed by substitution of the limits, leads to the final result:

$$\tau_{11} = 2G_0 \tau^2 \dot{\gamma}_0^2 \qquad \text{(ssc) (e)}$$

The values of τ_{22} and τ_{33} are both zero, as can be seen from the null values of the 2,2 and 3,3 terms in equation (d) of Example 8-4. Thus the conclusions about normal stresses for the LEF are

$$N_1 = \tau_{11} - \tau_{22} = 2G_0 \tau^2 \dot{\gamma}_0^2 \qquad \text{(ssc) (f)}$$

and

$$N_2 = \tau_{22} - \tau_{33} = 0 \qquad \text{(ssc) (g)}$$

Note that N_1 is positive (ssc), which results from the tensile character of τ_{11}.

The normal stresses found in Example 8-6 show a couple of very important aspects of the LEF. First and foremost is the finite value of N_1. Unlike the infinitesimal strain tensor, the LEF has resulted in a normal stress from the geometry alone. The second aspect of importance is the dependence of N_1 on the square of the shear rate. We know from previous discussions that this must be true from symmetry alone. The normal-stress coefficient Ψ_1 is thus independent of shear rate. So there is an aspect of linearity in the LEF formulation. A disappointing result is the absence of a second normal-stress difference. While experiments show that Ψ_2 is small, it would certainly be nice to have this feature arise naturally as well. And, indeed it does, as will be seen in the next section.

2. The behavior of the LEF in extension

It is fairly safe to say that the Finger relative-strain tensor would be a mere entry in a table if it weren't for its facile handling of normal stresses in shear, as illustrated in Example 8-6.   But the surge in popularity of extensional flow experiments in the 1970's brought even more excitement when the results of these experiments were compared with the predictions of the LEF.   The qualitative behavior of the LEF and that of an easily stretched LDPE melt were virtually identical. All this from mechanics alone!

Unfortunately, the derivation of the extensional result is complicated, even for the simplest single-exponential memory function and for constant extensional rate. (The derivation can be found in Appendix 8-2.) Instead, we will examine the result and discuss its behavior.

First of all, consider the data found in Figure 7-16. At low rates, the result follows the expected trend for a linear viscoelastic material: the stress rises exponentially to a constant value. At higher rates, the stress follows the linear trend for a ways, and then starts increasing and shoots a decade or more above the linear value before failing.

Now, we present the result of the analysis of this experiment using the LEF, but simplified by using a memory function with a single relaxation time. The constant extension rate $\dot{\varepsilon}_0$ is applied at time zero. Recall, that this means the sample length must be increased exponentially, as described in Problem 3-6. The method for calculating the key elements of the Finger tensor was illustrated in Figure 8-4 and equations (8-41) through (8-45). The rather complicated expression is

$$\eta_E^+(t) = \frac{2G_0\tau}{1-2\dot{\varepsilon}_0\tau}\left[1 - e^{-t(1-2\dot{\varepsilon}_0\tau)/\tau}\right] + \frac{G_0\tau}{1+\dot{\varepsilon}_0\tau}\left[1 - e^{-t(1+\dot{\varepsilon}_0\tau)/\tau}\right] \qquad (8\text{-}47)$$

where the relaxation modulus is the usual $G_0\exp(-t/\tau)$. This can be written in dimensionless form by dividing through by $G_0\tau$, which is the steady shear viscosity.

On examining this equation we note a couple of obvious facts. One is that there are two terms. Another is that there is a strong interplay between the relaxation time and rate. The second term has this modification, but otherwise is a well-behaved exponential increase with time to a constant value of $G_0\tau$. The first term, though, is anything but well behaved! In fact, it looks like it might go to infinity if the extension rate happened to be exactly $1/2\,\tau$. However, this is not the case, as the factor in square brackets also goes to zero. Expanding the exponential shows that the $1 - 2\dot{\varepsilon}_0\tau$ cancels completely. But there is something very, very important that does happen at an extension rate of

$1/2\tau$: the exponent changes sign from negative at low rates to positive at rates higher than $1/2\tau$. Thus the stress will start to increase exponentially.

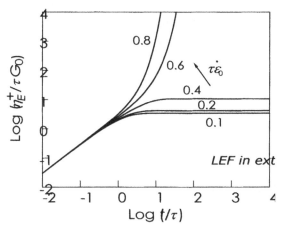

Figure 8-5. Prediction of the LEF in simple extension at constant rate. A single-term exponential is used for the memory function. At low extension rate $\dot{\varepsilon}_0$, the response is linear, leveling out to an extensional viscosity of $3G_0\tau$, following Trouton's rule. At higher rates ($\dot{\varepsilon}_0 > 1/2\tau$), the LEF model predicts that the stress should increase dramatically and not level off.

A plot of this important result is shown in Figure 8-5. The key features are the linear behavior at low extension rates and the positive exponential increase in extensional viscosity at high rates. Compare these with the experimental results in Figure 7-16. You will agree there is a strong qualitative resemblance.

F. THE CAUCHY STRAIN MEASURE

As shown above, the key aspect of the Finger tensor was the dependence of the relative strain on the square of the partial derivatives of the present positions of material points with respect to the past positions. What would be the result if the derivatives instead were the past positions of the material points with respect to the present or reference positions? In short, the result would be the Cauchy strain tensor, also called the Cauchy relative strain tensor.

Similar to the Finger tensor, we can write the components of the Cauchy tensor in rectangular coordinates as

$$C_{ij} = \frac{\partial x_1'}{\partial x_i}\frac{\partial x_1'}{\partial x_j} + \frac{\partial x_2'}{\partial x_i}\frac{\partial x_2'}{\partial x_j} + \frac{\partial x_3'}{\partial x_i}\frac{\partial x_3'}{\partial x_j} \tag{8-48}$$

For practical reasons that shall become apparent, we will define the stress calculated with this tensor in terms of the modified Cauchy tensor

$$\delta_{ij} - C_{ij} \tag{8-49}$$

The δ_{ij} serves the same purpose as with the Finger tensor; it eliminates the 1's from the diagonal. The change of signs, though, is somewhat of mystery. To see the reason for this, consider the case of steady, simple shear as outlined in Example 8-4. In particular, consider the 1,2 component of the Cauchy tensor. As describe in equation (a) of Example 8-4, the key displacement function is

$$x_1' = x_1 - x_2'\dot{\gamma}_0(t-t') = x_1 - x_2\dot{\gamma}_0(t-t') \tag{8-50}$$

where the steady shear rate $\dot{\gamma}_0$ is positive. The second equality recognizes that for simple shear, $x_2' = x_2$ at all times. Performing the derivatives for the 1,2 component of **C** according to equation (8-48) gives

$$C_{21} = C_{12} = -\dot{\gamma}_0(t-t') \tag{8-51}$$

The minus sign that appears as the result of the term $\partial x_1'/\partial x_2$ is most inconvenient, which explains the reason for the formulation shown in equation (8-49). The sign reversal then gives positive shear stresses (ssc). For (fsc), one might want the signs reversed so that there would be no need for a negative sign on the expression for stress, as is done by Bird et al.,[2] as well as several other authors.

1. Cauchy fluid behavior in shear

As was done with the Lodge elastic fluid, we will test the Cauchy fluid first in steady shear. The trick for any steady flow is to integrate the stress contributions from way, way back (i.e., $-\infty$), to zero time, which is also the reference position. Thus, for the shear stresses generated by the Cauchy fluid, we have the expression

$$\tau_{21} = \int_{-\infty}^{0} M(t-t')\left[C_{21}(t,t') - \delta_{21}\right]dt' \qquad \text{(ssc)} \tag{8-52}$$

The value of δ_{21}, of course, is zero. To find the 2,1 component of the Cauchy tensor, we start with the displacement functions, i.e., the positions at past time t' relative to their position at $t' = 0$ (the present). For simple shear with flow to the right (positive x_1 direction), the past positions are to the left at lower values of x_1, as was described with equation (8-50). The x_2 and x_3 values do not vary. The displacement functions, then, are the same as those in equations (a), (b) and (c) of Example 8-4. These are differentiated according to the formula in equation (8-48).

We want to pay special attention to the diagonal elements. For the 1,1 element, we have from equation (8-48)

$$C_{11} = \left(\frac{\partial x_1'}{\partial x_1}\right)^2 + \left(\frac{\partial x_2'}{\partial x_1}\right)^2 + \left(\frac{\partial x_3'}{\partial x_1}\right)^2 \tag{8-53}$$

As x_2' and x_3' depend only on x_2 and x_3, respectively, the second two terms are zero. For the first term, we refer to equation (8-50), which shows that this term will simply be 1. For C_{22} we have also 1 from the second term, but the first term is more interesting; equation (8-50) suggests it is not zero. Taking the derivative of equation (8-50) with respect to x_2 gives

$$C_{22} = \left(\frac{\partial x_1'}{\partial x_2}\right)^2 + \left(\frac{\partial x_2'}{\partial x_2}\right)^2 + \left(\frac{\partial x_3'}{\partial x_2}\right)^2 = 1 + \left(-\dot{\gamma}_0^2(t-t')\right)^2 \tag{8-54}$$

The 3,3 term also becomes 1, because the x_3 positions depend only on the present x_3 positions for all time.

The result for the Cauchy tensor, modified according to equation (8-49), is

$$\boldsymbol{\delta} - \mathbf{C} = \begin{pmatrix} 0 & \dot{\gamma}_0(t-t') & 0 \\ \dot{\gamma}_0(t-t') & -\dot{\gamma}_0^2(t-t')^2 & 0 \\ 0 & 0 & 0 \end{pmatrix} \tag{8-55}$$

Certainly a couple of aspects of this result should be discussed. First, the shear rate $\dot{\gamma}_0$ is constant and positive; also, it multiplies the time terms [i.e., don't read this element as $\dot{\gamma}_0$ of $(t - t')$]. As with the Finger tensor, the 1,2 and 2,1 terms of the Cauchy formulation will give a constant viscosity. Nothing special there. The 2,2 term is most interesting, as it looks like it predicts a positive first normal-stress difference N_1, which is good, but also a negative second normal-stress difference N_2. That's really something! Unfortunately, the magnitudes of the two will be the same, which is nowhere close to what is observed. Furthermore, there is nothing that can be done about it, as anything we do to reduce N_2 will also reduce N_1, or so it seems.

In spite of the gloomy outlook for the use of $\boldsymbol{\delta} - \mathbf{C}$, there is one thing that can be done, and that is to use $\mathbf{C}$ in conjunction with $\mathbf{C}^{-1}$. So, if we start with $\mathbf{C}^{-1}$ and feed in a little $\mathbf{C}$, we can have both normal-stress differences in about the right amounts. This idea has led to a class of models, which use the strain measure

$$a\left(\mathbf{C}^{-1} - \boldsymbol{\delta}\right) + (1-a)\left(\boldsymbol{\delta} - \mathbf{C}\right) \tag{ssc (8-56)}$$

with the parameter a having values around, say, 0.9 to match the steady shear results. One of these is the widely used K-BKZ model, in its simplest form.[†††] In more general form, a is dependent on strain.

Example 8-7: *Find the shear viscosity of a fluid with strain measure described in equation (8-56), based on the Cauchy relative strain tensor. Use an exponential memory function with a single relaxation time.*

As we are interested in the shear stresses only, the strain element of importance will be the 2,1 element (or the 1,2). The expression for steady shear stress at a shear rate $\dot\gamma_0$ is given by the equation

$$\tau_{21} = \int_{-\infty}^{0} (G_0/\tau)e^{-(t-t')/\tau}\{a[\dot\gamma_0(t-t')] + (1-a)[\dot\gamma_0(t-t')]\}dt' \qquad \text{(ssc) (a)}$$

where the { } is the strain measure from equation (8-56) but with only the 2,1 elements. The [ ] brackets contain the expressions for the 2,1 element for steady shear. These were described by equation (8-55) for the Cauchy tensor, and equation (d) in Example 8-4 for the Finger tensor.

It should be evident at this point that the two terms can be combined and both are linear in shear rate. Dividing through by $\dot\gamma_0$ gives the viscosity, and it is devoid of any dependence on shear rate. Once again, we find that while the combination of the two relative strain tensors $\mathbf{C}$ and $\mathbf{C}^{-1}$ leads to finite N_1 and N_2 predictions, the material constants—both η and Ψ—are independent of shear rate. To get realistic results in steady shear, we must think of some way of fixing things up.

G. FIXING UP INTEGRAL EQUATIONS BASED ON C AND C⁻¹

By now we should be convinced that finding shear-rate-dependent material functions using the nonlinear strain measures developed in Sections E and F alone is basically hopeless. On reflecting about this, we really shouldn't be too surprised, because the nonlinear strain measure doesn't reflect on the fluid itself; it is merely geometry. Polymeric fluids have special structure (entangled

---

[†††] The K-BKZ model was developed in the early 1960's by rheologists A. Kaye, B. Bernstein, E. A. Kearsley and L. J. Zapas, the first working independently. In their formulation, the parameter a is not arbitrary; it is instead associated with the stored work in the deforming melt and the dependence of this work on the invariants of $\mathbf{C}^{-1}$.

molecules) that leads to the shear-rate effects, and we need to imagine what shearing does to these structures.

1. Effect of shearing on the memory function

On thinking about the shear thinning of polymeric fluids, it is natural to picture highly entangled coils behaving like wool from a sheep. Wool, as it comes from the sheep, is highly entangled and must be carded to put it into a workable state. The term "workable" might be associated with a lower viscosity for the polymer melt or solution. This could translate into a modification of the parameters of the relaxation function according to either the amount of shear or the shear rate, or both.

We know that temperature has a strong influence on viscosity, especially near the glass transition temperature, T_g. We do not hesitate to write functions (Arrhenius, WLF) that describe this dependency (see Chapter 4). In most cases, the relaxation times are affected significantly more than the modulus parameters, which is consistent with t-T superposition. Why not do the same for shear rate, shear stress, or shear strain? As we have seen, shear rate is incorporated in the formulation of the GNF models. Can we do the same for the integral constitutive equations? Let's take a look.

Suppose we start with the LEF equation for shear and examine the effect of shear rate on the predicted viscosity. We will assume that only the relaxation time τ is affected, and in the following fashion

$$\tau(I_2) = \frac{\tau_0}{1 + a\sqrt{I_2/2}} \tag{8-57}$$

where I_2 is the second invariant of the rate-of-deformation tensor $\dot{\gamma}$. How will this influence the solution already presented in Example 8-5? As we know from that example, the shear viscosity of an LEF is simply $G_0\tau$, given a single-element Maxwell relaxation function $G(t; G_0, \tau) = G_0\exp(-t/\tau)$. We would like to have the result give a shear-rate-dependent steady shear viscosity.

For simple shear, the expression in equation (8-57) for the relaxation time becomes

$$\tau = \frac{\tau_0}{1 + a\dot{\gamma}} \tag{8-58}$$

as has been explained in Chapter 3, Example 3-1. The integral expression for the steady shear viscosity for a single-exponential LEF fluid is

$$\eta = \frac{1}{\dot\gamma_0} \int_{-\infty}^{0} \frac{G_0}{\tau} e^{t'/\tau} t' \dot\gamma_0 dt' \tag{8-59}$$

where $\dot\gamma_0$ is the steady shear rate. As the shear rate is constant over the entire range of the integral, we can treat τ as a constant. Thus the usual integration obtains (see Problem 8-17), giving the expression

$$\eta = G_0 \tau = G_0 \frac{\tau_0}{1 + a\dot\gamma_0} \tag{8-60}$$

Thus, at least for this problem, we have introduced shear-rate dependence in the viscosity. It is evident that the expression in equation (8-57) can be improved by fitting the steady shear data for a GNF model with three or more parameters, and using this expression to influence the relaxation time.

In a sense, we finally have reached a goal of providing a connection between the limited, but useful, world of the GNF models and a high-strain constitutive expression that can qualitatively predict the entire spectrum of viscoelastic response resulting from linear, extensional, shearing, biaxial, and transient deformations, in any combination.

By introducing a dependence of the relaxation time on an invariant of the rate-of-deformation tensor, the LEF can be given a shear-rate-dependent viscosity. However, this procedure introduces some serious concerns, such as:

a) Will the other predictions of the LEF (e.g., extensional viscosity, normal stresses) be improved, or affected adversely?

b) For transient measurements, will the integrals be impossibly complicated and require numerical methods?

c) Should the viscosity depend on strain as well as strain rate? If so, what is the proper invariant of the strain?

d) Similarly, would it make more sense to have viscosity depend on an invariant of the total stress tensor? Pressure is one of the invariants, and viscosity does depend on pressure.

In some respects, most of these questions have been addressed over the years. Diligent researchers have proposed new forms and generated nonlinear experimental data to test the results.[3] The results have been mixed, with some forms working well for shear, but not so well with extensional flows.

To simplify the math, a class of equations has been developed with separable memory functions.[4] This key assumption can be written in the form

$$M(t, I_1, I_2, I_2) = H(I_1, I_2, I_2) m(t) \tag{8-61}$$

where the Γ's are invariants of the strain tensors. At the most elementary level, this is equivalent to the modification of G_0 in equation (8-59), as opposed to τ. The function H has been coined the "damping function," as it modifies the relaxation behavior. The I_i are defined somewhat differently than those for the rate-of-deformation tensor; thus: $I_1 = \text{tr}\ (\mathbf{C}^{-1})$; $I_2 = \{I_1 - \text{tr}[(\mathbf{C}^{-1})^2]\}/2$; $I_3 = \text{det}\ (\mathbf{C}^{-1})$.[‡‡‡] For incompressible fluids, the latter is zero. Because of the dependency of the damping function on the strain invariants, the value of M is changing not only with time over the integration, but with strain. This complicates the determination of analytical expressions.

A key expectation from separability of this sort is that the relaxation modulus will show nonlinear behavior by a downward shift along the log $G(t)$ axis, rather than along the log t axis, as with temperature. With some polymer melts, the nonlinear stress-relaxation behavior has been observed to follow this type of shift remarkably well, even up to rather high strains.

While equations of this type have become very popular, there are dozens of other approaches. One related category that has been shown to be particularly capable is the set fashioned after the original K-BKZ model.[§§§] The Rivlin-Sawyers model is one example, and is described below:

$$\tau_{ij}(t) = \int_{-\infty}^{t} M(t - t')\big[\phi_1(I_1, I_2)(C_{ij}^{-1} - \delta_{ij}) + \phi_2(I_1, I_2)(\delta_{ij} - C_{ij})\big]dt' \quad \text{(ssc)} \quad (8\text{-}62)$$

In this equation, ϕ_1 and ϕ_2 are functions of the invariants I_1 and I_2 of the Finger tensor, as shown above.

The important aspect of this formulation is its flexibility, because the functional forms of ϕ_1 and ϕ_2 can be changed, along with the parameter values in these functions. To introduce a second normal-stress difference in shear, it is necessary to have ϕ_2 non-zero, as we have seen.

The exercise of finding useful forms for the damping functions has endured for many years. The hope, of course, is to find expressions that will reproduce much of the rheological behavior of polymeric fluids. The reasoning beyond simply having a nifty equation is that the parameters in the equations can perhaps be uniquely related to features of the molecular structure of the polymer. While there is progress in this regard, the hope has yet to be realized in any practical fashion. This fact prompts the search for molecular-scale theories of polymer dynamics that will suggest or even determine forms for the

---

[‡‡‡] The nomenclature $(C^{-1})^2$ means $(C^{-1}) \cdot (C^{-1})$.

[§§§] See footnote near Equation (8-56).

damping functions. These theories may even point out the hopelessness of describing response in all deformations with a single equation.

APPENDIX 8-1: THE RELAXATION FUNCTION

Being a linear property, changes to the form of the relaxation function do not change the nature nonlinear behavior unless the parameters are allowed to depend on invariants of the stress or rate-of-deformation tensors. Even then, these changes do not create different phenomena, such as the appearance of normal stresses. As might be expected, broadening the relaxation function broadens all rheological responses. Not surprisingly, polymers with broad molecular weight distributions tend to have broad relaxation functions.

The standard and easiest way to broaden the relaxation function is to use a Prony series, i.e., a sum of exponential decay functions. Thus

$$G(t) = \sum_{i=1}^{n} G_i e^{-t/\tau_i} \tag{a}$$

Recall that the Rouse theory results in a series of this form, but with $G_i = NkT$, where N is number concentration of chains, and $\tau_i = \tau_1/i^2$, where τ_1 is the maximum relaxation time. The total number of modes (terms), n, is proportional to the chain length. The widely space modes limit its usefulness for fitting actual data.

A popular three-parameter relaxation function is the stretched exponential

$$G(t) = G_0 e^{-(t/\tau)^{\beta}} \tag{b}$$

where the value of β determines the degree of broadening. As β decreases, the relaxation process is "stretched" over wider time scales. This function is also known as the Kohlrausch-Williams-Watts function. Narkis et al.[5] reported that $\beta = 0.5$ reproduced quite closely the measured relaxation of anionic polystyrene fractions with very narrow molecular weight distributions. These functions have proven to be very useful for examining the role of non-exponential relaxation on material response. One reason is that many of the transforms to the other viscoelastic functions have been worked out. Table 8-2 lists some of these.

Table 8-2. Viscoelastic functions based on the KWW equation.

Function	Formula	Notes	References
Steady viscosity	$\eta = G_0\tau\,\Gamma\!\left(1+\dfrac{1}{\beta}\right)$	Γ is the gamma function	5
Complex dynamic modulus	A variation of the Cole-Davidson function	for $\beta = 1/2$	6
Relaxation spectrum	$H(s) = \tfrac{1}{2}G_0\sqrt{\dfrac{s}{\tau}}\,e^{-s/4\tau}$	for $\beta = 1/2$	"

APPENDIX 8-2: CONSTANT-RATE EXTENSION OF THE LEF

In spite of its apparent simplicity, the solution of the single-relaxation-time LEF in "simple" uniaxial extension becomes algebraically complex, i.e., messy. The derivation follows the usual route: first and foremost we write down the displacement functions. A mistake here dooms the entire derivation. Then the definitions of the Finger strain are used to calculate the normal-stress difference, $\tau_{11} - \tau_{22} = \sigma_{11} - \sigma_{22} = \sigma_T$. Forgetting the stress in the 2 direction is also fatal. After this, the integration starts. Important in this step is to recognize that the integral must be divided into two parts: one from $-\infty$ to 0, and then from 0 on up to the present time t. The reason for this is the strain is constant (but not zero) for the $-\infty$ to 0 zone, while it increases (at rate of $\dot{\varepsilon}_0$) over the range 0 to t.

Toward the end, we will use a slightly indirect method of showing that the end result matches the form shown in equation (8-47). This illustrates a problem common in such derivations—it's often difficult to see easily that two equations are, in fact, the same, vs. recognizing that a mistake was made. This can be particularly annoying when using symbol manipulation programs such as Maple.

To simplify the algebra, it may help to express terms in dimensionless form. This maneuver can cut down on the number of symbols, reducing the clutter. This example will illustrate how this is accomplished.

1. Displacement functions

The key variable in the stretching process is the extension rate, $\dot{\varepsilon}_0$. This will be a positive number. In checking the displacement functions, it is very important to keep this in mind. Thinking in terms of a strain diagram such as Figure 8-2 can help to make sure the signs of the terms in the displacement functions are correct.

For $-\infty < t' \leq 0$, we have a fixed, negative strain. The width of the sample will be greater than that at the reference state ($t' = t$), while the length will be less. The displacement in the length direction is

$$x_1' = x_1 e^{-\dot{\varepsilon}_0 t} \tag{a}$$

with no dependence on x_2 or x_3 (if the flow is uniform). Note that the past positions (primed) are less than the reference positions (unprimed) for positive values of t, and the positions are independent of the variable t'. The sample, in addition to getting longer, is getting thinner with time, so the past positions will be larger than the present ones corresponding to the thicker sample. Thus, we can write by inspection the expressions

$$x_2' = x_2 e^{\dot{\varepsilon}_0 t/2} \tag{b}$$

$$x_3' = x_3 e^{\dot{\varepsilon}_0 t/2} \tag{c}$$

The reason for the two in the denominator of the exponent is that the velocity gradients in the thickness directions are both ½ that of the velocity gradient in the stretch direction ($\dot{\varepsilon}_0$). This is needed to preserve volume.

For $0 \leq t' \leq t$, the situation is a bit more complicated, because the positions will depend on the integration variable t'. For the stretch (1) direction, the past positions will still be less than those at the present, so we can write

$$x_1' = x_1 e^{-(t-t')\dot{\varepsilon}_0} \tag{d}$$

To make sure all the signs are correct, check the expression at $t' = 0$ and $t' = t$. The former should give the same expression as (a), while the latter should give the result $x_1' = x_1$. Both of these work out with the signs as shown in equation (d).

The displacement functions for the transverse directions follow the same pattern, and are recorded below:

$$x_2' = x_2 e^{\dot{\varepsilon}_0 (t-t')/2} \tag{e}$$

$$x_3' = x_3 e^{\dot{\varepsilon}_0 (t-t')/2} \tag{f}$$

Note that it helps to express the past positions in terms of the present, and keep everything in the exponent positive, letting the sign out front take care of the overall value of the exponent.

2. Finger strains

We are now in a position to calculate the Finger strains. As we are subtracting two stresses to get the total tensile stress, we do not need to worry about the 1's for the diagonal elements of the Finger strain, as the ones cancel out. For C_{11}^{-1}, the only active term in the expression shown in equation (8-40) is $\partial x_1 / \partial x_1'$, because there is no dependence of the positions on the 2 or 3 locations (assuming the flow is homogeneous). This applies also for the other directions. Solving for x_1 in equation (d) and taking the derivative gives

$$C_{11}^{-1} = e^{2\dot{\varepsilon}_0 (t-t')} \tag{g}$$

where the 2 comes from squaring the $\partial x_1 / \partial x_1'$, as prescribed by equation (8-40). Now with equation (g) in hand, we can easily get the relative strain for either $0 \leq t' \leq t$ or $-\infty < t' \leq 0$, the latter by setting $t' = 0$. For calculating the stress difference, we also need the 2,2 component of the Finger relative strain tensor. It is

$$C_{22}^{-1} = e^{-\dot{\varepsilon}_0 (t-t')} \tag{h}$$

for $0 \leq t' \leq t$. Again, the expression for $-\infty < t' \leq 0$ is obtained by setting $t' = 0$. Note that the ½ has disappeared due to the squaring. The negative sign corresponds to the shrinkage of the width as the sample is stretched.

3. Expressions for the stress

$\underline{-\infty < t' \leq 0}$

The tensile stress comes directly from $\sigma_T = \sigma_{11} - \sigma_{22} = \tau_{11} - \tau_{22}$ and equation (8-46), along with the strain values from the equations above. For $-\infty < t' \leq 0$, we have

$$\int_{-\infty}^{0} M(t-t') \left[C_{11}^{-1}(t,t') - C_{22}^{-1}(t,t') \right] dt' \tag{i}$$

which, on substituting for the memory function M and the strains, becomes

$$\int_{-\infty}^{0} \frac{G_0}{\tau} e^{-(t-t')/\tau} \left[e^{2\dot{\varepsilon}_0 t} - e^{-\dot{\varepsilon}_0 t} \right] dt' \tag{j}$$

This integral is easy to do, as the square-bracketed expression is constant over the range of integration. It, and the usual factor in the memory function can thus be pulled from the integral. These moves leave

$$\frac{G_0}{\tau} e^{-t/\tau} \left[e^{2\dot{\varepsilon}_0 t} - e^{-\dot{\varepsilon}_0 t} \right] \int_{-\infty}^{0} e^{t'/\tau} dt' \tag{k}$$

The integral itself becomes simply τ, which cancels the τ out front. The result is

$$G_0 e^{-t/\tau} \left[e^{2\dot{\varepsilon}_0 t} - e^{-\dot{\varepsilon}_0 t} \right] \tag{i}$$

Before moving on to the $0 \le t' \le t$ time interval, we want to take the simple expression in equation (i) and try to make it look something like the answer in equation (8-47). We will then transform the variables to dimensionless forms, which will help considerably in the $0 \le t' \le t$ time interval.

First, the terms in the square brackets are multiplied by the front factor and then the factor t/τ is removed in the exponent, as illustrated below:

$$G_0 \left(e^{2\dot{\varepsilon}_0 t - t/\tau} - e^{-\dot{\varepsilon}_0 t - t/\tau} \right) = G_0 \left(e^{t/\tau(2\dot{\varepsilon}_0 \tau - 1)} - e^{-t/\tau(\dot{\varepsilon}_0 \tau - 1)} \right) \tag{i}$$

What this does is to create rate and time terms that are both reduced by the time constant τ from the memory function. The parameter τ is a reasonable choice because it is invariant with time and rate. The combination $\dot{\varepsilon}_0 t$ would not have this quality, as $\dot{\varepsilon}_0$ will change from run to run; however, it will be fixed for a given run.

We will do two more steps. One is prompted by equation (8-47), which has negative exponents. Thus equation (i) will become

$$G_0 \left(e^{-t/\tau(1 - 2\dot{\varepsilon}_0 \tau)} - e^{-t/\tau(1 + \dot{\varepsilon}_0 \tau)} \right) \tag{j}$$

which is starting to look much like the answer. Our final step will be to make the following substitutions:

$$y = \sigma_T / G_0$$
$$x = t / \tau$$
$$u = \dot{\varepsilon}_0 \tau$$

With these, expression (j) becomes

$$y_1 = e^{-x(1 - 2u)} - e^{-x(1 + u)} \tag{k}$$

where the subscript 1 has been added to make it clear that the reduced stress is merely a contribution to the total stress.

$0 \le t' \le t$

The $0 \le t' \le t$ interval is where the algebra gets complicated, but our reduced variables will help. For the stress contribution from this interval, we have

$$\int_0^t \frac{G_0}{\tau} e^{-(t-t')/\tau} \left[e^{2\dot\varepsilon_0(t-t')} - e^{-\dot\varepsilon_0(t-t')} \right] dt' \tag{l}$$

Rather than fight through this integration, our first step will be to transform to dimensionless variables, as defined above:

$$y_2 = \int_0^x e^{-(x-x')} \left(e^{2u(x-x')} - e^{-u(x-x')} \right) dx' \tag{m}$$

where the variable $x' = t'/\tau$. A further substitution is now apparent, namely, $s = x - x'$. On making this substitution and switching the integral limits to eliminate the negative sign due to $ds = -dx'$, the expression below is obtained:

$$y_2 = \int_0^x e^{-s} \left(e^{2us} - e^{-us} \right) ds \tag{n}$$

This expression is beginning to look quite simple indeed, but we will do one more manipulation to give it the right look, namely, similar to equation (k) above with the two terms $(1 - 2u)$ and $(1 + u)$. Thus

$$y_2 = \int_0^x e^{-s(1-2u)} ds - \int_0^x e^{-s(1+u)} ds \tag{o}$$

Each of these integrals is easily evaluated to give

$$y_2 = \left[-\frac{1}{1-2u} e^{-s(1-2u)} \right]_0^x + \left[\frac{1}{1+u} e^{-s(1+u)} \right]_0^x \tag{p}$$

On substituting the limits, we obtain

$$y_2 = -\frac{e^{-x(1-2u)} - 1}{1-2u} + \frac{e^{-x(1+u)} - 1}{1+u} \tag{q}$$

As a final step aimed at easing the path to the answer in equation, we can switch the signs to yield

$$y_2 = \frac{1 - e^{-x(1-2u)}}{1 - 2u} - \frac{1 - e^{-x(1+u)}}{1 + u} \tag{r}$$

4. Combining the terms

At this point, we need to add y_1 and y_2. Another glance at equation (8-47) shows that we need to work on y_1 to get it in the right form. The equation

$$y_1 = \frac{(1 - 2u)e^{-x(1-2u)}}{1 - 2u} - \frac{(1 + u)e^{-x(1+u)}}{1 + u} \tag{s}$$

can be seen to be identical to equation (k). Now, combining y_1 and y_2 and canceling common terms in the numerator gives the expression

$$y = y_1 + y_2 = \frac{1 - 2ue^{-x(1-2u)}}{1 - 2u} - \frac{1 + ue^{-x(1+u)}}{1 + u} \tag{t}$$

which, unfortunately, looks nothing like equation (8-47). At this point, there are a couple of things one can do. The first item, as always, is to check the equation at known values. Two good ones are at $x = 0$, where the stress should be equal to zero (it is), and at $x = \infty$ for slow rates. The latter should lead to linear stress response and a steady-state viscosity (y/u) equal to 3 (it does, but see Problem 8-15). Another approach is to show numerically that equation (t) and equation (8-47) give identical results under all conditions. This is less work than it might seem. But the most satisfying is to spend a few hours trying to make equation (t) look like equation (8-47). Without going into detail, the trick is to subtract 1 from the first term and add the 1 back by subtracting 1 from the second term. Thus,

$$y = \frac{1 - 2ue^{-x(1-2u)}}{1 - 2u} - \frac{1 - 2u}{1 - 2u} - \left[\frac{1 + ue^{-x(1+u)}}{1 + u} - \frac{1 + u}{1 + u} \right] \tag{u}$$

Combining terms with like denominators and changing the sign of the last term gives the result

$$y = \frac{2u - 2ue^{-x(1-2u)}}{1 - 2u} + \frac{u - ue^{-x(1+u)}}{1 + u} \tag{v}$$

which now strongly resembles equation (8-47). We can see that it is identical on recalling that

$$\eta_E^+ \equiv \frac{\sigma_T}{\dot{\varepsilon}}$$

(ssc) (w)

The dimensionless form of viscosity for the notation used here is y/u. Transforming this back to familiar variables gives

$$\frac{y}{u} = \frac{\sigma_T(t)/G_0}{\dot{\varepsilon}_0 \tau} = \frac{\eta_E^+}{G_0 \tau}$$

(ssc) (x)

which explains the $G_0 t$ in the numerator of the terms on the right-hand side of equation (8-47).

A couple of conclusions can be drawn from this example. The first lesson is that long algebraic derivations can easily lead to errors. Symbolic manipulation programs can help, but may not match the forms being sought. For example, a popular on-line symbolic program[****] gives

$$e^{-(u+1)s}\left(\frac{e^{3us}}{2u-1} + \frac{1}{u+1} \right)$$

(y)

for equation (n). This answer is not wrong, but not in the desired form, and manipulation of this result to the desired form can still lead to errors. The second lesson is that the use of reduced (dimensionless) variables can simplify the algebra significantly, reducing the chances for error. Reduced variables also lead to simplified expressions that compress many possible answers for different parameter values into succinct expressions. For example, equation (v) shows that the reduced stressing viscosity (y/u) will depend parametrically only on the reduced deformation rate u and directly only on reduced time, x. There is no need to calculate many results for many values of rate, relaxation time, etc.

PROBLEMS

8-1. One depiction of the results of a transient extensional measurement and constant rate $\dot{\varepsilon}$ is to plot $\log \eta_E^+(t)$ vs. $\log t$, where $\eta_E^+(t)$ is the transient (or "stressing") viscosity, and t is time. Assuming the initial response is linear, predict the slope of such a plot at low times, assuming:

[****] http://integrals.wolfram.com/

(a) $G(t) = G_0 e^{-t/\tau}$

(b) $G(t) = G_0 e^{-(t/\tau)^\beta}$ with $\beta = 1/2$

8-2. For single-term stretched (slower than exponential) relaxation functions, one often sees for $G(t)/G_0$ the following:

(a) $1/(1 + x^n)$

(b) $\exp(-x^n)$

In these equations, $x = t/\tau$. What are the limits on n for liquid-like behavior, i.e., finite viscosity?

8-3. In extensional deformations, the material is subjected to hydrostatic tension of $-\sigma_T/3$. This tends to expand the material. The expansion amount is usually described in terms of the Poisson ratio ν, which is the negative of transverse strain ε_2 divided by the longitudinal strain, ε_1. Thus $\nu = -\varepsilon_2/\varepsilon_1$.

(a) Assuming that we can otherwise treat the material as a Hookean solid, show that Young's modulus E is given by $E = 2(1 + \nu)G$.

(b) The bulk modulus[††††] K is usually defined as the inverse of the compressibility, β, i.e.,

$$K = \frac{1}{\beta} = -\frac{\partial P}{\partial \ln V}\bigg|_T \qquad (8-63)$$

Assuming the material properties are uniform and insensitive to strain, relate the bulk modulus to the Poisson ratio and shear modulus, G.

8-4. The assertion was made just below equation (8-39) that, for incompressible materials, the deformation ratio in the transverse direction of a uniform sample stretched in simple extension by an amount $\lambda = L/L_0$ must be always by $1/\lambda^{1/2}$, no matter what the strain. Verify this assertion.

8-5. In equation (d) of Example 8-6, the bottom limit for $s = \infty$ is summarily ignored without comment, yet it contains the term $s^2 \exp(-s/\tau)$. Show, using l'Hôpital's rule, that this term goes to zero as s approaches infinity. If you have limited faith in this rule, program a spread sheet to examine the trend numerically.

8-6. Find the stress-strain function for a crosslinked, rubbery polymer with a shear modulus of $G_0 = 3$ MPa using:

(a) The Finger strain measure

(b) The Cauchy strain measure

Assume the stretching starts at time zero, and continues until time t. Which strain measure seems more appropriate? (Hint: do not neglect the uv term.)

[††††] The nomenclature for bulk modulus and compressibility may vary.

8-7. An LEF with a single-element Maxwell relaxation function is sheared suddenly at zero time and then held for various times t to investigate the relaxation of stress. Find the relaxation behavior as a function of time t, and strain γ_0.

8-8. (Challenging) The standard extensional experiment for polymer melts is to extend the sample with a constant rate of deformation. As discussed in Chapter 7, this requires special equipment. Suppose, instead, a sample is stretched in a common tensile tester, which moves one clamp away from the other at constant velocity V_{XH}. (XH stands for crosshead.) What is the expected response of an LEF sample in this test? Assume that the relaxation modulus is single exponential and the stretching is initiated at zero time.

8-9. (Challenging) One test of the behavior of the Cauchy strain measure at low strains might be to check the relaxation behavior as was done with the LEF in Problem 8-7 and see that it was independent of strain. Another possibility is to examine the dynamic behavior of the Cauchy material: the moduli G' and G'' should be independent of strain, at least at low strains, and the expressions should be identical to those derived using the Boltzmann equation. Check this out using a single-element Maxwell relaxation function.

8-10. Using an LEF model with a single relaxation time, predict and compare the relaxation of N_1 and σ_{21} after abruptly stopping the deformation of a material that has been continuously sheared for a long time.

8-11. One way of introducing nonlinearity in integral equations such as the LEF is to use a concept pioneered by Tanner and Simmons.[7] The argument is that an elastic fluid would deform to a certain strain, and then fail, with the failure corresponding to complete untangling of the polymer chains. Thus the integration, using the present position as the reference, would be carried from $t' = t$ back by an amount t_B, where t_B is the break time equal to $a/\dot{\gamma}_0$. In this expression, a is an adjustable parameter and $\dot{\gamma}_0$ is the applied shear rate. Thus the integral would take on the form

$$\tau(t) = \int_{t-t_B}^{t} M(t-t')\left[\mathbf{C}^{-1}(t,t') - \delta\right] dt' \qquad \text{(ssc)} \quad (8\text{-}64)$$

if we assume the Finger relative strain measure $(\mathbf{C}^{-1})$ is adequate. This is a statement of the Tanner-Simmons model.

(a) What is the relationship between a and the breaking strain γ_B?

(b) Investigate the behavior of an LEF with this modification. Use a steady-shear startup experiment, and a memory function with a single relaxation time.

(c) Plot the results using reduced variable such as τ_{21}/G_0 vs. t/t_B or $\dot{\gamma}_0 t$. Comment on the appearance.

8-12. It was stated in the text that the broadening of the relaxation behavior fails to introduce *nonlinear* shear behavior into the LEF model. Check this out using a double element Maxwellian relaxation modulus, i.e., $G(t) = G_1\exp(t/\tau_1) + G_2\exp(t/\tau_2)$.

8-13. (Computer) Romberg integration is a method for evaluating an integral numerically. The algorithm is very easy to remember: just use the trapezoidal rule with various (uniform) intervals and extrapolate the results to zero interval size. The trapezoidal rule is simply the average of equally spaced values of the integrand multiplied by the difference between the upper and lower integration limits. Obviously, a lower limit of $-\infty$ is a bit difficult to handle, but the exponential memory function means that $-\infty$ can be safely replaced by, say, 10 relaxation times back from the time at which you wish to find the stress.

(a) Set up and solve the problem with in algebraic form for an LEF model with a single relaxation time and N evaluation points spaced equally along the time axis. Assume a simple-shear startup experiment with the start of the deformation at zero time.

(b) Program the problem on a spread sheet using the "average" function (AVERAGE in Excel®) to do the integration. Assume a zero-time modulus of 1 GPa, a relaxation time of 1 s, a shear rate of 1 s^{-1} and a time of 1 s. Use 10 equally spaced times for a start. (Hint: space the points such that one lands on $t' = 0$, to facilitate dividing up the integral.)

(c) Increase the number of points N to 20, 30 and 40. Compile the results in a table. Extrapolate these linearly vs. $1/N$ to find a better estimate of the value of the integral. Check this value against the exact answer from part (a).

8-14. (Challenging) In the Tanner-Simmons model, the concept is a strain-induced breakup of the entanglement network. Thus, it seems reasonable that the breaking strain would be referenced to the unstressed state of the material. For the familiar simple-shear stress-growth experiment, the reference state would be zero time when the deformation starts.

(a) Explore what happens using the LEF model, but integrate over the range $-\infty < t' \leq t_B$, instead of over the usual range from $-\infty < t' \leq t$ or $t - t_B < t' \leq t$ used in Problem 8-11. The strains would still be referenced to positions at $t' = t$. Plot the result.

(b) Similarly, find out what happens with the time range $-\infty < t' \leq t_B$ but using the linear Boltzmann integral, equation (8-6). This assumption is identical to the Tobolsky-Chapoy approximation, equation (5-18).

8-15. In Appendix 8-2, the assertion was made that equation (t) would give a relative extensional viscosity of 3 at $t \to \infty$ and at slow rates. Demonstrate this clearly.

8-16. For Appendix 8-2, compare equation (y), after substituting in the limits, with equation (r). Show that the two results are the same.

8-17. Example 8-5 derives the steady shear viscosity of the single-exponential LEF via a substitution ($s = t - t'$), which may leave the reader somewhat uneasy because the usual method for deriving steady-state properties via integral equations is to integrate from $-\infty$ to 0. Show that the latter approach leads to the same result.

8-18. An often-used procedure for subjecting a material to biaxial extension is to squeeze a cylindrical sample of polymer melt between two large well-lubricated plates (see descriptions in Chapter 7). From this experiment, a biaxial stress-growth viscosity can be defined[8] as

$$\eta_B^+(t;\dot{\varepsilon}_B) \equiv \frac{\sigma_{11} - \sigma_{22}}{\dot{\varepsilon}_B} \qquad \text{(ssc)} \quad (8\text{-}65)$$

where the stretch rate

$$\dot{\varepsilon}_B = \frac{\partial V_1}{\partial x_1} \qquad (8\text{-}66)$$

is held constant after the start of the experiment. In the squeezing geometry, the 1 direction is radial, and the 2 direction is axial. If we ignore atmospheric pressure, the radial surface is free of stress.

(a) If the initial sample thickness is H_0, describe the thickness-vs.-time profile $H(t)$ that needs to be applied to hold the stretch rate $\dot{\varepsilon}_B$ constant.

(b) Write the relationship between the force F and the stresses produced by the stretching.

(c) (Challenging) Assuming a single-exponential relaxation modulus $G(t) = G_0 \exp(-t/\tau)$ describe the time dependence of the biaxial extensional viscosity η_B for an LEF. Plot the results using log-log scales and reduced variables. (See Appendix 8-2 for examples of reduced variables that might be useful.)

(d) Explain the signs of the stresses in terms of both (ssc) and (fsc).

8-19. Examine the development in Appendix 8-2 depicting the stress produced by the steady extension of an LEF.

(a) Relate the three values of C_{ii}^{-1} to the laboratory elongation ratio $\lambda(t;\dot{\varepsilon}_0) = L(t)/L_0$, where $L(t)$ is the sample length at time t and L_0 is the initial length, that is, when the extension process starts.

(b) Calculate the values of the first and second invariants of $\mathbf{C}^{-1}$ at the present time if the elongation ratio λ at the present time is 4.

(c) If the reference condition remains unchanged [i.e., $\lambda(t) = 4$], what are the values of the first and second invariants at $\lambda(t') = 1$, corresponding to the start of the experiment.

8-20. Find the memory function for a Rouse relaxation function. Comment on the strength of each term at zero values of the argument relative to the values of the partial moduli for the original Rouse function, and the spacing of each term.

8-21. (Challenging) The LEF can predict normal stresses in steady shear flow.

(a) Show that the LEF is also capable of predicting normal stresses in oscillatory shear deformations.

(b) Show that the normal stresses oscillate at twice the frequency ω of the shear stresses and their amplitude scales as the square of the shear strain, as in Figure 6-11.

(c) Is the shape of the normal-stress wave $\sin(2\omega t)$ or $\sin^2(\omega t)$?

8-22. One fix-up for the LEF equation that will lead to shear-rate-dependent viscosity is to have the relaxation times in the memory function depend on the magnitude of the rate of deformation. Suppose, for example that we state that

$$\tau = \frac{\tau_0}{1 + \dot{\gamma}/\dot{\gamma}_0} \tag{a}$$

where τ is the relaxation time in a single-exponential relaxation function $G = G_0 \exp(-t/\tau)$ and τ_0 and $\dot{\gamma}_0$ are parameters.

(a) Derive the expression for the steady-shear viscosity using the above information. Is the behavior in accord with that for typical GNF models? Show this by fitting the data with the Cross model.

(b) Repeat above for the first normal-stress-difference coefficient.

(c) Compare the characteristic times τ_0 and $1/\dot{\gamma}_0$ from equation (a) with $1/\dot{\gamma}_0$ from the Cross-model fit to the viscosity, and with $\Psi_{1,0}/2\eta_0$. Which comparisons seem to match the best?

REFERENCES

1. See, for example, M. T. Shaw and W. J. MacKnight, *Introduction to Polymer Viscoelasticity*, Wiley, New York, 2005, Chapter 6.

2. R. B. Bird, R. C. Armstrong and O. Hassager *Dynamics of Polymeric Liquids*, Vol. 1, 2nd ed., Wiley-Interscience, New York, 1987.

3. R. G. Larson, *Constitutive Equations for Polymer Melts and Solutions*. Butterworth, New York, 1988.

4. M. H. Wagner "Zur netzwerktheorie von Polymer-Schmelzen," *Rheol. Acta*, **18**, 33–50 (1979).

5. M. Narkis, I. L. Hopkins and A. V. Tobolsky, "Studies on the stress relaxation of polystyrenes in the rubbery-flow region," *Polym. Eng. Sci.*, **10**, 66–69 (1970).

6. C. P. Lindsey and G. D. Paterson, "Detailed comparison of the Williams–Watts and Cole–Davidson functions," *J. Chem. Phys.*, **73**, 3348–3357 (1980).

7. R. I. Tanner and J. M. Simmons, "Combined simple and sinusoidal shearing in elastic liquids," *Chem. Eng. Sci.*, **22**, 1803–1815 (1967).

8. J. M. Dealy, "Official nomenclature for material functions describing the response of a viscoelastic fluid to various shearing and extensional deformations," *J. Rheol.*, **39**, 253–265 (1995).

9
Molecular Origins of Rheological Behavior

As described in Chapter 8, much can be said about the nonlinear rheological functions using a strain measure that holds at high strains as well as low. A key to much of the success with fluids was the use of the present position of material elements as the reference state against which strain is calculated.

Nothing in the developments in Chapter 8 requires the presence of macromolecules; or, for that matter, any molecules at all. The models thus developed are referred to as *continuum models*. The goal of the *molecular models* is to bring in features of the molecular structure with the hope of connecting the molecular features (chain length, side groups, secondary forces) to the rheological behavior. Roughly, this is the path taken in the development of the kinetic or statistical description of elastomers, but with the addition of the motion of the chains as time goes on.

A. DESCRIPTION OF POLYMER MOLECULES

Before building models of chain dynamics, we need to know what reasonable descriptions of polymer chains might be. In a single-component melt, we know that the unperturbed polymer molecule is in a completely relaxed state, existing as a coil of chained segments that have no preference for interaction with the segments of other chains vs. those on the same chain. As a result, we expect the chains to be entangled, a state that is clearly demonstrated by the linear viscoelastic response of melts and concentrated solutions.

But what happens as the molecular assembly is strained to very high levels? With very dilute chains in a solvent, we might expect the chains to uncoil in the direction of the principal tensile stress, and drag on the solvent in a Rouse-like fashion, giving essentially linear response in shear. In simple extension, the chains would be interacting with the medium over long distances, which might introduce strong nonlinearities. In the melt, we might take the view that the assembly acts more like an elastomer, but with temporary crosslinks. As mentioned in Chapter 8, this approach leads to the models based on the Finger relative strain tensor.

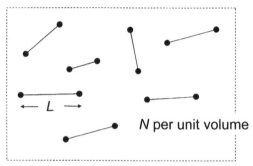

Figure 9-1. Illustration of a dilute suspension of dumbbell-shaped molecules in a Newtonian medium. Unlike a dilute suspension of spheres, this suspension is predicted to have a finite N_1 that depends on temperature. Why? Because Brownian motion tends to disorient the dumbbells, giving an elastic response during flow. Randomization goes faster at higher temperature.

There is no attempt in this chapter to go into detail concerning the derivation of constitutive equations derived from molecular theories. This is a hugely complex field. The resulting mathematical expressions can be complicated, even for the simplest model. For example, the general expression for stress developed on deforming a fluid comprising rigid dumbbells (Figure 9-1) suspended in a Newtonian fluid has a Newtonian term followed by an infinite series of terms containing multiple[*] time-dependent integrals.[1]

Is such a material a realistic molecular model for a polymer solution? Such seems unlikely, although possibly for dilute solutions of rigid-rod macromolecules.[†] Needed are molecular models for concentrated solutions and melts of highly entangled molecules executing complex motions.

---

[*] A multiple integral involves two or more integration variables, often spatial. An example of time-dependent multiple integral is $\int\limits_{-\infty}^{t}\int\limits_{-\infty}^{t'} K(t'';t',t)dt''dt'$, where K is the integrand containing time-dependent memory and strain information.

[†] Distributing the molecular volume along the rigid rod, rather than concentrating it at the ends, changes only the rotational time for the "molecule."

B. THE ROUSE CHAIN—A LIMITED DESCRIPTION OF POLYMER BEHAVIOR

The Rouse theory of polymer dynamics is a baseline molecular theory that is described in detail in most treatments of linear viscoelasticity.[2] The Rouse chain is represented as alternating beads and springs, which certainly seems to be an improvement over the rigid dumbbell described above. The beads are treated as spheres that are dragged through the fluid. The springs are considered to be Gaussian springs, i.e., identical to those used in the kinetic theory of rubber elasticity. Figure 9-2 is a depiction of this model.

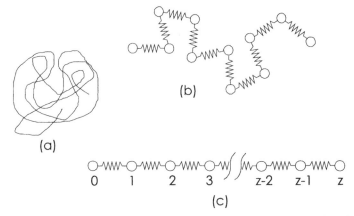

Figure 9-2. The coil shown in (a) is replaced by freely rotating beads and springs depicted in (b). In (c), the chain of beads and springs has been depicted in one dimension to simplify the equations of motion. (Reprinted with permission of John Wiley & Sons, Inc. from Shaw and MacKnight.[2])

An important aspect of the Rouse theory is the force produced by dragging the spheres through surrounding molecules. Rather than cope with the details of this medium, the theory simply replaces it by a continuum with a viscosity η. While it is perhaps not obvious, this choice means that the response of a Rouse fluid will be similar to the LEF model, but with a specified series of exponential relaxation terms. The classical expression for the Rouse spectrum is

$$G(t) = NkT \sum_{p=1}^{z} e^{-t/\tau_p} \qquad (9\text{-}1)$$

where N is the number density of polymer molecules, z is the number of physical segments in the chain, and kT has the usual meaning. (The polymer chains are assumed to all be of the same length.) The relaxation times τ_p are given by the equation

$$\tau_p = \frac{6M(\eta - \eta_s)}{cRT\pi^2 p^2}$$

(9-2)

for a polymer solution of concentration c in a solvent with viscosity η_s. For a melt, just change c to ρ, the melt density, and drop the solvent viscosity. M is the molecular weight of the polymer. Note that we are using RT instead of kT to correspond to the normal use of M as the mass per mole (g/mol) of polymer, not the mass per molecule.

For a working molecular rheological theory, we might combine the Rouse spectrum with the Finger strain tensor, the molecular aspect coming into the picture via the dependence of the relaxation times on molecular weight via equation (9-2) and the auxiliary viscoelastic relation

$$\eta = \int_0^\infty G(t)dt = NkT\sum_{p=1}^z \tau_p$$

(9-3)

By using the Finger tensor, we naturally endow the result with all the good and bad points of the LEF, including constant viscosity, strain-hardening extensional viscosity, finite first normal-stress difference, and no second normal-stress difference. The only impact of the distribution of relaxation times is to broaden the response.

A central problem with the Rouse model, and the related Zimm[‡] model,[3] is that neither deals with polymer entanglements in a consistent and effective fashion. We can appreciate without writing any equations that entanglements between polymer chains will be the primary cause of the high dependence of viscosity on molecular weight, as well as nonlinear effects such as shear thinning and normal stresses. But before leaving the Rouse model, we need to ask what kind of changes in bead-spring models might lead to nonlinear effects. For example, would springs with different characteristics be helpful? Or is a wholesale revision in thinking needed?

C. OTHER CHAIN-LIKE MODELS

As hinted above, there are many possible combinations of beads, springs and rods that have been investigated as dilute suspensions in a Newtonian or linear viscoelastic matrix. Their principal appeal, other than the mathematical

---

[‡] Bruno Zimm developed the Zimm modification (reference 2) a few years after the publication of the Rouse model. It recognizes that the isolated polymer coil will provide hydrodynamic shielding of a portion of the segments. The result is a modification of the Rouse times.

tractability, is that they convey, at first glance, a clear picture of the molecular dynamics that produce the extra stresses. For example, the uniaxial extension of a dumbbell suspension, if sufficiently intense, will tend to line up the dumbbells in the flow direction. On ceasing the flow, Brownian motion will dominate and randomize the directions of the dumbbells. The result is an exponential relaxation of stress with a relaxation time τ given by

$$\tau = \frac{\delta_s L^2}{12kT} \tag{9-4}$$

where δ_s is the Stokes' drag coefficient for the spheres at the ends. The drag coefficient for spheres is given by

$$\delta_s = \frac{F}{V} = 6\pi\eta r \tag{9-5}$$

where F is the force, V is the velocity, η is the viscosity of the medium and r is the radius of the spheres. The dimensions of δ_s are $[M][t]^{-1}$. SI units will be kg/s or N s/m. All this applies for one molecule.

Example 9-1: *Estimate the relaxation time for a dumbbell with a molecular weight of 20 kDa. Make necessary, but reasonable assumptions.*

We will assume all the molecular volume is concentrated into two spheres, one at each end of dumbbell, and these spheres have a density of 1 g/cm^3 = 1000 kg/m^3. Other assumptions include:
- $L = 30$ nm $= 3\times10^{-8}$ m
- $\eta = 10^{-3}$ kg/m s
- $T = 300$ K

The value of k in SI units is 1.38×10^{-23} J/K.

Some intermediate calculations include the radius of each sphere, r. The result of this is

$$r = \frac{1}{2}\left(\frac{3M/N_A}{\pi\rho}\right)^{1/3} \tag{a}$$

where N_A is Avogadro's number (6.022×10^{23}). Substituting the quantities above gives the result $r = 1.6$ nm $= 1.6\times10^{-9}$ m, which seems reasonable.

Using equation (9-5) to calculate the friction coefficient gives about $\delta_s = 3\times10^{-11}$ N s/m. Finally, we use equation (9-4) to calculate the time constant, obtaining the result $\tau = 5\times10^{-7}$ s $= 0.5$ µs. A time this short could be observed dielectrically.

Table 9-1. Concentration conversions for two components. [a]

Want [b]	Have [b]	Conversion [c,d]
c, g_p/dL_t	N, mol_p/m_t^3	$NM/10$
c, g_p/dL_t	ϕ_p	$\rho_p\phi_p/10$
ϕ_p	N, mol_p/m_t^3	NM/ρ_p
ϕ_p	c, g_p/dL_t	$10c/\rho_p$
N, mol_p/m_t^3	ϕ_p	$\phi_p\rho_p/M$
N, mol_p/m_t^3	c, g_p/dL_t	$10c/M$
n, particles/m_t^3	N, mol_p/m_t^3	NN_A

[a] Subscript p is for particles (polymer); m for medium (solvent); t for total system.

[b] n [=] particles/m_t^3; N [=] mol_p/m_t^3; M [=] kg_p/mol_p; m [=] kg_p/particle;
ρ_p [=] kg_p/m_p^3; c [=] g_p/dL_t

[c] N_A = Avogadro's number = 6.02×10^{23} particles/mol

[d] In these equations, volume fraction (ϕ) is based on specific or molar volumes of pure components, and not on partial volumes (e.g., volumes occupied in solution).

Example 9-1 demonstrates that a dumbbell of decent size is moved around very quickly by Brownian motion. The contribution of the dumbbells to the viscosity of a suspension is $NRT\tau$, where N is the moles of dumbbells per cubic meter, mol/m^3, and τ is the characteristic time. To convert this to the more palatable concentration (kg/m^3), N needs to be multiplied by molecular weight in kg/mol. Dividing this result by the dumbbell's density in kg/m will give the volume fraction ϕ_p of dumbbells. All this leads to the formula

$$\eta - \eta_m = nkT\tau = NRT\tau = \frac{\phi_p\rho_p}{M}RT\,\tau \qquad (9\text{-}6)$$

which can be used to estimate the impact of dumbbells on solution viscosity. In this equation, η_m is the matrix (solvent) viscosity, n is the number density of dumbbells, N is the molar density of dumbbells, ρ_p is the particle density, and M is the molecular weight of the particles.

Before trying this out, let's run through the units to make sure we are being consistent (a "reality" check) with respect to dimensions of the variables and the choice of units. When checking in this fashion, it is important to keep track of which component is being discussed. Thus we use sub p to designate the particle (polymer), sub m to designate the medium (solvent), and sub t to designate the total mixture.[§] Using the order shown in equation (9-6), we have

$$\text{Pa s} = \dfrac{\dfrac{m_p^3}{m_t^3} \dfrac{kg_p}{m_p^3} \dfrac{J}{mol_p K} K s}{\dfrac{kg_p}{mol_p}} \tag{9-7}$$

After considerable cancellation of units on the right-hand side, we can see that the two sides balance, on recalling that Pa $[=]$ J/m^3.

Example 9-2: *Calculate the viscosity increment for the solution discussed in Example 9-1 for a volume fraction of 0.01 (1%) dumbbells. Compare with the Einstein equation for spheres.*

First off, an examination of the Einstein equation prediction, which is given by the equation

$$\eta = \eta_m (1 + 2.5\phi_p) \tag{a}$$

at low particle concentrations, ϕ_p. With $\phi_p = 0.01$, the suspension viscosity η is predicted to be 1.025 times the medium (solvent) viscosity η_m.

The application of equation (9-6) requires reference to the assumptions in Example 9-1. The substitutions are shown below:

$$\eta - \eta_m = \dfrac{(0.01)(5\times10^{-7}\text{ s})(1000\text{ kg/m}^3)(8.314\text{ J/mol K})(300\text{ K})}{20\text{ kg/mol}} \tag{b}$$

Note the consistent use of SI units. Doing the arithmetic, gives a viscosity increment of 0.00062 Pa s, or a suspension viscosity of 1.6 times the matrix viscosity of 0.001 Pa s. This is considerably higher than the same volume distributed as spheres instead of dumbbells.

[§] This notation brings to mind suspensions, and the difference between a suspension and a solution. We like to think of particles as solid objects with fixed shape, whereas components of a solution as flexible molecules, all atoms of which are exposed to short-range interactions with the solvent. However, the boundary blurs with inflexible molecules of large size, e.g., Bucky balls, graphene sheets, polyphenylene rods, helical polypeptide rods. Thermodynamically, suspensions of such particles, or even of particles such as gold nanospheres, can act like solutions in every way, although with highly nonideal behavior.

In terms of intrinsic viscosity, commonly reported as dL/g, volume fraction can be substituted using $c/100$ (c [=] g/dL) if the density of the polymer is 1000 kg/m^3 and 100 denominator stands for 100 g/dL for the solution (Table 9-1). Thus, the intrinsic viscosity of a sphere is 0.025 dL/g. For the dumbbell used in this example, the intrinsic viscosity works out to be 0.62 dL/g, assuming a linear increase in solution viscosity with concentration.

These two examples illustrate the key difference between a molecular theory and a continuum theory. With the latter, the existence of molecules is not recognized or needed. With the molecular theory, we are able to estimate measurable rheological properties (e.g., intrinsic viscosity) using only information concerning the size and shape of the molecules. The prediction is as good as the model.

Models of this sort also suggest forms of continuum rheological equations that might be useful when suitably modified as illustrated in the Chapter 8 for some integral forms. Like the differential equation of motion of the Maxwell element,

$$\frac{d\gamma_{21}}{dt} = \frac{1}{G_0}\frac{d\tau_{21}}{dt} + \frac{G_0}{\eta} \qquad \text{(ssc) (9-8)}$$

which has the familiar solution

$$\tau_{21}(t) = \int_{-\infty}^{t} G_0 e^{-(t-t')G_0/\eta}\frac{d\gamma_{21}(t')}{dt'}dt' \qquad \text{(ssc) (9-9)}$$

the differential equation for a molecular model can be solved, at least in principle. Unfortunately, as mentioned above, these models tend to be mathematically complicated; for example, that for a Hookean dumbbell (rod replaced by a linear spring) has the differential form in extension

$$\tau + \tau_H\frac{\mathcal{D}\tau}{\mathcal{D}t} - \frac{\tau_H}{2}\left(\dot{\gamma}\cdot\tau + \tau\cdot\dot{\gamma}\right) = nkT\tau_H\dot{\gamma} + \eta_m\tau_H\left(\frac{\mathcal{D}\dot{\gamma}}{\mathcal{D}t} - \dot{\gamma}\cdot\dot{\gamma}\right) + \eta_m\dot{\gamma} \quad \text{(ssc) (9-10)}$$

where the $\mathcal{D}/\mathcal{D}t$ stands for the time derivative taken while moving and rotating with the fluid. Its definition, in terms of the fixed frame and using τ as an example tensor, is

$$\frac{\mathcal{D}\tau}{\mathcal{D}t} = \frac{\partial\tau}{\partial t} + v\cdot\nabla\tau + \frac{1}{2}\left(\omega\cdot\tau - \tau\cdot\omega\right) \qquad (9-11)$$

where ω is the vorticity tensor. The definition of vorticity in rectangular coordinates is

$$\omega_{ij} = \frac{\partial v_i}{\partial x_j} - \frac{\partial v_j}{\partial x_i} \qquad (9\text{-}12)$$

The term $v \cdot \nabla \tau$ is the most complex. This tensor comprises terms illustrated by the 1,2 term below:

$$(v \cdot \nabla \tau)_{12} = v_1 \frac{\partial \tau_{12}}{\partial x_1} + v_2 \frac{\partial \tau_{12}}{\partial x_2} + v_3 \frac{\partial \tau_{12}}{\partial x_3} \qquad (9\text{-}13)$$

For homogeneous flows, this and all the terms in $v \cdot \nabla \tau$ are zero.

Equations (9-10) through (9-13) are presented to show that things can get very complicated, especially compared to the equivalent integral expression for this model

$$\tau(t) = \int_{-\infty}^{t} \frac{nkT}{\tau_H} e^{-(t-t')/\tau_H} \left(C^{-1}(t,t') - \delta \right) dt' + \eta_m \dot{\gamma} \qquad \text{(ssc)} \quad (9\text{-}14)$$

which is an LEF with the relaxation time and zero-time modulus defined by the model. The relaxation time τ_H for the Hookean dumbbell is the same idea as the Rouse model, i.e., the time required for the spring, with spring constant H, to drag the beads through the viscous medium. The expression is

$$\tau_H = \frac{3\pi\eta_m r}{2H} \qquad (9\text{-}15)$$

The point of the comparison between the Hookean dumbbell and the Maxwell model is to suggest that the proper handling of a simple molecular model can lead to agreement with the continuum mechanical concept of a measure for high strain. The comparison also points out that the differential and integral expressions are connected.

A significant practical use of differential models is in computational fluid dynamics. With the differential model, the stress and stress gradients expected at a point are directly relatable to the conditions at neighboring points without integration over time. To have access to manageable differential expressions, simplified versions have been developed, e.g., the convected Maxwell model.

The problem with the simplified differential model, as with the simpler integral models, is that the predictions can fall short of real polymer behavior. Nevertheless, even the simplest models bring in viscoelastic features that can provide useful information about the flow. Needless to say, this area is beyond the scope of this book.

D. DEALING WITH ENTANGLEMENTS

The molecular models outlined in Sections B and C provide valuable insight but fall short of providing the relationships needed for dealing with polymer melts and concentration solutions. Molecular entanglement is the feature of these fluids that leads to a high viscosity and strong dependence of viscosity on shear rate.

1. Some early ideas

On picturing the behavior of an entangled melt in a velocity field with high velocity gradients, one might imagine that the entanglements simply knot up in spite of intense Brownian motion. Flow would then be realized by:

- Slip at the wall
- Crack propagation through the melt
- Breakage of chains
- Granular flow of lumps comprising the more highly entangled regions

While all of these mechanisms have been proposed and tested for some polymers, only "slip at the wall" is supported by firm experimental evidence, and then only at elevated stress levels. "Granular flow of lumps," sometimes referred to as "flow units," has some support for certain polymers, particularly poly(vinyl chloride), but is very difficult to exclude for others.

   The picture of rotating flow units can be extended to molecular levels, and is in concert with an early theory of shear thinning promoted mainly by F. Bueche (not to be confused with A. M. Bueche). The abstract to his key publication[4] is very clear, and reproduced below:

"The measured viscosity of bulk polymers and their solutions varies with the magnitude of the applied shearing stress. In this paper we show that this is a natural consequence of the fact that the molecules are caused to rotate by the shearing action. This rotation gives rise to sinusoidal forces which alternately stretch and compress the molecules. The viscous behavior of such a system is shown to be analogous to that of a spring immersed in a viscous medium and being acted upon by sinusoidal viscous forces. We have treated the problem quantitatively using the method of a previous paper. Good agreement is obtained between theory and experiment without the aid of adjustable parameters. It is found that, if the molecule is considered shielded, the theory accurately represents the experimental data for dilute solutions. The behavior of the bulk polymer is well represented by means of the free draining approach. It is pointed out that the variation of the viscosity of dilute polymer solutions with the rate of shear may provide a convenient method for determining molecular weights of coiling type polymers."

"Shielded" and "free draining" are concepts borrowed from spring-bead viscoelastic theories of Rouse and Zimm, and control the distribution of relaxation times. The analogy mentioned reminds one of the Cox-Merz rule, and suggests that entanglements are, in a sense, ignored. As with the Rouse theory, the Bueche relaxation time is often written in terms of the zero-shear-rate viscosity η_0 as

$$\tau_1 = \frac{12(\eta_0 - \eta_m)M}{\pi^2 cRT} \tag{9-16}$$

where η_m is the viscosity of the matrix. For melts, the density ρ is substituted for c, and the η_m is dropped. The expression for the shear-rate-dependent viscosity is

$$\eta(\dot{\gamma}) = \eta_m + (\eta_0 - \eta_m)\left[1 - \frac{6}{\pi^2}\sum_{p=1}^{N}\frac{\dot{\gamma}_R^2}{p^2(p^4 + \dot{\gamma}_R^2)}\left(2 - \frac{\dot{\gamma}_R^2}{p^4 + \dot{\gamma}_R^2}\right)\right] \tag{9-17}$$

where $\dot{\gamma}_R = \tau_1\dot{\gamma}$ and N is the total number of modes. The sum, as it is set up, has $N \to \infty$, but the rapid convergence requires only a moderate number of terms to produce reasonable results.

While attractive because of explicit algebraic form, we must regard this approach as more interpolative than predictive because we need to know η_0 before viscosity predictions can be made.

Example 9-3: *A polymer is known to have a molecular weight of 100 kDa. The zero-shear-rate viscosity is measured at 190 °C and found to be 10 kPa s. Estimate its viscosity at 100 s^{-1} using the Bueche expression. Assume the density of the melt is 1 g/cm^3.*

The first step is to calculate the characteristic time τ_1 using equation (9-16). This is mostly a matter of keeping the units straight. Using SI units consistently gives

$$\tau_1 = \frac{(12)(10000\text{ Pa s})(100\text{ kg/mol})}{(\pi^2)(1000\text{ kg/m}^3)(8.314\text{ J/mol K})(190 + 273.2\text{ K})} \tag{a}$$

Recall that Pa [=] J/m^3, so all the units save the seconds in the viscosity cancel out very nicely. Doing the arithmetic gives the result $\tau_1 = 0.32$ s, which is reasonable.

The next job is to tackle the sum. The value of $\dot{\gamma}_R$ is 32, so the index p will need to be run quite high to get reasonable precision. With a spread sheet, set up the index p in the first column. In the second column, calculate the term corresponding to p. In the third column, run a sum of the terms. Keep adding terms until the sum doesn't change much, perhaps less than 1 in 10^4. This could be checked in a fourth column.

The result of this is shown below:

Constants	Comment	p	Term	Sum	Relative Change
32	Reduced shear rate	1	1.0000	1.00000	
1024	Square of above	2	0.2499	1.24994	0.24994
		3	0.1105	1.36045	0.08842
		4	0.0600	1.42045	0.04410
		5	0.0343	1.45471	0.02411
		6	0.0191	1.47382	0.01314
		7	0.0104	1.48420	0.00704
		8	0.0056	1.48982	0.00379
		9	0.0031	1.49293	0.00209
		10	0.0018	1.49470	0.00119
		11	0.0010	1.49575	0.00070
		12	0.0006	1.49638	0.00043
		13	0.0004	1.49679	0.00027
		14	0.0003	1.49705	0.00017
		15	0.0002	1.49722	0.00012
		16	0.0001	1.49734	0.00008

In all, 16 terms are required to bring the change in the sum below our target of 1 part in 10^4. The reason for the large number of terms is the high value of the reduced shear rate. The sum, when entered into equation (9-17), yields a viscosity of 1000 Pa s, which is $1/10^{th}$ of η_0.

Is the result shown in Example 9-3 realistic? See Problem 9-7 or 9-8 for example using real data.

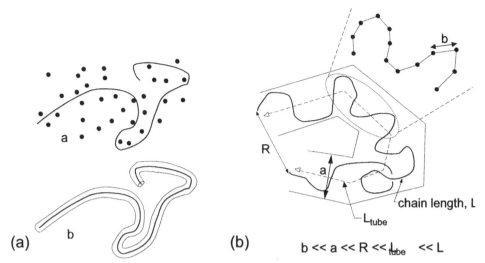

Figure 9-3. Handling of entanglements by the tube model: the tube schematic in (a) and the geometric variables defined in (b). All interactions of the two intersecting chains are treated equally, and there are no attractive forces. Extra impediments to end-to-end diffusion include branches on the diffusing chain. If sufficiently long, a branch must first diffuse out of its constraints before the main chain can move. Two branches or more create a very difficult situation, something akin to a fly trying to get off flypaper. [Reprinted with permission of John Wiley & Sons, Inc. from Shaw and MacKnight (2005). [2]]

A more predictive approach was developed by William Graessley[5] and collaborators during the late 1960s. Their theory introduces explicitly an entanglement, which occurs when two coils impinge. The entanglement will persist as long as the center point of one coil is closer to the other than the sum of their two radii of gyration, assuming that the center of masses of each coil stay on their respective streamlines. This concept leads to an increase of the persistence time with molecular weight and concentration, and a decrease with shear rate. Rotation of the coils in concert with the fluid vorticity, a feature of the Bueche theory, is not needed. Instead, this source of energy dissipation is replaced by the extra dissipation produced by the doublet as it rotates, because now the solvent must go around the entire assembly, assuming the coils are not "free-draining."

The details are somewhat involved and only the end results will be presented here. The Graessley expression for the viscosity is

$$\frac{\eta}{\eta_0} = [g(\theta)]^{3/2} h(\theta) \qquad (9\text{-}18a)$$

where the two functions of θ are

$$g(\theta) = \frac{2}{\pi}\left(\cot^{-1}\theta + \frac{\theta}{1+\theta^2}\right) \qquad (9\text{-}18\text{b})$$

and

$$h(\theta) = \frac{2}{\pi}\left(\cot^{-1}\theta + \frac{\theta(1-\theta^2)}{(1+\theta^2)^2}\right) \qquad (9\text{-}18\text{c})$$

The variable θ is where the trouble starts, as it contains the relative viscosity as well as the expected dependence on shear rate, as can be seen below:

$$\theta = \frac{\eta}{\eta_0}\frac{\dot{\gamma}\tau_1}{2} \qquad (9\text{-}18\text{d})$$

The τ_1 is the Bueche time constant, i.e., twice the Rouse maximum relaxation time, and is where the molecular weight information enters. Because of the implicit nature of the final equation, and other limitations, the Graessley approach as shown here has not been as fully tested as the Bueche theory.

Both the Bueche and the Graessley approaches are descriptions for a polymeric material with a single molecular weight. Distributions must be handled by some form of averaging according to the amounts and relaxation properties of each component in the blend. Neither theory suggests exactly how this should be done, at least in their original forms.

2. The "tube diffusion" concept

A completely different picture is one where entanglements are considered as temporary impediments to long-range displacements of the macromolecules. Intense Brownian motion keeps these entanglements "loose" enough, so they do not become permanently knotted. How can such an irregular structure be modeled to produce realistic results?

The key concept, as pioneered by de Gennes, Doi and Edwards, and others is that the duration of the entanglement is not determined by the geometry of the coil, but by the time is takes for the molecule to work itself loose from the impediments by end-to-end Brownian diffusion (reptation). The assembly of impediments can be visualized as a tube, but as the polymer molecule pokes out from one end, a similar length of tube at the other end disappears. Auxiliary ideas are that the tube deforms affinely with the bulk fluid, and the force developed by the deformed tube relaxes linearly with its remaining length. Later concepts included the realization that the molecules making up the tube wall also diffuse, and when they diffuse away from the wall, the tube radius expands. This process speeds up the relaxation of stress. Various terms have been associated with this important feature, including "tube expansion" and "constraint release."

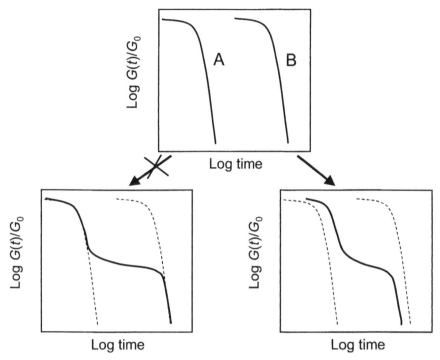

Figure 9-4. Schematic illustration of the effect of tube expansion on the relaxation of polymer molecules in a blend. The behavior of the two is not additive, as in the lower left; instead, the behavior depicted in the lower right is observed.

A simple but important outcome of the tube expansion concept is that it predicts the relaxation behavior of two components of different molecular weight in a binary blend will both be severally modified relative to their behavior in the pure state. Thus, the relaxation modulus of the blend will not be a simple weighted sum of the moduli of the two components; instead, the relaxation of the low-molecular-weight molecules will be slowed, while the high-molecular-weight molecules will relax faster. This is pictured in Figure 9-4. The relaxation in this case is often expressed as a quadratic weighting with interaction terms, e.g.,

$$[F(t)]^{1/2} = w_1[F_1(t)]^{1/2} + w_2[F_2(t)]^{1/2} \tag{9-19}$$

where w_i are the weight fractions of each of two components 1 and 2. The symbol F stands for the reduced relaxation modulus, i.e., $G(t)/G_0$.

While the ideas and scaling results associated with the tube models have enjoyed wide recognition and acceptance, practical results are somewhat difficult to find.   Attempts have been made to express these often complex descriptions in terms of familiar continuum models.   An example is one expressed in terms of the K-BKZ model, which may be written in factorized form[**] as

$$\tau(t) = \int_{-\infty}^{t} M(t - t')\{\phi_1[\mathbf{C}^{-1}(t,t') - \boldsymbol{\delta}] - \phi_2[\boldsymbol{\delta} - \mathbf{C}(t,t')]\}dt' \qquad \text{(ssc) (9-20)}$$

where ϕ_1 and ϕ_2 are derivatives an unknown strain energy potential function, often called $W(I_1,I_2)$.   The derivatives are taken respect to strain invariants I_1 and I_2, respectively, of $\mathbf{C}^{-1}$.   Recall that with this equation, we can predict any stress combination for any deformation; however, it looks hopelessly complicated.   For this reason, a host of simplified forms have been proposed, including ones with $\phi_2 = 0$, which gets rid of $\mathbf{C}$ altogether.   A popular example is the Wagner model with $\phi_1 = \exp(-a_0\sqrt{a_1 I_1 + (1 - a_1)I_2 - 3})$, where a_0 and a_1 are parameters.

Other needed information from theory is the nature of the memory function; theory suggests that

$$M(x) = \frac{8G_0}{\pi^2 \tau_D} \sum_{p=1,3,...}^{N} \exp\left(-xp^2 / \tau_D\right) \qquad (9\text{-}21a)$$

corresponding to the relaxation function

$$G(x) = \frac{8G_0}{\pi^2} \sum_{p=1,3,...}^{N} \frac{\exp(-xp^2 / \tau_D)}{p^2} \qquad (9\text{-}21b)$$

where τ_D is the diffusion time associated with the reptation of the macromolecule out of its tube.[6]   We will examine τ_D below.

The form of equation (9-21) deserves comment.   First of all, the p^2 spacing of the relaxation modes is very Rouse-like.   However, the spacing of the modes is increased by including only the odd values of p.   This makes $M(x)$ very bumpy; the longest and second longest times will be spaced by a ratio of 9:1, or almost a whole decade.   This would be very obvious for a narrow-MWD polymer.   Note that the partial moduli drop quickly with mode number,

---

[**] "Factorized" refers here to the separation of the integrand into strain- and time-dependent parts.   Needless to say, the strain-dependent part usually depends on time, but the time-dependent part often does not depend on strain, at least at low strains.   It is simply the memory function $M(t)$.

whereas the Rouse values are constant at NkT, where N is the chain concentration.

As for the strain-dependent functions, these are expressed in terms of a strain-energy function $W(I_1, I_2)$, where I_1 and I_2 are the first and second invariants of the Finger tensor, $\mathbf{C}^{-1}$. A form of the strain energy function is[7]

$$W(I_1, I_2) = 5\ln\frac{I_1 + \sqrt{4I_2 + 13} - 1}{7} \tag{9-22}$$

Certainly, the exact fractions give the impression that this expression derives from theory, but it is only an approximation. The expressions for I_1 and I_2 have been given previously; see Chapter 8, Section G-1.

Returning to the principal relaxation time, we need to explore briefly its origin. Figure 9-3 shows the state of affairs for a polymer molecule confined by a tube made up of many other polymer molecules. It's only direction of escape is out the ends of a tube, where it enters the media in a relaxed configuration. The rate at which it does this is approximated by the Nernst-Einstein equation

$$D = kTm \tag{9-23}$$

where D is the diffusivity and m is the mobility of the molecule. Diffusivity has the usual dimensions $[L]^2/[t]$, with the SI units m^2/s, and is related to the distance diffused x by the expression

$$D = x^2/2t \tag{9-24}$$

If we associate the time with our characteristic time τ_D, and further assume that the mobility m decreases as the diffusion distance x increases with the length of the polymer chain, then the well-known proportionality

$$\tau_D \sim N^3 \tag{9-25}$$

obtains, where N is the number of physical segments in the chain.

This easily derived scaling relationship was the first molecular explanation for the strong dependence of viscosity on molecular weight.

E. SUMMARY OF PREDICTIONS OF MOLECULAR THEORY

Molecular theory of deformation and flow of macromolecules is not a new area of endeavor. Many have worked long and hard toward the goal of explaining phenomena in terms of molecular structural models, and the results have been quite remarkable. Some of the relationships between the parameters of molecular models and the sizes and shapes of actual molecules are obvious in the simplest cases, but obscure in the more complicated ones. Branched structures are very important commercially, and recent theory has finally been able to make some reasonable predictions.

Rather than assume a model for the polymer molecule, the molecular modeler can work with known chemical structure and interaction potentials, and examine the motions of the individual atoms in the chain. This area of endeavor, called variously molecular dynamics or *ab initio* molecular dynamics, is still stymied by computer limitations. The fundamental assumption is that the spatial average of an assembly of molecules at an instant in time can be simulated by the time-average behavior of a very limited set of molecules constrained spatially by periodic boundary conditions and "thermostated" to keep the system's temperature constant. The big issue in what appears to be a perfectly general approach is the accuracy of the unbounded pair potentials between atoms in the chains and the atoms in surrounding medium. Pair potentials can and have been modified to include the effects of neighboring atoms. Electrostatic interactions are also needed for charged species.

Imposing deformation on the system is not difficult; the atoms within the system are simply moved affinely in increments that exceed by far the time associated with the averaging process. The motion raises the energy of the system, with the "thermostat" bringing the kinetic energy of the atoms into line with the fixed system temperature. The energy lost to the thermostat is proportional to the viscosity.

Obviously, molecular dynamics does not present the properties of the system as neat analytical constitutive expressions that can be used to find the response of the system in arbitrary deformations. This shortcoming may be regarded as severe or trivial, depending upon your point of view.[8]

PROBLEMS

9-1. A well-known problem with the Rouse model is the zero-time modulus. As written, it is clear that the zero-time modulus $G(0)$ will be $zNkT$, where z is the number of physical segments in the chain and N is the number density of polymer chains. In contrast, the empirical evidence suggests that not much happens to $G(0)$ as the molecular weight is changed, in stark contrast to the viscosity. Does the Rouse model give a more realistic prediction for viscosity? Explain.

9-2. In Example 9-2, the intrinsic viscosity of a rigid dumbbell was worked out for a specific example. Derive a general expression for dumbbells with spherical ends of radius r and separated by a distance L.

9-3. In Examples 9-1 and 9-2, the viscosity of the medium was used to calculate the drag coefficient for the spheres at the ends of the dumbbell. Suppose, instead, that the total viscosity were used, implying some hydrodynamic interaction between dumbbells in the suspension. Use the same numbers as those in the Examples 9-1 and 9-2. (Hint: this is an iterative calculation; set up on a spread sheet.)

9-4. If the volume of the spheres at the two ends of the rigid-rod dumbbell is distributed along the length of the rod, a lumpy dumbbell (a.k.a., shish kebab[1]) results. For such a model, which might be considered closer to a rod-like polymer structure than is the simple dumbbell, the rotational relaxation time τ_N is given by the expression

$$\tau_N = \frac{\pi r \eta_m L^2}{12kT} \frac{N(N+1)}{N-1}$$

where N is the number of lumps of equal radius r. The expression for the viscosity is the same as for the simple dumbbell, but with τ replaced by τ_N in equation (9-6)

(a) Check to see if this equation is general, i.e., it will work for a simple two-sphere dumbbell as well.

(b) What dependence of viscosity of molecular weight does one expect for the distributed dumbbell? State clearly any assumptions.

9-5. Derive a general expression for the intrinsic viscosity of a dumbbell.

9-6. (Computer) Find the power-law slope predicted by the Bueche expression, equation (9-17). Use the suggested modification for melts, and work with the reduced variables, $\dot{\gamma}_R$ and η/η_0.

9-7. Repeat Example 9-3 using either the data from Problem 6-1 or the NIST SRM 2481 data.

9-8. (Challenging) One criticism of the Bueche expression is that is, in effect, based on an argument that dynamic dissipation via Rouse-like motions is the source of viscous dissipation during flow. Compare the flow viscosity predicted by the Bueche expression of equation (9-17) with the complex dynamic viscosity magnitude as predicted by the Rouse expression. For the latter, use equation (9-1) plus the relationship for the Maxwell model

$$G'(\omega) = G_0 \frac{(\tau\omega)^2}{1+(\tau\omega)^2}; \quad G''(\omega) = G_0 \frac{\tau\omega}{1+(\tau\omega)^2}$$

applied to each exponential term. Then use the Cox-Merz relationship to approximate the flow viscosity.

9-9. Repeat Problem 9-8 using the Tobolsky-Chapoy approximation, equation (5-18).

9-10. (Computer) Prepare a graph of the Bueche expression using reduced variables on log scales. Is the transition from Newtonian to power-law behavior sharper or narrower than that predicted by the Cross model? (Hint: while best done using a general high-level programming language, the Bueche expression can be evaluated using a spread sheet. Put the reduced shear rates in the first column, and the p values in the first row, leaving the A1 cell blank. For each reduced shear rate, assemble the terms in the sum across the second row using a least 15. Make sure the p values are treated as constants by either naming them, or using $ in the cell addresses (e.g., A2 for $p = 1$). Do the summing and combining with the constants in the expression in the last column to yield the reduced viscosity. Then copy this row into the remaining rows.

9-11. Reportedly, the power-law slope for the Graessley model is $-3/4$. Assuming that this is true, develop the right-hand side of equation (9-18) and see if it is consistent with the left-hand side being expressed as $a(\dot{\gamma}_R)^{-3/4}$, where $\dot{\gamma}_R = \dot{\gamma}\,\tau_1$. (Note: the function $\cot^{-1}x = \pi/2 - \tan^{-1}x$, so at high x, this term will approach zero.)

9-12. (Computer) The mixing rules for blends of polymers of the same structure are very diverse, but predictions of mixture behavior will often start with equation (9-19). Using the following viscoelastic data for each component, predict the relaxation behavior of a 50:50 blend of the two. Plot your results similarly to the schematics shown in Figure 9-4. (Hint: it may be helpful to describe the data with a KWW relaxation function.)

M_w = 93 kDa		M_w = 315 kDa	
Log (t, s)	Log [$E(t)$, Pa]	Log (t, s)	Log [$E(t)$, Pa]
-0.146	5.981	-0.064	6.032
0.026	5.904	0.263	5.962
0.214	5.834	0.541	5.917
0.394	5.751	0.860	5.860
0.598	5.674	1.179	5.821
0.778	5.611	1.473	5.777
1.023	5.515	1.775	5.732
1.268	5.419	2.094	5.694
1.530	5.311	2.356	5.655
1.767	5.209	2.658	5.611
2.029	5.113	2.969	5.553
2.225	5.017	3.304	5.502
2.413	4.915	3.770	5.413
2.609	4.794	4.048	5.343
2.773	4.685	4.301	5.266
2.969	4.532	4.767	5.087
3.083	4.404	4.988	4.985
3.222	4.251	5.176	4.883
3.345	4.053	5.519	4.640
3.451	3.836	5.650	4.538
3.516	3.645	5.789	4.404
3.582	3.447	5.920	4.251
3.623	3.294	6.116	3.951
		6.181	3.804
		6.247	3.657
		6.345	3.332

9-13. In the article "Molecular Dynamics" in Wikipedia, there currently is a movie of a two-dimensional molecular dynamics "experiment" involving deformation. In this movie a copper atom is hurled at an atomically flat copper crystal. The movie shows the motion of the atoms on the surface as the energy of the moving atom is dissipated.

(a) If all the potentials are conservative, no energy should be lost and the atoms in the lattice would continue to vibrate wildly, spreading throughout the lattice; however, the motions die away. Where does the kinetic energy go?

(b) At equilibrium, the motions of the atoms are small. Why do they keep vibrating?

(c) Observe carefully the spacing of the added copper atom relative to those in the lattice. It is somewhat larger. Why?

9-14. (Computer) Consider the memory function derived from simple reptation theory as describe by equation (9-21).

(a) Graph the relaxation function described by this equation using log-log scales and reduced variables $G(t)/G(0)$ and t/τ_D using 100 modes ($N = 100$). Cover the reduce time scale over five decades from 0.01 to 100.

(b) Using the data generated from part (a), describe the general shape of the relaxation function using a stretched-exponential (KWW) function.

9-15. Show that equation (9-21a) is consistent with equation (9-21b).

REFERENCES

1. R. B. Bird, R. C. Armstrong and O. Hassager, *Dynamics of Polymeric Liquids*, Vol. 1, 1st ed. (Wiley, New York, 1977), pp. 80–84.

2. M. T. Shaw and W. J. MacKnight, *Introduction to Polymer Viscoelasticity*, 3rd ed., Wiley, New York, 2005.

3. B. H. Zimm, "Dynamics of polymer molecules in dilute solution: viscoelasticity, flow birefringence and dielectric loss," *J. Chem. Phys.*, **24**, 269–278 (1956).

4. F. Bueche, "Influence of rate of shear on the apparent viscosity of A—Dilute polymer solutions, and B—Bulk polymers," *J. Chem. Phys.*, **22**, 1570–1576 (1954); F. Bueche, J. K. Wood, and J. R. Wray, "Variation of viscosity of coiling polymers with shear rate," *J. Chem. Phys.*, **24**, 903 (1956); F. Bueche and S. W. Harding, "New absolute molecular weight method (linear high polymers)," *J. Chem. Phys.*, **27**, 1210 (1957).

5. W. W. Graessley, "Molecular entanglement theory of flow behavior in amorphous polymers," *J. Chem. Phys.*, **43**, 2696–2703 (1965).

6. R. G. Larson, T. Sridhar, L. G. Leal, G. H. McKinley, A. E. Likhtman and T. C. B. McLeish, "Definitions of entanglement spacing and time constants in the tube model," *J. Rheol.*, **47**, 809–818 (2003).

7. R. G. Larson, *The Structure and Rheology of Complex Fluids*, Oxford University Press, New York, 1999, pp. 156–162.

8. S. C. Glotzer and W. Paul, "Molecular and mesoscale simulation methods for polymer materials," *Ann. Rev. Mater. Res.*, **32**, 401–436 (2002).

10

Elementary Polymer Processing Concepts

The subject of polymer processing is huge, as indicated by the large list of textbooks, monographs and journals devoted to the topic.[1]   In one chapter, the coverage will necessarily be limited.   Those processing methods that are discussed will be done in a fashion that acquaints the reader with the concepts, as opposed to detailed analysis and lists of processing problems faced by commercial users.   The focus is on methods commonly used in the laboratory to prepare test specimens from small amounts of polymer.

The idea of processing polymers is to take a polymer, usually present in a bulk form (powder, granules, pellets, solution, emulsion, melt), and shape it into an object—hopefully, a useful object.   Thus, inherently, processing involves deformation at some point in the process.

A. SIMPLE LABORATORY PROCESSING METHODS

1. Spin casting

As an example, consider the standard laboratory process of spin casting, starting with a polymer solution. The procedure, at least on a laboratory scale, is to dribble a fixed amount of the solution onto a horizontal disk that is or will rotate at a high rate. Centrifugal force moves the solution outward in all directions, with excess flying off the edge of the disk (Figure 10-1). The objective of this process is to form a very thin film or coating that is uniform in thickness and smooth, after the solvent evaporates. Polymer films of molecular thickness have been thus formed.

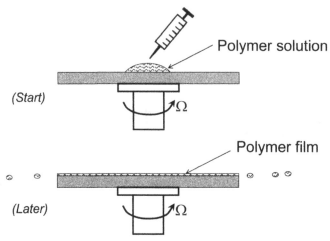

Figure 10-1. Spin casting of a polymer solution, an elementary but important polymer processing operation. The thickness is limited only by the dilution, rotation speed, rheological characteristics of the solution, and the interfacial stability of the thin film.

How can such a process be analyzed? One approach is to assume that the film forms a uniform coating, as shown in lower schematic of Figure 10-1. Then the question becomes one of deciding if uniform thinning is consistent with the mass balances and rheological properties. Along the way, we might find some information on how the rate of thinning will scale with rotation speed, viscosity, etc. This will be explored in Example 10-1.

Example 10-1: *A polymer solution with viscosity η forms a film of uniform thickness on a flat, horizontal disk that is rotating at an angular velocity Ω. Describe the change in thickness with time, and (possibly) with radius. The solution has a mass density ρ.*

As with other problems, our first step is to consider the conservation of mass. Because the density is assumed to be constant, conservation of mass is equivalent to the usual incompressibility assumption. Next we define "bulk velocity" as $V_B(t, r) = Q(t, r)/2\pi r H(t, r)$, where Q is the flow rate across a circle of radius r and H is the film thickness. For the moment, we will consider the case where H depends only on time. Q is most definitely a function of r, as well as t, and is equal to the material fed from the inner portion of the disk, i.e., $Q(t, r) = \pi r^2 |dH(t)/dt|$. Thus, we see that the flow rate must increase as the square of the radius if the film is to remain uniform, which means the bulk velocity increases linearly with radius.

So, is $V_B(r) \propto r$ compatible with the equation of motion?   Without examining the equation itself, we can safely surmise that the shear rate, and the thus the shear stress must increase linearly with radius.   However, what is providing the force to generate this increasing stress?   It's not a pressure drop, as the pressure is most certainly uniformly atmospheric everywhere.   The source is the centrifugal acceleration, which is $r\Omega^2$.   Thus, we can see with this simple argument, that a uniform film of a Newtonian fluid on a spinning disk does not violate either the mass or the momentum balances.   Furthermore, we expect, but haven't demonstrated, that a non-uniform film will tend to become more uniform as time goes on.

In practice, of course, there are many complicating factors such as the evaporation of solvent, the flow of air over the film's surface, and non-Newtonian effects.

Before leaving this, a glance at the equation of motion would be appropriate. We pick the radial component in cylindrical coordinates from Appendix 2-1. Discarding terms that are zero due to symmetry, we are left with

$$\rho\left(\frac{\partial V_r}{\partial t} + V_r \frac{\partial V_r}{\partial r}\right) = -\frac{\partial p}{\partial r} + \left(\frac{1}{r}\frac{\partial}{\partial r}(r\tau_{rr}) - \frac{\tau_{\theta\theta}}{r} + \frac{\partial \tau_{rz}}{\partial z}\right) + \rho g_{rr} \quad \text{(ssc)}^* \quad \text{(a)}$$

Now, some important assumptions need to be made.   These are quite common in many polymer-processing analyses.   First of all, we assume that time changes are slight and can be ignored.   This is called a *quasi-steady-state* assumption.   This gets rid of the first term on the left-hand side, which represents the acceleration of the fluid.   The other density-based term we can't ignore right away.   This term is the "fire hose" term—the momentum imparted by high- velocity fluid hitting the differential volume on one face.   But let's compare it with the last term on the right-hand side, the centrifugal acceleration $r\Omega^2$.   The bulk velocity from above is

$$V_B = \frac{r|\dot{H}|}{2H} \tag{b}$$

so the term in question is roughly

$$V_r \frac{\partial V_r}{\partial r} = \frac{r|\dot{H}|}{2H}\frac{|\dot{H}|}{2H} = \frac{r}{4}\left(\frac{\dot{H}}{H}\right)^2 \tag{c}$$

---

* See Appendix 3-3 for definitions of (ssc) and (fsc).

In the face-off between these two terms, the critical difference boils down to Ω vs. $\frac{1}{2}(d\ln H/dt)$. Typical rotational speeds are 5000 rpm $\approx$ 500 rad/s. The thinning process, on the other hand, can take minutes, at least when the film is getting thin. For example, a 10-μm film might take a minute to thin to 5 μm, giving a value for $\frac{1}{2}(d\ln H/dt)$ of $1/120 \approx 0.01$ s^{-1}. Thus, the radial acceleration dominants by far the density-dependent terms.

The next assumption is involved with the relative magnitude of the various velocity-gradient terms, and thereby the magnitudes of the extra stresses. In this geometry, we have velocity gradients in both the r and z directions. Examination of the components of the rate-of-deformation tensor for the same geometry is needed to decide which are important. The relevant ones, according to equation (a) are rr, $\theta\theta$ and rz. These are displayed below:

$$\dot{\gamma}_{rr} = 2\frac{\partial V_r}{\partial r} \tag{d}$$

$$\dot{\gamma}_{\theta\theta} = 2\left(\frac{1}{r}\frac{\partial V_\theta}{\partial \theta} + \frac{V_r}{r}\right) = 2\frac{V_r}{r} \tag{e}$$

$$\dot{\gamma}_{zr} = \dot{\gamma}_{rz} = \frac{\partial V_z}{\partial r} + \frac{\partial V_r}{\partial z} \tag{f}$$

Equation (d) expresses the stretching of the material as it goes outward. Using the bulk velocity to estimate the velocity gradient gives us $d\ln H/dt$. Equation (e) gives the same result for stretching in the circumferential direction. For the rz component, equation (f), the first term will be zero identically if we assume the thinning is uniform. The second term $\partial V_r/\partial z$, on the other hand, will be something like $V_B/2H = (r/4H)\ d\ln H/dt$. Thus, this term is many times larger than the others because r is millimeters, while H is micrometers.

With this in mind, we can fairly safely drop all but the gradients in the velocity in the z direction. This very important process is known as the *lubrication approximation* after the solution for the flow of oil in a journal bearing. There the major gradient is in the r direction with flow in the θ direction. Because the axle is slightly off center, there is also a velocity gradient in the θ direction, but it is safe to ignore, at least for a Newtonian fluid.

Putting all this together gives the differential equation

$$0 = \rho r\Omega^2 + \eta\frac{r}{4H}\frac{d\ln H}{dt} \tag{g}$$

which can be solved to give us an approximate solution

$$\frac{1}{H} = \frac{1}{H_0} + \frac{4\rho\Omega^2 t}{\eta} \tag{h}$$

where H_0 is the initial thickness. This solution is independent of r.

It is important to realize that in this derivation, we have made many serious and possibly limiting assumptions that may well restrict the usefulness of the result. First and foremost, we have assumed that the polymer solution cast onto the spinning disk is Newtonian, with time-independent viscosity (in spite of the evaporation of solvent). We have then made a lubrication approximation, neglecting terms that will be small when the film is thin. We have also approximated the shear stresses using an average velocity gradient, whereas in fact we need to incorporate the actual velocity gradient (parabolic) into the calculation.

Two important approximations are used in Example 10-1: *quasi-steady state* and *lubrication*. Respectively, these change transient problems into steady ones, and reduce the dimensionality of the problem. Obviously, they are both wrong, but there application to a problem can often give useful results that provide a baseline against which data can be compared.

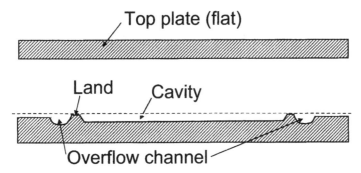

Figure 10-2. Laboratory cavity mold for pressing flat sheets. The combination of land and overflow channel allow the full force of the press to be applied to the sample. The slight flow of melt over the land maintains a high pressure in the cavity. Under high pressure, air bubbles trapped in the melt are forced into solution, where they stay as the sample is cooled.

2. Compression molding

Compression molding is one of the earliest methods of forming a lump of polymer into a flat film of precisely controlled thickness. For making a film in the laboratory, the mold can be simply two flat plates. The film thickness can be controlled by using shims. Alternatively, a cavity can be machined into one of the plates to give a thicker sample of controlled thickness. Figure 10-2 shows a cross section of a typical cavity mold. Other simple shapes can be made as well by machining a more complex cavity. Clearly, the complexity of the shape is geometrically limited.

The basic flow in the simple pressing of a film between flat plates is squeezing, similar to squeezing of ketchup out from a hamburger. The flow is in the radial direction, and the pressure drop that develops is primarily due to the shear deformation of the melt as it flows outward between the closely spaced plates. The volumetric flow rate at each value of r, the radial coordinate, is determined by the closing rate, $-dH/dt$, which we will call $\dot{H}$, a positive number. The flow rate is then $Q(r) = \pi r^2 \dot{H}$. Note that the flow rate increases quadratically, whereas the cross-sectional area that this melt must flow through ($2\pi rH$) increases only linearly with radius. Thus the average velocity must increase linearly with radius.

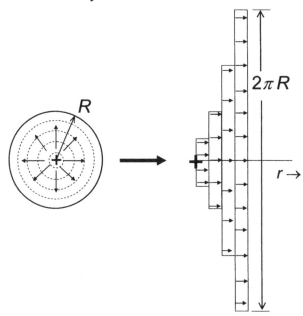

Figure 10-3. Simplification of squeezing flow used to calculate the pressure profile. The straightened-out rings on the right will be treated as slit dies in series with increased flow rate at each position. This treatment is equivalent to lubrication approximation, as it removes from consideration all velocity gradients except in the z (axial) direction due to flow in the r direction.

These simple facts mean trouble. First of all, the flow will be complicated, comprising shear between the disks (mainly) but distorted by extensional flow due to stretching in the radial direction, and compression in the z direction. The rates of these deformations will change with radial position, which may change material properties. A changing velocity profile means that material will be flowing in the z direction as well as the radial direction simply due to the rearrangement of the velocity profile. Obviously a situation like this calls for assumptions, and the quasi-steady and lubrication approximations seem appropriate if the squeezing is fairly slow and the gap is small compared to the radius. Furthermore, we will start with a Newtonian fluid, and see if we can expand the solution to GNF fluids.

With these assumptions in mind, we can approximate the flow between the disks as a series of straight slits, as depicted in Figure 10-3. The formula for the pressure gradient of a Newtonian fluid flowing through a wide slit is well known; it's

$$\frac{dP}{dx} = -\frac{12\eta Q}{WH^3} \tag{10-1}$$

where x is the flow direction, η is the viscosity, Q is the flow rate, and W and H are the width and gap of the slit, respectively.

Using the construct illustrated in Figure 10-3, we can see that $W(r) = 2\pi r$, $x = r$, and $Q(r) = \pi r^2 \dot{H}$. Substituting these into equation (10-1) yields the differential equation

$$\frac{dP(r)}{dr} = -\frac{12\eta\pi r^2 \dot{H}}{2\pi r H^3} = -\frac{6\eta \dot{H}}{H^3} r \tag{10-2}$$

which shows immediately that the pressure profile will be quadratic.

Now, after pursuing this rather crude path, we might ask if this result agrees with the prediction of the radial component of the differential equation of motion, as modified for a Newtonian fluid (Stokes equations). Good question, so let's check this. The equation of motion modified in the usual fashion for slow flow and symmetry is

$$\frac{\partial P(r)}{\partial r} = \eta \left[\frac{\partial}{\partial r}\left(\frac{1}{r}\frac{\partial(V_r r)}{\partial r}\right) + \frac{\partial^2 V_r}{\partial z^2} \right] \tag{ssc} \tag{10-3}$$

which only bears a faint resemblance to equation (10-2). Because we are dealing with a Newtonian fluid, we can safely assume the velocity profile is quadratic, as given below:

$$V_r(r,z) = \frac{3}{2}\bar{V}(r)\left[1 - \left(\frac{z}{H/2}\right)^2\right] \tag{10-4}$$

where $\bar{V}$ is the bulk (average) velocity and the z coordinate starts at the plane halfway between the two plates. Clearly $\bar{V}(r) = Q(r)/2\pi r H = r\dot{H}/2H$, and

$$V(r,z) = \frac{3r\dot{H}}{4H}\left[1 - \left(\frac{z}{H/2}\right)^2\right] \tag{10-5}$$

where $\dot{H}$ is considered positive. However, the only important part of this discussion is that while velocity V depends upon z as well as r, it depends in the same fashion (linearly) at each z on the variable r. So, when the radius doubles, the bulk velocity doubles, and the velocity in each layer of fluid also doubles. This means that rV_r will be quadratic in r, and the derivative with respect to r will be linear. Dividing by r then gives a constant, which produces zero when the derivative is taken again with respect to r. (The constant does depend on z.) Thus the first term in the square brackets of equation (10-3) is zero. The second term requires equation (10-5), which gives the result

$$\frac{\partial P(r)}{\partial r} = \eta\left[-\frac{3r\dot{H}}{4H}\frac{2}{(H/2)^2}\right] = -\frac{6\eta\dot{H}}{H^3}r \tag{ssc} \tag{10-6}$$

The pressure drops with radius, as we expect. Because the pressure depends only with radius, equation (10-6) can be integrated easily to give the pressure profile

$$P(r) = \frac{3\eta\dot{H}}{H^3}\left(R^2 - r^2\right) = \frac{3\eta R^2\dot{H}}{H^3}\left[1 - \left(\frac{r}{R}\right)^2\right] \tag{10-7}$$

where R is the maximum radius, which will vary with time. The positive force required to move the plates together at a velocity $\dot{H}$ is given as the integral of the pressure over the entire area of the plates. The general form is

$$F = \int_0^R P(r)2\pi r\, dr \tag{10-8}$$

Why are we not using σ_{zz} instead of pressure? The simple reason is that for Newtonian fluids, they are the same for this type of flow. Integration of this equation using the pressure profile from equation (10-7) gives the final result

$$F = \frac{3\pi\eta R^4}{2H^3}\dot{H} \tag{10-9}$$

This equation can be expressed in an alternative form using the initial charge to the mold V_s, which is related to R and H by the simple equation $V_s = \pi R^2 H$. Substituting this in for R gives

$$F = \frac{3\eta V_s^{\,2}}{2\pi H^5}\dot{H} \tag{10-10}$$

This form points out vividly the difficulty of preparing thin films by simple compression molding between two flat plates: the closing velocity drops rapidly if a constant force is applied to the plates. One can get around this by using a disk mold with a fixed radius instead of using large, flat plates to squeeze a sample of fixed volume. The excess is allowed to flow out of the mold. The fixed radius case is just equation (10-9), which shows the velocity depends on H^3 instead of H^5. From a processing point of view, though, it is messy to have material spilling out, so one designs an overflow channel something like that described in Figure 10-2. A simplification that can be made on the spot is illustrated in Figure 10-4. This geometry can be used to make films thin enough for transmission infrared spectroscopy.

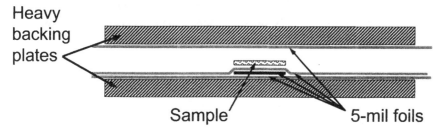

Figure 10-4. Simplified mold for forming a thin film from a melt using constant-radius squeezing flow. The dark insert is disk-shaped. To avoid damaging the platens of the press, it is critical to protect the platens with heaving plates, as the normal stress over the disk insert can easily rise above the yield point of steel. (Note: a "mil" is 0.001 in. = 25.4 μm.)

3. Vacuum-compression molding

Bubbles are a problem often encountered with laboratory compression molding, as there is little one can do to eliminate air (water, solvent) from samples that are available only in the form of a powder or a precipitate from a solvent/non-solvent purification procedure. Certainly one can do a preliminary pressing and remove all air and solvent by vacuum drying. As described above, one can also juggle the temperatures and pressure to force all vapor into solution in the sample. In an early study, researchers described a process whereby a powdered sample was molded under vacuum, using a piston-cylinder mold shown schematically in Figure 10-5.

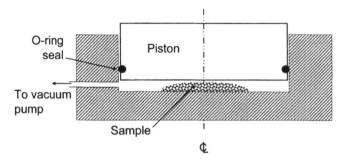

Figure 10-5. Classical design of a vacuum-compression mold. In some designs, the bottom is removable so the piston and sample can be pushed out of the mold.

The advantages of molding under vacuum were found to be numerous.[2] Tensile strength, for example, was increased and the distributions of failure stresses and strains were markedly narrowed. While most compression molding leaves remnants of the boundaries between the original grains in the sample, the vacuum encouraged complete fusion of these weak interfaces. Optical properties were also improved, probably for the same reason. Molding pressures were greatly reduced; in fact, it was reported that atmospheric pressure acting on the cylinder was enough. Consequently, the moldings were completely isotropic, with no orientation. Oxidative degradation was reduced or absent, depending on the quality of the vacuum.

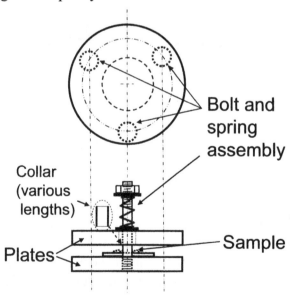

Figure 10-6. Schematic of vacuum-compression mold for use in a vacuum oven. The collars are optional, and are sized to fit over the bolts and take the spring force off of the sample to control the thickness of the final film. The vacuum oven must be very clean to avoid contaminating the sample, and must be capable of holding a high vacuum.

The negative aspects are quite apparent from the drawing. While in principle, one can isolate the polymer from the mold using release films along the bottom of the cylinder and piston, in practice the mold often sticks together due to stray powder. Also, one must pull a vacuum while the sample is cold so no air is trapped, and heating the thick, cold mold can be very tedious. Cooling is also slow, which encourages the growth of larger crystals in semicrystalline polymers. Large crystals can reduce tensile strength and toughness.

Those lacking a press, but equipped with a heated vacuum oven, can bypass the machining and sticking problems of the vacuum mold design shown in Figure 10-6 by using flat plates forced together by strong springs. Stops can be incorporated into the design to produce films of set thickness. The entire assembly is inserted into the vacuum oven and left overnight to mold. As most vacuum ovens can't heat much beyond 200 °C, this design is limited to polymers with low softening points. The bottom plate needs to have good thermal contact with the bottom of the oven; otherwise, heating can be slow.

B. ELEMENTARY EXTRUSION CONCEPTS

The combination of an extruder attached to a shaping device opens up the possibility of continuous processing. Shapes made by these devices, which are very common in industry as well as in laboratories, are termed *profiles*. For preparing laboratory specimens, common extruded shapes include tubes, tapes, rods, films and fibers. The drawback of any continuous device is that it generally requires a large amount of material to reach steady operating conditions. This has been somewhat averted by small extruders that recirculate the mix until steady conditions are reached. The charge can then be extruded through a tiny tape or fiber die in a quasi-continuous fashion.

1. Forming dies

Continuous extrusion is mostly confined to the fabrication of objects with an unchanging cross section.[†] Such are generated by forming dies with a huge variety of shapes and sizes. Clearly, only the simplest can be discussed in any detail.

[†] The "mostly" comes from the fact that extruders can be hooked up to molds to make three-dimensional objects. Blow molding and injection molding machines are examples, although these two devices also use axial motion of the screw to control the discharge rate.

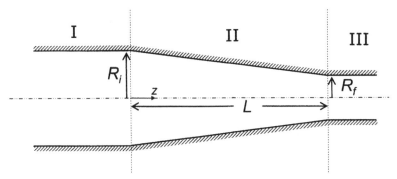

Figure 10-7. Taper die for forming monofilament or other cylindrical profiles. Design variables include the length of the taper (Section II) and the land length (Section III).

Consider the design shown in Figure 10-7. Section I is the delivery channel and would be hooked to the melt source, an extruder or gear pump. Section II is the *forming* section; here, the melt stream is reduced in diameter as a result of the taper. While clearly there is an axial velocity gradient in Section II, we will find that it is small compared to the velocity gradients in the radial direction, i.e., $\partial V_z/\partial r > \partial V_z/\partial z$. Section III is termed the *die land*, and is designed to allow the melt to relax to reduce die swell. In some designs, this section may even have a reverse taper.

To reduce die swell and residual orientation, Section II can be made longer. The Deborah number associated with the relaxation will be the maximum relaxation time of the melt τ_1 divided by the residence time t_R. As the volume of truncated cone in Section II increases linearly with length, the residence time will likewise increase linearly. The penalty for doing this is higher pressure drop in that section. Will the pressure drop also increase linearly with length? Or possibly faster?

Example 10-2: *Find the pressure drop for a GNF flowing in the tapered section of the die depicted in Figure 10-7. Use the lubrication approximation.*

If the taper is not too severe, so $\partial V_z/\partial r > \partial V_z/\partial z$, the pressure drop in the tapered section will be due mainly to the shearing flow. The only aspect that changes is the radius, which is be given by

$$R(z) = R_i - \frac{R_i - R_f}{L} z \qquad (a)$$

where R_i and R_f are the initial and final radii, respectively; z is the axial distance measured from the beginning of the tapered section; and L is the total length of Section II. According to the lubrication approximation, we will consider only the pressure drop due shearing, which is tantamount to breaking the tapered section into a series of cylindrical tubes of decreasing radius. Each tube will be dz in length, and have a pressure drop given by the expression

$$-\frac{dP}{dz} = \frac{2\tau_R}{R} = \frac{2\eta(\dot{\gamma}_R)\dot{\gamma}_R}{R} \qquad \text{(ssc) (b)}$$

for any fluid. In this equation, τ_R and $\dot{\gamma}_R$ are the shear stress and rate magnitudes at the wall, respectively, and $\eta(\dot{\gamma}_R)$ is the viscosity at $\dot{\gamma}_R$. The problem is relating $\dot{\gamma}_R$ or τ_R to Q, the flow rate through the tube. Why Q? Because Q is the observable and constant, whereas τ_R and $\dot{\gamma}_R$ are not. The connection is via the familiar [Appendix 7-2, equation (e)] expression

$$Q = \pi \int_0^R \dot{\gamma}(r) r^2 dr \qquad \text{(c)}$$

This equation assumes no slip at the wall. For a Newtonian fluid, we have the familiar result that $\dot{\gamma}_R = 4Q/\pi R^3$, while for a power-law fluid the result is

$$\dot{\gamma}_R = \frac{(3n+1)Q}{n\pi R^3} \qquad \text{(d)}$$

where n is the power-law exponent. GNF expressions other than the power-law expression are much more complicated and the algebraic equation obtained on integrating the velocity profile to get Q cannot, in general, be solved explicitly for $\dot{\gamma}_R$ or τ_R. This means a numerical solution is usually required.

There are special cases, however. One is the GNF expression

$$\eta = \frac{\eta_0}{1+\tau/\tau_0} \qquad \text{(e)}$$

where τ is the shear stress and τ_0 is a parameter. (This is the Ferry model, which in turn is a special case of the Ellis fluid with $n = 1/2$; see Table 5-1.) This expression can be integrated quite easily to give an expression for τ_R in terms of Q and R that can applied to equation (b). However, the integration of this result after substituting in (a) for R vs. z, gives an exceedingly complicated result.

Returning to the Newtonian case, we have the expression

$$-\frac{\partial P}{\partial z} = \frac{8Q\eta}{\pi}\frac{1}{R^4} = \frac{8Q\eta}{\pi}\frac{1}{\left[R_i - (R_i - R_f)\dfrac{z}{L}\right]^4} \qquad \text{(f)}$$

This equation can be integrated to give

$$\Delta P = \frac{8Q\eta L}{3\pi(R_i - R_f)}\left[\frac{1}{R_f^3} - \frac{1}{R_i^3}\right] \tag{g}$$

where the signs are chosen to give a positive pressure drop.

If we guessed that we could use the average diameter $(R_f + R_i)/2$ in the Poiseuille equation, we would get the incorrect expression:

$$\Delta P = \frac{8Q\eta L}{\pi\left((R_i + R_f)/2\right)^4} \tag{h}$$

See also Problems 10-13, 10-14 and 10-15.

In answer to the question concerning the influence of length, for a given initial and final radius, equation (g) of Example 10-2 indicates that the pressure drop increases linearly with length.

Although more complicated cases can be approached using the lubrication approximation, the cross-sectional shapes of the extrudate must be quite simple to have any hope of developing an algebraic expression. Simple shapes are either axisymmetric, e.g., rods and tubes or planar, e.g., wide sheets. These problems, which are two dimensional in nature, are transformable to one-dimensional problems via the lubrication approximation.

2. Drag flow

The flow through a die with stationary walls is pressure-driven. While it easy to visualize developing pressure using a piston in a cylinder, how does one establish a continuous source of polymer solution or melt? Some kind of pump is needed. In addition, for melts, it is necessary to melt the solid as well as pump the resulting melt continuously and in a fashion that excludes air and smoothes out temperature gradients and pressure pulsations.

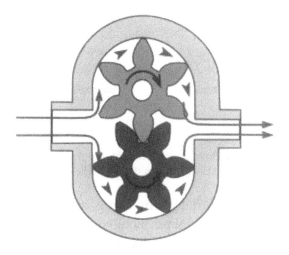

Figure 10-8. Illustration of a gear pump. Note that this pump is positive displacement, meaning the delivery flow rate is independent of the delivery pressure. To avoid dangerously high pressures, a relief valve on the delivery side must be installed. (From Wikipedia. This image may be freely copied according to the GNU license agreement. See http://commons.wikimedia.org/wiki/Commons:GNU_Free_Documentation_License for more details.)

First of all, most pumps used for low-viscosity fluids are not suitable for high-viscosity polymer melts. Can a centrifugal pump be used, however, for solutions? Certainly, but at the risk of mechanically degrading the polymer because of the high velocity of the blade tips. The one traditional pump that is used is the gear pump, illustrated in Figure 10-8. This device is useful in that it delivers very precise flow rates needed for spinning. The delivery rate is virtually independent of upstream pressure and fluid viscosity. However, gear pumps are expensive and difficult to clean. They are also not suitable for melting solid polymer; if pellets are dumped in the inlet, pellets will emerge from the outlet.

The premier device for handling melting and delivery jobs is the screw extruder.[‡] These pumps come in many configurations and sizes. For simplicity we will consider only the simplest device.

[‡] While both use screws, the single-screw extruder and the Archimedes screw work on completely different principles. The latter requires gravity, whereas the extruder does not. See http://en.wikipedia.org/wiki/Archimedes_screw for a description of the Archimedes screw. (The Wikipedia article is currently misleading in that it equates the two in one place.)

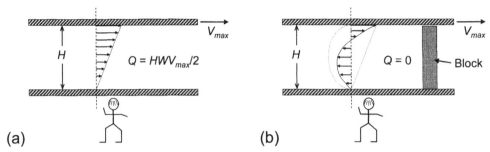

Figure 10-9. On left, simple drag-flow device; on right, the flow is blocked. The device on the left produces the maximum possible flow rate; on the right, the maximum possible pressure.

To understand how a screw extruder works, it is necessary to understand *drag flow*. Imagine yourself as the observer pictured in Figure 10-9a. You and the bottom plate are stationary, whereas the top plate is moving to the right. Polymer melt sticks to the top plate and is dragged to the right at a velocity that is proportional to its distance from the stationary bottom plate. Thus the total flow rate past you is

$$Q = V_{max}WH/2 \qquad (10\text{-}11)$$

where V_{max} is the plate velocity, and W and H are the width and height, respectively, of the channel. Of course if both plates were moving (see Problem 10-9), the flow rate would be twice this value. So far, though, our device has not produced any pressure, and certainly it is not going to be a good mixer.

To pressurize the melt in the simple drag-flow device shown in Figure 10-9 it is necessary to introduce a restriction. The restriction is usually the die, but we can block off the end completely, as shown in Figure 10-9b. Unlike the gear pump, the pressure is limited because the pressurized melt simply flows backward. In fact, the entire volume simply circulates: forward at the top, backwards near the bottom. The device accomplishes no pumping, but there clearly is some mixing action.[§] This type of flow becomes important in understanding the screw geometry.

Can we calculate the pressure? Well, somewhat crudely, but for isothermal Newtonian fluids, the results are quite accurate. Proceeding, we recall the equation for flow of a Newtonian fluid in a slit given in equation (10-1), and solve it for backward flow rate Q_2 as

---

[§] In fact, the mixing is rather poor, as the motion is mainly laminar and the velocity gradients are relatively mild. More complicated designs are used to provide intense velocity gradients both perpendicular and parallel to the flow.

$$Q_2 = \frac{WH^3}{12\eta} \frac{dP}{dx} \qquad (10\text{-}12)$$

As the situation is nearly constant along the flow direction x, we can assume the pressure gradient is constant and thus equal to P/L, where L is the length of the extruder. Because the net flow is zero, we can boldly equate the pressure-driven backflow rate Q_2 with the forward drag flow rate Q_1, thus

$$Q = Q_1 - Q_2 = \frac{WHV_{max}}{2} - \frac{WH^3 P}{12\eta L} = 0 \qquad (10\text{-}13)$$

Solving this for P gives the desired result

$$P = \frac{6\eta V_{max} L}{H^2} \qquad (10\text{-}14)$$

In spite of its simplicity, this equation carries some important information. First of all, we can see that the viscosity η is critical for building pressure. Drag flow is not going to work well for low-viscosity fluids. Also note that a large extruder with a high flow capacity is not going to produce as much pressure as a smaller device. The geometrical factors are clear: increase the length L and decrease the channel depth H to develop high pressure.

Of course, both designs in Figure 10-9 are useful only in establishing the extremes of what one can expect from this simple, planar extruder. In between these would be a more practical case where we introduce a slit die at the discharge end to make a film. This is illustrated below

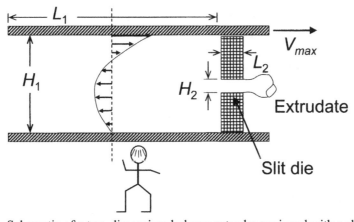

Figure 10-10. Schematic of a two-dimensional planar extruder equipped with a slit die.

The analysis of this setup follows a similar path except that the die geometry counts in determining the flow rate vs. pressure characteristics. While the drag flow contribution remains the same, the developed pressure pushes material through the die as well as back down the channel. So, we have for the material balance

$$\frac{H_1 W V_{max}}{2} - \frac{W H_1^3 P}{12 \eta L_1} = \frac{W H_2^3 P}{12 \eta L_2} \tag{10-15}$$

The terms on the left-hand side represent the net flow forward as provided by the extruder, and this amount balances the flow through the die, as given by the term on the right-hand side.

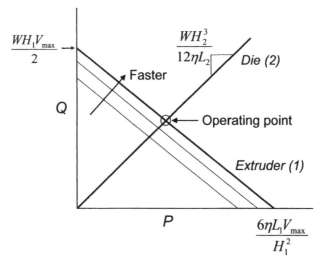

Figure 10-11. Characteristic curves for extruder and die. Those pictured are for the simple example resulting in equations (10-15), (10-16) and (10-17). The axes on this traditional plot are the opposite of those commonly used for fluid pumps, such as centrifugal pumps.[3]

Everything in this equation is known except for the pressure, and we can solve for P to give

$$P = \frac{6 H_1 V_{max} \eta}{\dfrac{H_1^3}{L_1} + \dfrac{H_2^3}{L_2}} \tag{10-16}$$

The corresponding flow rate is

$$Q = \frac{WH_1 V_{max}}{2\left[\dfrac{L_2}{L_1}\left(\dfrac{H_1}{H_2}\right)^3 + 1\right]} \qquad (10\text{-}17)$$

Note that the flow rate is independent of the viscosity, because the viscosity impacts both the flow rate in the die and in the channel equally.

These equations are for isothermal flow in a very specific geometry and also, importantly, for Newtonian fluids. These results are often pictured in terms of the *characteristics* of the extruder and die plotted on linear Q vs. P scales. For our simple planar two-dimensional device, the plots are straight lines, as shown in Figure 10-11.

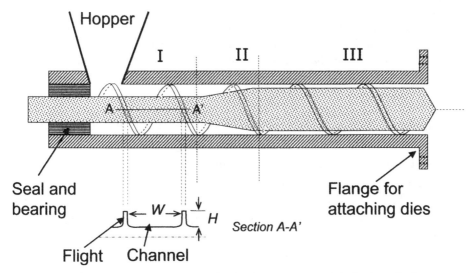

Figure 10-12. Schematic of single-screw plasticating extruder with the barrel slice opened to look inside. Section A-A' shows the general shape of the channel. Sections I, II and III are the feed, compression and metering zones, respectively.

Of course, actual screw extruders processing real polymers have some significant differences, but the principles are similar. Figure 10-12 depicts, very schematically, what a single-screw extruder looks like. As can be seen, the feed and compression zones are dealing mainly with squeezing and melting the pellets fed to the hopper. The metering zone is responsible for most of the pressurization of the molten polymer. Typically, it occupies from 5 to 15 turns of the screw.

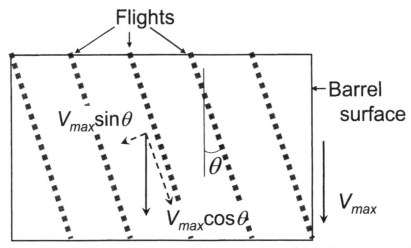

Figure 10-13. Unrolled screw and barrel. The unwrapped barrel is sliding on top of the flights. The velocity component $V_{max} \cos \theta$ drags the melt along the channel, while the component $V_{max} \sin \theta$ causes rotation in the channel, which can help with the melting and mixing. Many screws are designed such that the pitch (distance between adjacent flights) is one diameter, i.e., $\theta = \tan^{-1}(1/\pi) \sim 17°$.

The resemblance between our simple planar extruder and the single-screw extruder become more evident if we do two imaginary things. First of all, we hold the screw stationary and rotate the barrel, the view we would enjoy if we were sitting on the screw. This assumption has no effect on the outcome. We then unwind the channel in the screw from its core and flatten it out. On top of this we place the barrel after slicing and flattening it. The result is depicted in Figure 10-13.

Application of the equations developed for the planar extruder can be applied directly to the screw, albeit with several assumptions. These include no leakage over the flight, no friction at sides of the channel, no slip at the bottom surface of the channel or at barrel surface, isothermal conditions, and narrow flights. Obviously, these may limit the usefulness of the results, but they do represent the best possible performance. On working through the geometry and assuming a pitch equal to the screw diameter, we find the screw characteristic

$$Q = \frac{N\pi^2 D^2 H}{2} \sin \theta \cos \theta - \frac{\pi D H^3}{\eta} \frac{P}{L_{III}} \sin^2 \theta \qquad (10\text{-}18)$$

where N is the speed of the screw in rad/s, H is the channel depth, and L_{III} is the axial length of the metering zone (Zone III in Figure 10-12). If the screw is of "square" design, then the factors can be calculated. The expression becomes then

$$Q = 0.145 N \pi^2 D^2 H - 0.0077 \frac{\pi D H^3}{\eta L_{III}} P \qquad (10\text{-}19)$$

where the coefficients are dimensionless. Note that the screw characteristic is still linear, and dependencies on channel depth are the same. The advantage of going to larger-diameter screws is clear; the drag contribution goes up as the square of the diameter, whereas the "back-flow" term increases only linearly with diameter. However, if the channel depth H and metering section length L_{III} are scaled with diameter, then Q scales as the diameter cubed.

Needless to say, this expression is a very crude approximation to the performance of a screw extruder. The single-screw extruder might be the most investigated and modeled piece of processing equipment there is. Commercial services are available to design screws for optimal performance (high throughput with minimal heating) for a given resin and extruder size. Beyond this, there are numerous refinements and modifications of the basic design, including twin-screw extruders of all sorts.

C. A DOWNSTREAM PROCESS—SPINNING

A die of some configuration is normally found immediately downstream of the melt delivery system, be it an extruder or a gear pump. Downstream of the die, a number of operations might take place. In a laboratory setting, often the extruded shape, e.g., a tape, is merely cooled down and wound up. If the extruder is being used as a mixer, then the extrudate will probably be chopped into pellets for later use. Other downstream operations include crosslinking, foaming, stamping, embossing, surface treatment, etc. One operation of interest to polymer scientists is orientation of the extrudate by stretching in one or two directions to make, for example, fibers or film.

As an example of such a process, we will consider melt spinning. The goal will be to predict the shape of the extrudate as its being drawn and the force required for the operation. As expected, there will be a number of simplifying assumptions.

The geometry is illustrated in Figure 10-14. The major assumptions are that the velocity profile is flat at each point along the thread, the velocity gradient in the radial direction is negligible (in spite of the fact that the thread is getting thinner), inertial terms are negligible, the velocity profile is flat coming out of the spinneret, and the melt is Newtonian. Other more minor assumptions include the absence of air friction and the negligible surface tension.

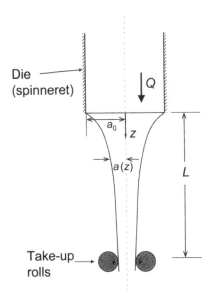

Figure 10-14. Schematic of melt spinning, with definition of geometric variable symbols.

Cylindrical coordinates are the appropriate choice for this problem. The equation of continuity simplifies to

$$\frac{1}{r}\frac{\partial}{\partial r}(rV_r) + \frac{\partial V_z}{\partial z} = 0 \tag{10-20}$$

This equation merely restates what we already know: locally the flow is simple extension.

With the inertial, surface and gravitational terms out of the picture, we can state that the force is constant along the thread line. Referring to Figure 10-14 for nomenclature, we can write this assumption as

$$\frac{d(a^2\sigma_{zz})}{dz} = 0 \tag{10-21}$$

where σ_{zz} is the total stress along the spin line and $a^2\sigma_{zz}$ is, of course, the tension T. Knowing that the fluid is Newtonian, we can connect the force to the dynamics using

$$\sigma_{zz} = \tau_{zz} - p = 2\eta\frac{dV_z}{dz} - p = 3\eta\frac{dV_z}{dz} \tag{ssc} \tag{10-22}$$

Note that we have used the total derivative based on the assumption that the axial velocity depends only on z. This assumption also leads to

$$\frac{dV_z}{dz} = \frac{d(Q/\pi a^2)}{dz} = \frac{Q}{\pi}\frac{d(1/a^2)}{dz} \qquad (10\text{-}23)$$

where Q is the flow rate, which is constant at every point along the thread line.[**] Since the applied stress σ_{zz} is simply $T/\pi a^2$, where T is the constant (positive) tension, we can use equations (10-22) and (10-23) to give

$$\frac{T}{\pi a^2} = \frac{3\eta Q}{\pi}\frac{d(1/a^2)}{dz} \qquad \text{(ssc)} \quad (10\text{-}24)$$

Substituting $u = 1/a^2$ and combining the constants gives the simple differential equation

$$\frac{du}{dz} = \frac{T}{3\eta Q}u \qquad \text{(ssc)} \quad (10\text{-}25)$$

which is easily solved to give the final result

$$a(z) = a_0 \exp(-Tz/6\eta Q) \qquad \text{(ssc)} \quad (10\text{-}26)$$

Thus the thickness of the thread decreases exponentially along the line. If Q and T can be measured, then the viscosity can be extracted from a picture of the thread profile. Likewise, if the viscosity is known from other measurements, we can predict the tension in the line associated with the reduction of diameter from a_0 to $a(z)$ at position z. On solving equation (10-26) for the tension we find

$$T = \frac{6\eta Q}{L}\ln\frac{a_0}{a_L} = \frac{3\eta Q}{L}\ln DR \qquad \text{(ssc)} \quad (10\text{-}27)$$

where a_L is the diameter at the end of a spin line of length L, and DR is the draw ratio. The draw ratio is equivalent to the familiar stretch ratio λ for transient extensional experiments. In other words, if we take a Lagrangian point of view by sitting on the thread as it travels down the spin line, it looks

---

[**] While we can assume the flow rate is constant in space and time, there are well-documented violations of this, even when the feed flow rate and take-up velocity are constant. The phenomenon is known as "draw resonance." Roughly what happens is a chance thickening of the thread hits the constant-velocity take-up roll and thus the take-up flow rate increases momentarily. This causes a thin spot in the thread somewhere upstream; when the thin spot hits the take-up roll, the flow decreases. And so on.

like the material we are sitting on is being stretched in a transient experiment by an amount $\lambda = DR$.

D. SUMMARY

As suggested earlier, this chapter contains some discussion of only a few of the thousands of process methods and their variations. Its main purpose is to demonstrate how rheology and processing interact strongly. Introduced is the important simplifying assumption: the lubrication approximation. On applying this approximation, we can progress using only one component of the equation of motion, at least for fairly simply geometries.

For more complex geometries and for conditions that clearly violate the lubrication approximation, the situation becomes exceedingly complex and numerical methods are the only hope. One can purchase (unfortunately, at very high cost) software packages that approach solutions for complex geometries using finite-element methods for tracking the changes in velocity and stress through a channel, in a mold, or even for flows with free surfaces. Clearly, the energy equation (heat balance) must also enter the calculation.

While progress with computational methods has been very significant, there are still many aspects that could be better. Viscoelastic fluids and materials with yield stresses are sometimes not handled completely. Conditions at the wall near sharp corners are still numerically challenging. Slip at the walls and elastic instabilities for high-stress flows are often ignored. In short, the bridge between rheological response and processing behavior is still under active construction.

APPENDIX 10-1: DENSITIES OF MELTS AT ELEVATED TEMPERATURES

All polymer scientists know that the density of a very pure HDPE is around 0.965 g/cm^3 at room temperature.[4] But what is the density at 220 °C, a typical processing temperature? And what happens if the pressure is raised to 20,000 psi? Such information is needed for conversion of mass flow rate to volume flow rate Q, as well as for more fundamental questions concerning the role of free volume on flow properties.

Fortunately, the pressure-volume-temperature (*PVT*) relationships for many polymer melts have been characterized (not an easy measurement). The observations are often given in terms of equations of state developed for polymers. Thus, in place of a huge table of data filling many pages, one needs only the form of the equation and the values of the parameters.

An example of a popular equation of state is that due to Flory, Orwoll and Vrij;[5] it's know as the FOV equation. Another is that due to Sanchez and Lacombe.[6] Parameters for these equations, and others, are tabulated in the *Polymer Handbook* and many articles. Parameters for various polyolefins have been included in an article by Han et al.[7] It is particularly important to note the temperature range for the correlation, as none of the equations is a perfect description of the behavior of polymer melts. While it is fully expected that the parameters will also depend upon molecular weight, the dependence is slight at molecular weights typical of processing.[††]

To illustrate the application of an equation of state, we pick the Sanchez-Lacombe (S-L) equation for a metallocene LLDPE with 10 wt% hexene comonomer and a room-temperature density of 0.923 g/cm^3. The S-L parameters are listed as $P* = 436.8$ MPa, $V* = 1.114$ cm^3/g, and $T* = 662.9$ K. We need the density at 220 °C, at low absolute pressures. Obviously, 0.923 g/cm^3 would be a very poor choice.

To progress, we write a statement of the S-L equation. It's

$$\tilde{P} + \frac{1}{\tilde{V}^2} + \tilde{T}\left[\ln\left(1 - \frac{1}{\tilde{V}}\right) + \frac{1}{\tilde{V}}\right] = 0 \tag{a}$$

where the symbols with the tilde are reduce variables, e.g., $\tilde{V} = V/V*$. As this equation can be solved algebraically only for P and T, we use a numerical solution to get V.

The first step is to calculate the known reduced variables $\tilde{P}$ and $\tilde{T}$. We will assume $P = 0.1$ MPa (1 atm) and $T = 493$ K (220 °C). The reduced variables are then

$$\tilde{P} = \frac{P}{P*} = \frac{0.1}{436.8} = 0.000229 \tag{b}$$

and

$$\tilde{T} = \frac{T}{T*} = \frac{493}{662.9} = 0.7437 \tag{c}$$

The S-L equation now reads

[††] The Sanchez-Lacombe model has an adjustment for relative molecular size, but leaves undetermined the cell size, given a molecular structure.

$$0 = 0.000229 + \frac{1}{\tilde{V}^2} + 0.7437 \left[\ln\left(1 - \frac{1}{\tilde{V}}\right) + \frac{1}{\tilde{V}} \right] \tag{d}$$

after putting all terms on the right-hand side. There are lots of ways of finding the solution(s) to this equation, but the easiest is via computer, e.g., "goal seek" with Excel®. The resulting value is $\tilde{V} = 1.213$, which gives a specific volume of 1.351 cm³/g and thus a density of 0.74 g/cm³. (For this resin and temperature, the FOV equation gives essentially the same answer, but this will not always be the case.) Clearly, using the room-temperature density of 0.923 g/cm³ in calculations of flow and shear rates could be a potentially serious mistake.

PROBLEMS

10-1. One way of achieving spin casting by gravity is to switch the axis of the disk to the horizontal direction and rotate the disk slowly such that $g \gg R\Omega^2$, where R is the radius of the disk and Ω is the rotation speed in rad/s.

(a) If $R = 2$ cm, what must Ω be in rpm to have the acceleration of gravity dominate, e.g., 10 times larger than $R\Omega^2$?

(b) Using the assumptions of Example 10-1, test the gravity-driven casting for film uniformity.

10-2. Another common film-making process is dip coating wherein a flat substrate (usually glass) is dipped into a polymer solution and hung vertically to drain and dry.

(a) What properties of the solution will determine the final film thickness?

(b) Develop an expression for the shear stress in the draining film assuming steady uniform downward flow on a vertical substrate.

(c) (Open end) Using the result for (b), assume a microscope slide is used as the substrate. Choose typical but convenient numbers for other properties to find the value of the shear stress. Judging from polymer solution viscosity functions displayed in the literature, is this stress likely to be in the Newtonian or non-Newtonian regime?

(d) (Open end) Describe the factors that determine the initial film thickness as the substrate is pulled out of the solution. What would happen if a high acceleration were applied, i.e., the substrate was yanked out of the solution?

10-3. Casting a film from a polymer solution often results in bubbles in the final product, depending upon the film thickness, solvent volatility, and the presence of dissolved or entrained air in the initial solution. One casting method that can largely eliminate bubbles and make a uniform film is depicted in Figure 10-15. This is a specialized variation of a general processing method referred to as *rotational molding*, and is also related to *spin casting* in that the dominant body forces acting on the material are produced by rotation. The success of this method depends greatly on having a highly concentric mold that can be rotated at a high speed.

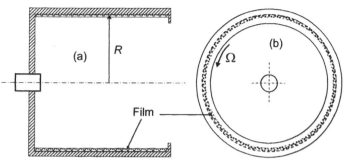

Figure 10-15. Rotational molding of a film from a polymer solution. The mold, shown in section in (a) and end on in (b), must be very carefully machined to obtain a uniform film. At a high speed, the radial acceleration is strong enough to rid the film of bubbles and suppress instabilities.

(a) Develop an expression for the buoyant force on a bubble of radius R_B in terms of the radius of the mold and the radial spinning velocity Ω.

(b) Derive an expression for the maximum time needed for the bubble to "float" to the surface in terms of the relevant film properties. Assume Stokes' law.

10-4. A researcher is attempting to prepare by compression molding a polymer film for transmission infrared analysis. The IR detector has a dynamic range of 100, i.e., the lowest transmissivity I/I_0 it can detect is 0.01 and thus the tolerable absorbance $A = -\log(I/I_0)$ for good results is 2.

(a) Assuming that the most highly absorbing band has an absorption coefficient[‡‡] of 200 mm^{-1}, what sample thickness should be used to obtain a spectrum with all bands on scale?

(b) The sample comprises 1 pellet weighing 10 mg, which will be squeezed between two very flat plates in a small press capable of safely exerting a force of 5000 lb$_f$. How long will it take to squeeze the pellet to a film that is thin enough for transmission spectroscopy, assuming the viscosity is 10 kPa s? Assume an initial thickness of 1 mm and a density of 1 g/cm^2.

(c) (Open end) Explain why a modification of the mold similar to that shown in Figure 10-4 would be useful. Suggest some design parameters for the insert.

10-5. In Section A-2 of Chapter 10, it was asserted that the pressure gradient in a slit was merely $12Q\eta/WH^3$. Starting with the equations of motion, show that this equation is in fact correct for a Newtonian fluid.

10-6. Equation (10-4) was presented without proof. This equation relates, for Newtonian fluids, the velocity profile in slit flow to the bulk velocity. The bulk velocity is defined as the flow rate Q divided by the cross-sectional area for flow. In the case of plane slit flow in a slit of width W and gap H, we have $\overline{V} = Q/WH$.

[‡‡] The nomenclature for absorption spectroscopy comprises a complicated mixture of terms developed by specialists using the various spectroscopic techniques. The vocabulary used here is widely used in infrared spectroscopy.

(a) With these conditions in mind, demonstrate the validity of equation (10-4) if $W \gg H$.

(b) Calculate the shear rate at the wall in terms of Q, W, and H.

10-7. One partial fix for the problem of pressing thin films from melts is to use a mold that is disk-shaped instead of flat. In this case, the squeezing flow will be constant radius instead of constant volume, once the charge starts to spill over the edges of the disk. Repeat Problem 10-4 (b), assuming the disk is 20 mm in radius.

10-8. (Challenging) A power-law fluid has a potentially bothersome property; its viscosity goes to infinity at zero shear rate. In compression molding, the very center of the mold has a very low deformation rate and thus possibly a very high viscosity.

(a) Starting with the lubrication concept, work through to the solution for compression molding of a power-law material.

(b) As the molding gets thinner, the shear rate drops and the viscosity increases. Does the compression process lock-up at a certain point?

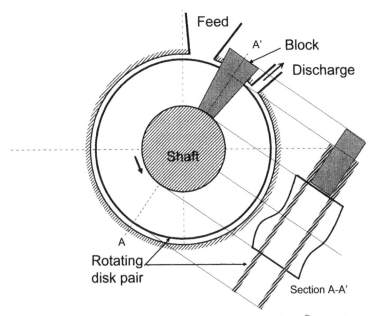

Figure 10-16. Schematic of an element of a Disc-Pack® extruder.

10-9. An element of a Disc-Pack® extruder[§§] comprises two disks mounted to a rotating shaft, all in a closely fitting cylinder. Melt is located between the disks and is dragged circumferentially by them. To remove the melt, either to another element or

---

[§§] The Disc-Pack® extruder was developed at the Farrel Corporation (Ansonia, CT) in the early 1970's. Farrel Corporation is perhaps best known for its Banbury mixers, which are a widely used batch compounding machine available in capacities up to 1000 lb_m. Continuous versions of the Banbury (FCM) are available in capacities up to nearly 100,000 lb_m/h.

to a die outside the cylinder, a block is inserted between the disks and an escape channel is provided through the wall of the cylinder.

(a) What is the maximum capacity of the element?

(b) Assuming laminar flow, develop an expression for maximum pressure increase that can be expected from such a device, assuming a Newtonian fluid of viscosity η. Define all geometric variables carefully.

10-10. The application of equation (10-8) was strange in that the pressure, instead of the normal stress, was used to calculate the force acting on the surface. Starting with the equations of motion, show that for laminar slit flow of a Newtonian fluid, the normal stress acting against the slit surfaces is equal to the pressure.

10-11. Flow through a tapered die looks somewhat similar to inward radial flow of fluid to a point sink. Starting with the equation of motion in spherical coordinates, investigate what terms have been eliminated by the lubrication approximation used in Example 10-2.

10-12. Show that the expression in equation (g) of Example 10-2 reduces to the Poiseuille equation as the initial and final radii approach each other.

10-13. The expression for a power-law fluid in a tapered tube is

$$\Delta P = \frac{2mL}{3n(R_i - R_f)} \left(\frac{Q(1+3n)}{\pi n} \right)^n \left[\frac{1}{R_f^3} - \frac{1}{R_i^3} \right] \tag{10-28}$$

Show that this expression reduces to that for Newtonian fluids.

10-14. Develop an expression for the difference between the predictions in equation (h) vs. equation (g) in Example 10-2.

10-15. (Challenging) The expression for the Ferry GNF was shown in Example 10-2. Carry this forward to develop an expression for $\tau_R (Q; R)$ for flow in a tube, where τ_R is the magnitude of the shear stress at the wall. From this and equation (b) in Example 10-2, develop an expression for the pressure drop in a tapered tube. Use integration software to help with the integration. Check you expression by examining the result at very low flow rates.

10-16. Using the lubrication approximation, develop an expression for the pressure drop for flow of a Newtonian fluid through a tapered slit die.

10-17. (Challenging) Repeat the development of the thread line profile and tension using a power-law in place of the Newtonian model.

10-18. Very stretchable polymer melts like branched, high-pressure LDPE can stand a tensile stresses of about 100 kPa without breaking. If the viscosity of the melt under these conditions is 4.7 kPa s (see Figure 5-4), describe what conditions should be used to achieve a draw ratio (DR) of 20.

10-19. Instabilities are known to occur in melt spinning, leading to thickness variation and eventually failure. One partial solution is to cool the thread line as it thins. Explain how this might help stabilize a thin spot in the thread.

10-20. Derive an expression for stretch ratio vs. time as a small volume of polymer travels downstream in a spin line. Is the deformation rate constant?

10-21. Estimate the impact of each of the density-dependent terms in the equation of motion for a Newtonian material with a viscosity of 1 kPa s and a density of 900 kg/m^3 during a spinning operation with a flow rate of 0.1 cm^3/s, an initial diameter of 7 mm, and a draw ratio of 10.

10-22. Between equations (10-25) and (10-26), there is one of those obnoxious "easily solved" phrases. Check the integration in a careful, step-by-step fashion.

10-23. (Challenging) Large pelletizing extruders can have capacities upward of 40,000 lb$_m$/h. How large a single-screw extruder is needed to reach this capacity? The metering section of a general-purpose polyethylene screw typically has a channel depth around 5% of the screw diameter and occupies the last eight turns of the screw. Assume a viscosity of 2 kPa s and a density of 0.9 g/cm^3. Also assume the extruder is run with a head pressure of 50% of the maximum pressure.

REFERENCES

1. Authors of a few of the more widely cited texts on polymer processing are: J. M. McKelvey (1962), S. Middleman (1977), D. G. Baird and D. I. Collias (1995), T. A. Osswald (1998), and Z. Tadmor and C. G. Gogos (2006).

2. C. E. Rogers, W. I. Vroom and R. F. Westover, "Getting the most from a polymer by vacuum-compression molding," *Modern Plastics*, **45**(13), 200, 203, 205, 206 (1968).

3. S. Mukesh, "Basics Concepts of Operation, Maintenance, and Troubleshooting, Part I," The Chemical Engineers' Resource Page, www.cheresources.com.

4. R. P. Quirk and M. A. A. Alsamarraie, in *Polymer Handbook, 4th Edition*, (J. Brandrup and E. H. Immergut, eds.), John Wiley & Sons, New York, 1989, p. V/20.

5. P. J. Flory, R. A. Orwoll and A. Vrij, "Statistical thermodynamics of chain molecule liquids. I. An equation of state for normal paraffin hydrocarbons," *J. Am. Chem. Soc.*, **86**, 3507–3514 (1964).

6. I. C. Sanchez and R. H. Lacombe, "An elementary molecular theory of classical fluids. Pure fluids," *J. Phys. Chem.*, **80**, 2352–2362 (1976).

7. S. J. Han, D. J. Lohse, P. D. Condo and L. H. Sperling, "Pressure-volume-temperature properties of polyolefin liquids and their melt miscibility," *J. Polym. Sci. Part B Polym. Phys.*, **37**, 2835–2844 (1999) (Note: there is an error in equation 16 describing the FOV equation; it should have a divide sign between the first two terms on the right-hand side.)

11

Quality-Control Rheology

While polymer scientists are typically interested in synthesis and structure-processing-property relationships of polymers, they often need to interact with the organizations responsible for polymer production. An important aspect of production is quality control (QC), i.e., testing to make sure the product is meeting specifications. Often such testing includes one or more flow tests, the variety of which is stunning. For the rheologist, these tests often present challenges in analysis and interpretation. For the QC engineer, the quest is always in the direction of faster, simpler, foolproof procedures using inexpensive but rugged equipment. But beyond this is the challenge of obtaining certification, verification, and acceptance of a test by a recognized standards organization such as American Society for Testing and Materials (ASTM).[*] This organization lists Standard Specifications for materials aimed at certain applications, and Standard Test Methods, which are designed to show if the material meets a Specification. ASTM methods relevant to the flow of polymers or materials with a polymeric component are listed in Appendix 11-1. This list, which is not at all exhaustive, illustrates the variety of methods and, to some extent, gives one a feeling for the motivation behind the development of these tests.

---

[*] The ASTM is now known as simply ASTM International. PDF copies of Specifications and Test Methods are available at ASTM.org. Current costs range from $30 to $50.

There are many other standards organizations, including ISO (International Standards Organization).   The idea of ISO was to provide consistency throughout the world.   As there are many interests—often conflicting— involved in advancing any standard procedure, it is clear that the job of ISO is not easy.   Whereas an ASTM method can be proposed, tested and adopted in a few years, many ISO standards and methods appear to be taking decades to finalize.[†]   For this reason, there are relatively few ISO rheology tests; most were developed by the TC-35 committee (paints and varnish).

Aside from meeting standards for a product, QC tests are often used to detect drift in the manufacturing process.   To avoid scrap, drift is best caught well before the product is "out of spec."   Clearly, the test must be very precise and free from drift itself.   The role of a standard material is to control against variance introduced by differences in the instrument, operator or location.   As mentioned in Chapter 1, a limited number of rheological standards are available from NIST.   For example, SRM (standard reference material) 1475 is a HDPE resin with melt index of 2.07 and a standard deviation of the mean[‡] of 0.006. (Needless to say, this level of precision is not common in most labs.)   Melt index will be explained below.

A. EXAMPLES OF METHODS USED BY VARIOUS INDUSTRIES

As mentioned above, most test methods are very specific to an industry or even a polymer type.   Our main reason for exploring some of these methods is that they are often quoted and thus part of a rheology conversation.   Furthermore, they can provide ideas for instrument design and operation.   An example of the latter was the so-called Monsanto gas rheometer, which was a widely used capillary rheometer that was essentially a laboratory version of the melt indexer.   Instead of a weight to provide pressure, an easily regulated, high-pressure gas supply was used.   What could be simpler?

[†] ISO standards can be purchased on-line at iso.org. Prices, currently denominated in Swiss Francs, range between 40 and 70 CHF. Languages are principally English and French. An examination of the table "International harmonized stage codes" can explain the complications of establishing an ISO standard.

[‡] Standard deviation of the mean is a measure of the precision of the mean. The standard deviation of the mean is defined as the estimate population standard deviation (the usual standard deviation) divided by the square root of the sample size.

1. Melt index

Perhaps the most-quoted ASTM Standard Test Method is D1238, "Standard Test Method for Melt Flow Rates of Thermoplastics by Extrusion Plastometer," a.k.a. Melt Flow Index or simply Melt Index (MI). This test, which varies somewhat with the test resin, was primarily promoted by a growing polyolefin industry in the early 1950's. As with many ASTM tests, D1238 specifies everything: sampling of the resin, the instrument, the conditions, and the procedure. Any deviation from these specifications means that the result will not be the Melt Index.

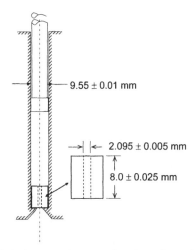

9.55 ± 0.01 mm

2.095 ± 0.005 mm

8.0 ± 0.025 mm

Figure 11-1. Geometry of the melt indexer. The extrudate is collected over a period of 10 min, and weighed, thereby giving the result in units of g/10 min. The higher the MI, the lower the viscosity. For polyethylene, the temperature is generally set at 190 °C.

The melt index test has been roundly criticized by many for good reasons, but it is so widely used that it is unlikely to disappear for some time. There have been many very careful analyses of the melt index in terms of fundamental rheological material functions, primarily $\eta(\dot{\gamma})$.[1]

In appearance, the melt indexer looks like a capillary rheometer, but the specifications do not allow changes in the geometry, which is depicted in Figure 11-1. The force on the piston is applied with a 2000-g mass, which gives a total mass of 2160 g. A high-load condition is also specified; it is delivered using 22 kg, which is ten times the normal load. A mass of 22 kg is equivalent to nearly 50 lb_m, which is a bit cumbersome to perch on top of the piston. Instruments that elevate the mass automatically are available.

A major shortcoming of melt indexer from a rheometry point of view is the low aspect ratio of the capillary. The reason for this design choice is to push the stress up to a realistic level with the use of weights only. (See Example 11-

1.) Somewhat fortuitously, it also introduces a significant contribution due to the build up of extensional stress in the entrance.

Example 11-1: *Approximate the nominal shear stress in a melt indexer with the total low-load mass (2160 g). Find the shear rate corresponding to this stress for the resin depicted in Figure 5-5, and describe the region of the flow curve in which this shear rate falls. Use the dimensions in Figure 11-1.*

As the apparent shear stress is called for, we use the equation

$$\tau_W = \frac{R\Delta P}{2L} = \frac{\Delta P}{4(L/D)} \tag{a}$$

where τ_W is the shear stress at the capillary wall; R, D and L are the capillary radius, diameter and length, respectively; and ΔP is the pressure. The aspect ratio is $L/D = 8.00/2.095 = 3.82$. The pressure, in SI units, is given by the quick calculation $\Delta P = 9.81$ N/kg $\times$ [2160 g/1000 g/kg]/[π(9.55 mm/1000 mm/m)^2/4] = 296 kPa. The nominal shear stress, via equation (a), is then $\tau_W =$ 19.4 kPa.

To find the corresponding shear rate, one can find the line of constant shear stress corresponding to 19.4 kPa. This line will have a slope of -1 and an intercept of $\log_{10}(19400 \text{ Pa}) = 4.3$ on the plot shown in Figure 5-5. Alternatively, one can use the Cross GNF model in equation (5-1) to fit the data and find the shear rate by solution of the Cross equation for the corresponding shear rate. This can be done by trial and error, or by an equation solver (e.g., "goal seek" on Excel®).

The shear rate turns out to be 2.33 s^{-1} (log shear rate = 0.37), which is almost in the power-law region.

As the data in Figure 5-5 are referenced to 150 °C instead of the standard 190 °C, this number will be lower than the result at the latter temperature, but still not into the range typical for extrusion. Additionally, the shear stress is sufficiently high such that the melt index is not likely to be a valid substitute for the zero-shear-rate viscosity.

A far more thorough analysis of the type shown here is provided by Gleissle.[2]

With skilled operators running the same melt indexer, ASTM reports[3] that the melt index of a 2-MI polyethylene can be reproduced with a standard error of 0.04 g/10 min. This increases to 0.094 g/10 min when comparing results on the same resin at different laboratories (different operator, different instrument). Information on error for other resins is listed in Table 11-1.

2. Rossi-Peakes flow [§]

Testing the melt flow of a thermoset resin in a melt indexer would definitely be a risky procedure. If the thermoset were to cure before extruding completely, the result would be a plugged instrument and a nasty clean-up job.[**]

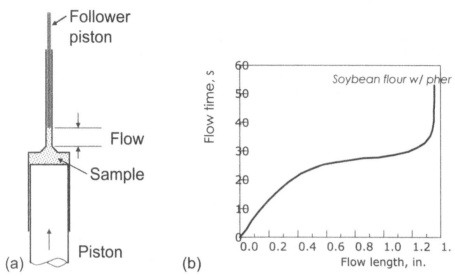

Figure 11-2. (a) Schematic of the geometry used for the Rossi-Peakes flow test and (b) data gathered with a formaldehyde-modified soybean flour mixed with a phenolic resin. Using the flow length as the independent variable was customary for this test. (Adapted with permission from G. H. Brother and L. L. McKinney, *Ind. Engr. Chem.*, **32**, 1002-1006.[4] Copyright 1940, American Chemical Society.)

For this reason, alternatives for examining the flow and curing of thermosets were developed. The Rossi-Peakes tester is one example.[5] The key part of this device, shown in section in Figure 11-2, is a capillary and a piston driven by a weight system. The capillary can be split apart to facilitate cleaning. While this geometry looks much like that for the melt indexer, the procedure is quite different. The idea is not to establish a steady flow, but to examine the penetration of the melt into the capillary as a function of time, using the upper piston. Temperature and load are the two parameters that can be changed.

[§] ASTM D569-90 for this test was withdrawn in 1995. The geometry is mentioned because of its provisions for thermosetting materials.

[**] For cleaning, the capillary in the melt indexer is normally pushed out of the top of the instrument using a brass rod. Not shown in Figure 11-1 is a method of removing the capillary out of the bottom of the barrel: the flange that holds the capillary in place can be unscrewed and removed. This facilitates the cleaning of the barrel, as well as the removal of the capillary if the melt starts to crosslink. The bottom plate is covered by an insulating layer.

Two things happen as the melt travels into the capillary. First of all, the stress drops because the path over which the polymer must flow is increasing. Secondly, the viscosity of the melt increases because of the advancement of the cure. As a result of these two factors, the melt front comes to a rather sudden stop. The point where the melt stops is the Rossi-Peakes flow.

Because the Rossi-Peakes flow tester is somewhat cumbersome to use, QC labs often use similar but less complicated tests. A popular one is the spiral flow test, which uses a mold with a spiral cavity.[6] The melt is pushed from a low-temperature cavity into an opening in the center of the spiral, which is held at a higher temperature. The distance the resin travels before curing is recorded.

3. Mooney viscosity

The Mooney test, used primarily by the rubber industry, is codified by ASTM D1646 and ISO289. It is also called the *shearing disk viscometer*, which describes the geometry quite well: a rotating disk immersed in a sample that is compressed in a disk-shaped cavity mold. This arrangement is depicted in Figure 11-3. Two disks are specified: a small (30.48 mm diameter) and a larger one (38.10 mm diameter). To prevent slip, all surfaces on the disk are grooved.

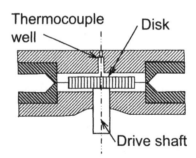

Figure 11-3. Schematic of the geometry for the Mooney shearing-disk viscometer. The polymer sample files the entire white area around the disk, and is confined by the dark-hatched seals on the edges of the cavity.

Oddly enough, the Mooney viscosity has units of torque instead of viscosity, because the geometry does lend itself to analysis to get the viscosity.[7] Thus a Mooney viscosity (often called simply "Mooney") of 1 corresponds to a steady torque of 0.083 N-m on the large rotor turning at 2 rpm. The gap on each face of the disk is 2.52 mm, while the clearance between the disk edge and the wall of the cavity is 6.4 mm for the large disk and 10.2 mm for the small disk.

The so-called Monsanto Curometer has a similar geometry, but with a bicone instead of disk. The cone surfaces are grooved to prevent slip. Operation is generally in the oscillatory mode. The deformation field using the bicone is nearly uniform, an advantage. As the name suggests, the Monsanto Curometer is used mainly for following the cure of thermosets.

Table 11-1 Results of round-robin testing for melt flow index (MFI) as listed in ASTM D1238.

Test	Mean MFI, g/10min	Precision[a] Within lab	Between labs
PE, 190 °C, 2.16 kg	0.27	0.009	0.014
	0.40	0.016	0.027
	2.04	0.040	0.094
	43.7	0.997	1.924
PP, 220 °C, 2.16 kg	2.25	0.052	0.214
	7.16	0.143	0.589
	32.6	0.693	0.945

[a] Precision is given as the estimated standard deviation of the mean. Most simply, this is the common "standard deviation" divided by square root of the number of observations in the sample.

B. TEST PRECISION

1. Estimating error

Precision refers to the variation in observed test results, given test units that are the same except for purely random fluctuations. As was described with the melt index test, the precision of test varies greatly depending upon the operator, the equipment and the details of the testing procedure. In spite of having a strict testing protocol, the operator-to-operator variation is always larger than the run-to-run variation when drawing test units from a uniform population.[††] A summary of ASTM results for "round-robin" melt-index measurements of test units drawn from uniform population is provided in Table 11-1.

Example 11-2: *Using error propagation, estimate the instrument-to-instrument variability for the melt indexer. Assume that a standard resin is used that has*

[††] There is, of course, no such thing as a perfectly uniform population. For example, if our population is a retained bag of production resin, it is well known that pellets within the bag can differ measurably. Usually, however, individual pellets are not tested; of sample of, say, 5 g is drawn. In this sample there may be 500 pellets, which would reduce the pellet-to-pellet variance by a factor of $1/\sqrt{500}$. However, dangers still lurk. For example, with handling, the larger pellets tend to sift to the top of the bag. Their large size might be due to different thermal exposure during pelletizing, which could well translate into differences in rheological properties.

an intrinsic variability of 0.5% in its viscosity due to compositional nonuniformity.

Sources of variance in the melt index test will include uncertainties in the viscosity $\eta(\sigma, T)$, the applied shear stress and the mass extruded. The melt index **MI** will be given approximately by the relationship

$$\mathbf{MI} = \frac{MgD_c^4 \rho(T)}{32\eta(\sigma,T)D_b^2 L_c} 10^3 t, \text{ in g} \tag{a}$$

where the subscript c stands for capillary and b for barrel. The 10^3 converts kilograms to grams; all the units for the variable symbols must be SI, including the time t in seconds, to get the right answer. Note that to some extent the temperature variation of η and ρ will compensate, although viscosity is generally the most sensitive. [Review the Doolittle expression, equation (6-24), as an example of the connection between density and viscosity.]

The error (variance) in the melt index will involve a weighted sum of the variances in all the variables. The weighting for each is given by the general expression

$$\sigma_y^2 = \sum_{i=1}^{m} \left(\frac{\partial y}{\partial x_i} \right)^2 \sigma_{x_i}^2 \tag{b}$$

where y is, in our case, the melt index, and the x_i are the m variables in the equation. The errors in each variable are designated by their variances $\sigma_{x_i}^2$. These must be either estimated or measured. Unfortunately, this process is sometimes very approximate and brings in questions concerning the population. For example, the weight with mass, M, can be calibrated against a standard mass, which will reduce the error relative to that expected by picking out any mass from all the available masses in the stockroom.

According to equation (a), there are eight variables that influence **MI**; these are listed in the table below, along with estimates of the error in each.

Symbol	Definition	Estimated error, σ_i
M	Mass, including piston rod, ~ 2.160 kg	0.0001 kg
g	Gravitational constant, ~ 9.81 m/s^2	0.001 m/s^2
D_c	Capillary diameter, ~ 0.0021 m	0.00001 m
ρ	Melt density, ~ 900 kg/m^3	0 [a]
η	Melt viscosity, ~ 10000 kg/m s	100 kg/m s
D_b	Barrel diameter, ~ 0.00955 m	0.00005 m [b]
L_c	Capillary length, ~ 0.008 m	0.00002 m
t	Measurement time, ~ 600 s	2 s

[a] Set to zero because viscosity changes with T will be in concert with and overwhelm density.
[b] Set at 50 μm because of wear and leakage. The latter will reduce the force on the melt because of viscous drag, and is probably a major source of error.

The derivatives in expression (b) are very important to the process. For example, the derivative of melt index with respect to the capillary diameter D_c is:

$$\frac{\partial \mathbf{MI}}{\partial D_c} = 4\frac{MgD_c^3\rho}{32\eta D_b^2 L_c}10^3 t = \frac{4\mathbf{MI}}{D_c} \tag{c}$$

It is evident from this expression that all the derivatives will be of similar form, e.g., for D_b, the derivative is $-2\mathbf{MI}/D_b$. (When squared, all weighting factors will be positive, which leads to the sad fact that errors accumulate; they do not cancel.) A common factor to all is $(\mathbf{MI})^2$, which is conveniently removed to the left-hand side of the equation. Similarly, all terms will be the square of the respective exponents (4 for D_c, 2 for D_b and 1 for the others) multiplied by the square of the relative error s_i/x_i.

Completing the arithmetic using the numbers in the table gives an estimated relative error in $\mathbf{MI}$ of 0.024, or 2.4%.

Precision can be estimated by examining the precision of the variables that in turn lead to the observation. Of great importance is the impact each variable has on the final result. This is demonstrated in Example 11-2.

In spite of carefully controlled geometry and conditions, the test results can drift due to subtle changes in procedure, temperature drift due to changes in heat transfer characteristics of the insulation or changes in room temperature, or barely detectable changes in geometry. As mentioned in Chapter 7, Section A, slight wear of the capillary can have a large effect on the results. For this reason, capillaries are generally made from tungsten carbide. Capillary radius can get smaller due to the accumulation of deposits of oxidized polymer that are difficult to remove. The role of standard materials is to guard against drift.

All that said, the impact of test protocol and operator skill on precision is often greater than problems with the test equipment.

2. Resin A vs. Resin B

Polymer solution or melt viscosity, and the related QC values, can vary for a number of reasons, the primary one being molecular weight. A rule of thumb is that the zero-shear-rate melt viscosity will be a function primarily of the weight-average molecular weight, M_W. As the stress associated with the MFI measurement is often well into the non-Newtonian region, the MFI will reflect both M_W and the molecular-weight distribution. With copolymers, changes of composition, including co-monomer amount and distribution, can change the results, especially if there is a possibility of microphase separation. Generally, a high-molecular-weight tail on the molecular-weight distribution has only a secondary effect; it will show up principally in the elastic response of the melt.

With this in mind, what changes or differences in MFI or other QC result should be considered important? If one supplier's resin (Resin A) is working and another supplier's resin (Resin B) is not, will the MFI reveal this in advance?

We can start to answer these questions by considering what would happen if we ran a blind test with two identical samples. This could be done, for example, by taking a portion of Resin A and labeling this as Resin B. These two would be submitted for testing without revealing the origin of the two samples. The operator would be asked to run each N times, being careful to draw test units at random, and running tests on A and B in random order for N times.

Table 11-2. Critical values of the Student t at the 0.05 probability level. [a]

d.f. [b]	1	2	3	5	10	∞
t critical [c]	12.71	4.3	3.18	2.57	2.23	1.96

[a] The 0.05 probability level means that there is a 5% chance that $|t|$ will be as large or larger than that listed even though the samples are drawn randomly from the same population (e.g., same bag of resin).
[b] Degrees of freedom. Usually N_A+N_B-2, where N_A and N_B are the number of observations for samples A and B, respectively.
[c] Critical value of t. The listed magnitude of t can be exceeded 5% of the time on comparing sample averages drawn from the same population.

Of course, no significant difference between the averages is expected; however, on repeating the tests over and over again, we will find that the averages occasionally can differ by an alarming amount. This phenomenon was studied years ago by Gosset in 1907,[8] and resulted in what's called the Student t test, available on almost every spread sheet including Excel®. The statistic t, is a ratio the difference in the two averages as compared to the error associate with these averages. If t is large, it is likely that the two samples are different enough to matter. We can be more specific about this by checking a table that gives the probability, roughly speaking, of being incorrect by stating that the two are indeed different.[‡‡] Note again from Table 11-2 that the observed t value must be large if the number of observations is low.

The general formula for calculating the t statistic is

$$t = \frac{\overline{y}_A - \overline{y}_B}{s_{AB}} \tag{11-1}$$

where the $\overline{y}$ are the averages and is an average standard deviation of the two samples. If the two samples have essentially equal variances, then the usual formula for s_{AB} is

$$s_{AB} = \sqrt{\frac{(n_A - 1)s_A^2 + (n_B - 1)s_B^2}{n_A + n_B - 2}\left(\frac{1}{n_A} + \frac{1}{n_B}\right)} \tag{11-2}$$

where the symbol s stands for standard deviation and n for the number of observations.

The t test is not the only way of comparing Resin A with Resin B; there are many other methods,[9] some of which might have better resolution or which examine the details of the distributions of observations in more detail. The Student t test looks only at the averages. It is, however, easy to apply and resistant to some deviation of the observations from the basic assumptions.[§§]

More details about this and other tests can be found in any text on statistics or experimental design. It is important to keep in mind the general principle: one can never be absolutely sure that two samples are different, and one can never, never assert that they are the same. With respect to the latter, one can set a value of t below which one can assume the resins will behave identically, but again there is a probability that this could happen just by chance. The tests help

---

[‡‡] The correct and incorrect ways of wording this statement have attracted much discussion, as has the designation of the probability level for deciding the difference between two samples is significant. Many will pick a probability of 5% as an acceptable probability of making a wrong decision; however, one must decide what is the risk associated with even this small value. The answer for borderline cases is often to gather more observations.

[§§] The assumptions behind the usual t test are normal distribution of values for each base population, and equality of the variances. If the latter is not the case, then a test for unequal variances is available, which is considerably more conservative. See, e.g., Press et al.[9]

to keep one honest, and point out emphatically the need for a sufficient number of independent repetitions.

APPENDIX 11-1: ASTM TESTS METHODS FOR RHEOLOGICAL CHARACTERIZATION

Table 11-3 list most of the ASTM flow tests relevant to polymers or polymer-containing materials. Note that most of the tests are specific for a class of materials or even a specific polymer. Many use a custom-made instrument of specified design, although some flexibility is allowed.

From a research point of view, only a few of the Standard Test Methods provide fundamental rheological properties. For this reason, many researchers eschew ASTM tests, or use them only in part. Measuring a quantity that cannot be interpreted in terms of molecular structure or used for design of processing equipment seems like a waste of time. However, many readers of research reports will be thankful for information regarding a new polymer that can be readily compared with common commercial materials. Very few manufacturers will advertise or even supply fundamental rheological data on their products (although they all have such data).

Table 11-3. ASTM tests for the flow and deformation of polymeric or polymer-containing materials.

ASTM Method	Title
D1238-10	Standard Test Method for Melt Flow Rates of Thermoplastics by Extrusion Plastometer
D1646-07	Standard Test Methods for Rubber-Viscosity, Stress Relaxation, and Pre-Vulcanization Characteristics (Mooney Viscometer)
D1823-95(2009)	Standard Test Method for Apparent Viscosity of Plastisols and Organosols at High Shear Rates by Extrusion Viscometer
D1824-95(2002)	Standard Test Method for Apparent Viscosity of Plastisols and Organosols at Low Shear Rates
D2084-07	Standard Test Method for Rubber Property-Vulcanization Using Oscillating Disk Cure Meter
D2162-06	Standard Practice for Basic Calibration of Master Viscometers and Viscosity Oil Standards
D2171-07e1	Standard Test Method for Viscosity of Asphalts by Vacuum Capillary Viscometer
D2196-05	Standard Test Methods for Rheological Properties of Non-Newtonian Materials by Rotational (Brookfield type) Viscometer
D2396-94(2004)	Standard Test Methods for Powder-Mix Time of Poly(Vinyl Chloride) (PVC) Resins Using a Torque Rheometer

ASTM Method	Title
D2538-02	Standard Practice for Fusion of Poly(Vinyl Chloride) (PVC) Compounds Using a Torque Rheometer
D2556-93a(2005)	Standard Test Method for Apparent Viscosity of Adhesives Having Shear-Rate-Dependent Flow Properties
D2639-08	Standard Test Method for Plastic Properties of Coal by the Constant-Torque Gieseler Plastometer
D2983-09	Standard Test Method for Low-Temperature Viscosity of Lubricants Measured by Brookfield Viscometer
D3056-05	Standard Test Method for Gel Time of Solventless Varnishes
D3123-09	Standard Test Method for Spiral Flow of Low-Pressure Thermosetting Molding Compounds
D3346-07	Standard Test Methods for Rubber Property-Processability of Emulsion SBR (Styrene-Butadiene Rubber) With the Mooney Viscometer (Delta Mooney)
D3364-99(2004)	Standard Test Method for Flow Rates for Poly(Vinyl Chloride) with Molecular Structural Implications
D3531-99(2009)	Standard Test Method for Resin Flow of Carbon Fiber-Epoxy Prepreg
D3532-99(2009)	Standard Test Method for Gel Time of Carbon Fiber-Epoxy Prepreg
D3795-00a(2006)	Standard Test Method for Thermal Flow, Cure, and Behavior Properties of Pourable Thermosetting Materials by Torque Rheometer
D4016-08	Standard Test Method for Viscosity of Chemical Grouts by Brookfield Viscometer (Laboratory Method)
D4040-10	Standard Test Method for Rheological Properties of Paste Printing and Vehicles by the Falling-Rod Viscometer
D4217-07	Standard Test Method for Gel Time of Thermosetting Coating Powder
D4242-07	Standard Test Method for Inclined Plate Flow for Thermosetting Coating Powders
D4243-99(2009)	Standard Test Method for Measurement of Average Viscometric Degree of Polymerization of New and Aged Electrical Papers and Boards
D4287-00(2010)	Standard Test Method for High-Shear Viscosity Using a Cone/Plate Viscometer
D446-07	Standard Specifications and Operating Instructions for Glass Capillary Kinematic Viscometers
D4473-08	Standard Test Method for Plastics: Dynamic Mechanical Properties: Cure Behavior
D4539-09	Standard Test Method for Filterability of Diesel Fuels by Low-Temperature Flow Test (LTFT)
D4603-03	Standard Test Method for Determining Inherent Viscosity of Poly(Ethylene Terephthalate) (PET) by Glass Capillary Viscometer
D4640-86(2009)	Standard Test Method for Determining Stroke Cure Time of Thermosetting Phenol-Formaldehyde Resins

ASTM Method	Title
D4683-09	Standard Test Method for Measuring Viscosity of New and Used Engine Oils at High Shear Rate and High Temperature by Tapered Bearing Simulator Viscometer at 150 °C
D4741-06	Standard Test Method for Measuring Viscosity at High Temperature and High Shear Rate by Tapered-Plug Viscometer
D4957-08	Standard Test Method for Apparent Viscosity of Asphalt Emulsion Residues and Non-Newtonian Bitumens by Vacuum Capillary Viscometer
D4989-90a(2008)	Standard Test Method for Apparent Viscosity (Flow) of Roofing Bitumens Using the Parallel Plate Plastometer
D5-06e1	Standard Test Method for Penetration of Bituminous Materials
D5099-08	Standard Test Methods for Rubber—Measurement of Processing Properties Using Capillary Rheometry
D5125-97(2005)	Standard Test Method for Viscosity of Paints and Related Materials by ISO Flow Cups
D5289-07a	Standard Test Method for Rubber Property-Vulcanization Using Rotorless Cure Meters
D5422-09	Standard Test Method for Measurement of Properties of Thermoplastic Materials by Screw-Extrusion Capillary Rheometer
D5478-09	Standard Test Methods for Viscosity of Materials by a Falling Needle Viscometer
D5481-04	Standard Test Method for Measuring Apparent Viscosity at High-Temperature and High-Shear Rate by Multicell Capillary Viscometer
D5581-07ae1	Standard Test Method for Resistance to Plastic Flow of Bituminous Mixtures Using Marshall Apparatus (6 inch-Diameter Specimen)
D562-01(2005)	Standard Test Method for Consistency of Paints Measuring Krebs Unit (KU) Viscosity Using a Stormer-Type Viscometer
D6049-03(2008)	Standard Test Method for Rubber Property-Measurement of the Viscous and Elastic Behavior of Unvulcanized Raw Rubbers and Rubber Compounds by Compression Between Parallel Plates
D6103-04	Standard Test Method for Flow Consistency of Controlled Low Strength Material (CLSM)
D6204-07	Standard Test Method for Rubber-Measurement of Unvulcanized Rheological Properties Using Rotorless Shear Rheometers
D6601-02(2008)	Standard Test Method for Rubber Properties—Measurement of Cure and After-Cure Dynamic Properties Using a Rotorless Shear Rheometer
D6606-00(2005)	Standard Test Method for Viscosity and Yield of Vehicles and Varnishes by the Duke Viscometer
D6616-07	Standard Test Method for Measuring Viscosity at High Shear Rate by Tapered Bearing Simulator Viscometer at 100 °C
D6648-08	Standard Test Method for Determining the Flexural Creep Stiffness of Asphalt Binder Using the Bending Beam Rheometer (BBR)
D6821-02(2007)	Standard Test Method for Low Temperature Viscosity of Drive Line Lubricants in a Constant Shear Stress Viscometer

ASTM Method	Title
D6927-06	Standard Test Method for Marshall Stability and Flow of Bituminous Mixtures
D7042-04	Standard Test Method for Dynamic Viscosity and Density of Liquids by Stabinger Viscometer (and the Calculation of Kinematic Viscosity)
D7109-07	Standard Test Method for Shear Stability of Polymer Containing Fluids Using a European Diesel Injector Apparatus at 30 and 90 Cycles
D7175-08	Standard Test Method for Determining the Rheological Properties of Asphalt Binder Using a Dynamic Shear Rheometer
D7226-06	Standard Test Method for Determining the Viscosity of Emulsified Asphalts Using a Rotational Paddle Viscometer
D7271-06	Standard Test Method for Viscoelastic Properties of Paste Ink Vehicle Using an Oscillatory Rheometer
D7279-08	Standard Test Method for Kinematic Viscosity of Transparent and Opaque Liquids by Automated Houillon Viscometer
D7312-07	Standard Test Method for Determining the Permanent Shear Strain and Complex Shear Modulus of Asphalt Mixtures Using the Superpave Shear Tester (SST)
D7346-07	Standard Test Method for No Flow Point of Petroleum Products
D7394-08	Standard Practice for Rheological Characterization of Architectural Coatings using Three Rotational Bench Viscometers
D7395-07	Standard Test Method for Cone/Plate Viscosity at a 500 s^{-1} Shear Rate
D7405-10a	Standard Test Method for Multiple Stress Creep and Recovery (MSCR) of Asphalt Binder Using a Dynamic Shear Rheometer
D7483-08	Standard Test Method for Determination of Dynamic Viscosity and Derived Kinematic Viscosity of Liquids by Oscillating Piston Viscometer
D7496-09	Standard Test Method for Viscosity of Emulsified Asphalt by Saybolt Furol Viscometer
D7552-09	Standard Test Method for Determining the Complex Shear Modulus (G*) of Bituminous Mixtures Using Dynamic Shear Rheometer
D7605-10	Standard Test Method for Thermoplastic Elastomers— Measurement of Polymer Melt Rheological Properties and Congealed Dynamic Properties Using Rotorless Shear Rheometers
D926-08	Standard Test Method for Rubber Property—Plasticity and Recovery (Parallel Plate Method)

PROBLEMS

11-1. (Open end) Explore the internet, including the ISO and VAMAS websites, to investigate the status of a melt-flow test equivalent to melt index, ASTM D1238.

Report the "stage" of development of the test. (Hint: Be sure to check resources describing melt index for clues.)

11-2. (Open end) Repeat Example 11-1, but with the high-load condition.

11-3. From the PP melt viscosity data below, gathered with a helical-barrel rheometer, estimate the melt index with the low load (2.16 kg). Check this against the measured value of 37.1 g/10 min. Offer a reason why the two differ, including in your discussion the relative order of the two values.

Log (shear rate, s^{-1})	Log (viscosity, Pa s)
1.496	2.185
1.957	2.093
2.179	2.009
2.316	1.954
2.427	1.907

Data from Wan et al.[10]

11-4. (Open end) The zero-shear-rate viscosity η_0 is known to be uniquely related to the weight-average molecular weight of the polymer. Judging from the data available in Chapter 5 and elsewhere, what load would be likely to bring the melt indexer into the range where it would reflect well the η_0 of typical polymer melts? Describe some possible practical problems with this testing condition.

11-5. Assuming the disk and chamber for the Mooney viscosity test produces viscometric flow, find the viscosity of a 45 Mooney elastomer, assuming Newtonian response. Neglect the contribution due to the edges (cylindrical surface) of the disk. Find also the maximum shear rate.

11-6. Again ignoring the contribution of stress on the outer cylindrical surface of the disk, relate the Mooney viscosity to the viscosity of the polymer using the Cross-Kaye[11] ¾-th's rule.

11-7. Often the Mooney torque is depicted as a summation of that due to the faces of the rotating disk, acting like two sets of parallel-disk fixtures, and the contribution from Couette flow between the disk edges and the chamber walls.

(a) Is this analysis likely to give a result that is too high, or too low? Give qualitative reasons for your response.

(b) Derive the torque in the Mooney geometry for a power-law fluid with power-law constants m and n. Assume the Couette contribution has a shear rate at the disk surface of

$$\dot{\gamma}_i = \frac{2\Omega}{n(1 - \kappa^{2/n})}$$

where Ω is the rotational rate in rad/s, and $\kappa = R_i/R_o$. The subscripts i and o refer to the inner and outer cylindrical surfaces, respectively, that confine the Couette flow.

(c) Is it possible to get the power-law parameters m and n from the Mooney rheometer? Explain.

11-8. The variation of the flow in the Rossi-Peakes apparatus has been applied to a method for measuring viscosity at very low stresses.[12] According to this method, a glass capillary tube is attached to a vacuum pump, and the open end is stuck into the polymer melt or solution. The length of tube L containing the fluid is measured after a certain time.

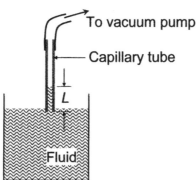

(a) Ignoring the density of the fluid, derive a general expression for the filled length vs. time for a Newtonian fluid.

(b) As the length of fluid in the tube increases, the shear stress drops. For a power-law fluid, this means that the viscosity increases, which in turn drops the flow rate even more. Will the power-law fluid simply stop moving? [Hint: assume that for a power-law fluid, $\tau_w = k(4Q/\pi R^3)^n$, where k and m differ by a multiplicative constant.]

11-9. Cleaning of the apparatus is a constant issue for experimentalists and QC personnel, especially the technician stuck with the job. The one relatively simple flow test that does not require cleaning is the parallel-plate plastometer (PPP), as the plates can be covered with a protective film.

(a) Find the relevant ASTM test covering the parallel-plate plastometer, and describe the nature of the flow in this device.

(b) Using the specifications given in the ASTM protocol, compare the stress levels in the PPP with those in the melt indexer.

(c) Leider and Bird[13] have described a test method whereby the time to reach ½ the original gap is the primary output. Assuming Newtonian behavior, relate the reciprocal of this time to melt-flow index.

(d) Repeat (c) for a power-law fluid.

(e) (Open end) Assuming the plate spacing vs. time is recorded on a computer, suggest a way to use this data in a fashion that will mimic the melt index test. Back up your suggestion with appropriate equations.

11-10. Review the error calculation for melt index (MFI) provided in Example 11-2.

(a) What is the predicted MFI, given the numbers provided in the example?

(b) Verify the error calculation. (Hint: use a spread sheet to help keep track of the many terms.)

(c) Which term is the major source of error?

(d) Compare the estimated error with the observations listed in Table 11-1.

11-11. (Open end) Suggest methods for detecting melt slip during a flow test when using the parallel-plate plastometer (squeeze flow).

11-12. ASTM 5289 concerns a cure test using a "rotorless" rheometer, which is very curious indeed. Investigate this standard, and report on the geometry.

11-13. (Computer) The MFI observations for two resins are tabulated below. Using the t tests for equal means, assuming equal and unequal variances, find the probabilities of being incorrect by rejecting the notion that these two resins are equivalent.

Resin A	Resin B
1.05	1.07
1.01	1.09

11-14. The piston of a melt indexer fits quite tightly into the barrel with an allowed clearance of only 38 μm. The skirt of the piston head is 6.35 mm long. Using the polymer described in Figure 5-5 and Example 11-1, calculate the steady-state drag on the piston if this annular gap were filled with melt. Use the piston speed as if the drag were absent.

REFERENCES

1. R. Longworth and E. T. Pieski, "Melt index as a measure of newtonian melt viscosity," *J. Polym. Sci. Part B Polym. Lett.*, **3**, 221–226 (1972).

2. W. Gleissle, "Influence of the measuring apparatus and method on the value of melt index," *Rheology*, **94**, March 1994, pp. 133–22.

3. ASTM International, D1238-10.

4. G. H. Brother and L. L. McKinney, "Protein plastics from soybean products," *Ind. Engr. Chem.*, **32**, 1002–1006 (1940).

5. S. Tonogai and S. Seto, "Monohole flow test with high-frequency preheating for evaluating flowability of thermosetting compounds," *Polym. Eng. Sci.*, **21**, 301–306 (1981).

6. U. F. González, S. F. Shen and C. Cohen, "Rheological characterization of fast-reacting thermosets through spiral flow experiments," *Polym. Eng. Sci.*, **32**, 172–181 (1992).

7. E. Ehabé, F. Bonfils, C. Aymard, A. K. Akinlabi and J. Sainte Beuve, "Modelling of Mooney viscosity relaxation in natural rubber," *Polym. Test.*, **24**, 620–627 (2005).

8. An interesting story concerning W. S. Gosset and the Student t distribution is at
http://en.wikipedia.org/wiki/William_Sealy_Gosset

9. W. H. Press, B. P. Flannery, S. A. Teukolsky and W. T. Vetterling, *Numerical Recipes*, 2nd ed., Cambridge University Press, Cambridge, UK, 1992.

10. C. Wan, C. Lu, D. B. Todd, L. Zhu, V. Tan, M.-W. Young, C. G. Gogos, "Determining melt flow index on-line using a helical barrel rheometer," *SPE ANTEC Proc.*, **66**, 223–227 (2008). [For more details, see A. M. Kraynik, J. H. Aubert, R. N. Chapman, and D. C. Gyure, "The helical screw rheometer," *SPE ANTEC Proc.*, **48**, 403–406 (1984).]

11. M. M. Cross and A. Kaye, "Simple procedures for obtaining viscosity-shear rate data from a parallel disk viscosimeter," *Polymer* **28**, 435–440 (1987).

12. R. M. McGlamery and A. A. Harban, "Two instruments for measuring the low-shear viscosity of polymer melts," *Mater. Res. Std.*, **3**, 1003–1007 (1963). See also ASTM D2171-07e1.

13. P. J. Leider and R. B. Bird, "Squeezing flow between parallel disks. I. Theoretical analysis," *Ind. Eng. Chem. Fund.*, **13**, 336–341 (1974).

12

Flow of Modified Polymers and Polymers with Supermolecular Structure

Rarely are polymers used commercially in their pure or neat state. The list of additives used for modifying polymer properties is long and varied. General categories include stabilizers, process aids, lubricants, plasticizers, fillers, pigments, flame retardants, impact modifiers, and fibrous reinforcements. A somewhat newer category includes nanoparticles and nanoneedles. All of these have rheological consequences, some more than others.

In addition to additives, polymers often feature morphology that is a direct consequence of the polymer structure. The most common example is crystallinity brought about by structural regularity. A less common one is the presence of liquid-crystal phases that persist at temperatures above the crystalline melting point. Liquid-crystal polymers (LCPs) have very special rheological behavior.

A. POLYMERS FILLED WITH PARTICULATES

Polymer properties can be modified at low cost by adding solid fillers. Aside from pigmentation, solid fillers are most often added to increase the modulus and heat-distortion temperature[*] of the resin. As might be expected, the fillers also increase the melt viscosity of the resin. Is this a serious problem, or is the effect minimal? We shall explore this question below.

Fillers come in a variety of sizes and shapes, both of which have rheological consequences. But they all have one characteristic in common: they are solid. This means that the surrounding fluid cannot flow through the particles; instead, it must flow around the particle. Consequently, the viscosity increases. As the particle concentration increases, the particles become more closely spaced and the flow patterns become increasingly tortuous. Attractive forces may cause the particles to agglomerate, which changes there effective shape.

1. Influence of particle concentration on viscosity

As we have seen in Chapter 9, the baseline influence of widely spaced spherical particles on suspension viscosity is given analytically by the Stokes-Einstein[†] law, written as

$$\eta = \eta_m \left(1 + \frac{5}{2} \phi_p \right) \qquad (12\text{-}1)$$

where η_m is the viscosity of the suspending fluid and ϕ_p is the volume fraction of particles. Note that nothing is said about particle size, which is quite reasonable, because the fluid has no metric against which to gauge the size. Equation 12-1 is important as it sets a lower bound for suspension viscosity; there is nothing one can do to reduce the viscosity below this value (although some tricks will be described below).

In contrast, there are plenty of effects that lead to higher viscosities than those predicted by the Stokes-Einstein equation. The most important is particle interactions as the concentration of filler is increased. If the particles are hard spheres with no attractive forces, then the interactions are purely hydrodynamic. For this situation, and for moderate concentrations, the Batchelor equation[1] is often quoted; it is

$$\eta = \eta_m \left(1 + 2.5 \phi_p + 6.2 \phi_p^2 \right) \qquad (12\text{-}2)$$

---

[*] Heat distortion temperature (HDT) is an important viscoelastic property of the resin. In short, a bar of polymer is put into three-point bending with a fixed load. The bar is then heated until the deflection exceeds a specified amount. ASTM D648 gives the specifications.

[†] In addition to Einstein, and around the same time, W. Sutherland and M. Smoluchowski started with the Stokes equation and developed expressions relating diffusivity and viscosity.

This equation introduces a quadratic term, implying hydrodynamic interaction between particles; but, as stated above, the coefficient for this term is restricted to spherical, non agglomerating particles.

Table 12-1. Summary of popular equations for suspension viscosity [a]

Name	Formula [b]	Reference
Stokes-Einstein	$\eta = \eta_m(1 + 2.5\,\phi_p)$	
Mooney	$\eta = \eta_m \exp\left[2.5\,\phi_p/(1 - \phi_p/\phi_{max})\right]$ $\eta = \eta_m \exp\left[K_E\,\phi_p/(1 - \phi_p/\phi_{max})\right]$	M. Mooney, *J. Colloid Sci.*, **6**, 162–170 (1951).
Maron-Pierce	$\eta = \eta_m/(1 - \phi_p/\phi_{max})^2$	S. H. Maron and P. E. Pierce, *J. Colloid Sci.*, **11**, 80–95 (1956)
Roscoe	$\eta = \eta_m(1 + \phi_p)^{2.5}$ $\eta = 1/(1 - 1.35\,\phi_p)^{2.5}$	R Roscoe, *Br. J. Appl. Phys.*, **3**, 267–269 (1952).
Frankel-Acrivos [c]	$\eta = \dfrac{9}{8}\eta_m \dfrac{(\phi_p/\phi_{max})^{1/3}}{1-(\phi_p/\phi_{max})^{1/3}}$	N. A. Frankel and A. Acrivos, *Chem. Eng. Sci.*, **22**, 847–853 (1967).
Krieger-Dougherty	$\eta = \eta_m\left(1 - \phi_p/\phi_{max}\right)^{-K_E\phi_{max}}$	I. M. Krieger and T. J. Dougherty, *Trans. Soc. Rheol.*, **3**, 137–152 (1959).
Hatschek	$\eta = \eta_m/\left(1 - \phi_p^{1/3}\right)$	E. Hatschek, *Proc. R. Soc. A*, **163**, 330–334 (1937).

[a] Apply to Newtonian response of suspensions of non-interacting hard spheres.
[b] η_m = matrix viscosity; ϕ_p = particle volume fraction; ϕ_{max} = maximum packing fraction; K_E = Einstein coefficient. For a more general equation, replace the constant 2.5 with K_E, a parameter.
[c] Valid only for high concentrations.

Empirical and semi-theoretical relationships covering much of the concentration range abound. Examples are listed in Table 12-1. Note that several of these equations diverge at an upper limited $\phi_{max} < 1$ for the volume fraction of solid particles in the suspension. In practice, this bound can be a bit elusive, although it can be estimated from the concentration of particles in the sediment at the bottom of a well-settled suspension. For randomly packed, smooth spheres of uniform size, $\phi_{max} \approx 0.64$. In most applications of these equations, both K_E and ϕ_{max} are retrieved by fitting the viscosity data at intermediate concentrations. Used in this fashion, the equation can be used to describe the viscosity of suspensions of moderately non-spherical particles.[2] As with any set of similar empirical equations, the one that works the best[‡] will depend upon the materials chosen for the test.

[‡] The concept of "best" in this context can be a rather complex statistical issue. Criteria included measures of the distance of the points from the line (the residuals), the randomness of the residuals, etc.

An important question is how to squeeze more particles into a polymer matrix or to reduce the viscosity at a given concentration. The easiest solution is to change the particle size distribution such that small particles into spaces left by the large ones. Going from a strictly unimodal size distribution to a suspension of large and optimally sized smaller particles has the greatest impact on viscosity. Figure 12-1 displays the results of theoretical calculations using multimodal distributions of spherical particles and assuming a volume fraction of 0.66, which is close to the maximum packing fraction. The reduction of viscosity by using multimodal distributions is large and very important.

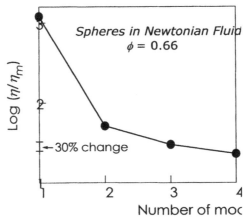

Figure 12-1. Illustration of the optimal reduction in viscosity by using a bimodal, trimodal and tetramodal distributions of spherical particles at a fixed volume fraction of 0.66. The sizes of the modes are optimized to fit into the spaces between the larger particles. The reduction is most dramatic on going from a unimodal to a bimodal distribution. (Data from Metzner.[2])

Currently, there is a great deal of interest in very small fillers for polymers. These are often referred to as nanoparticles because one or more dimensions of the particle is within the nanometer size range.[§] An example of the unique properties possible with of nanoparticle suspensions, consider that such particles, being much smaller than the wavelength of light, can modify the refractive index of a polymer without imparting opacity. As the contact of particle with polymer will be enhanced with decreasing size,[**] it is not

---

[§] Particles are considered as nanoparticles if at least one of its dimensions is less than 100 nm. At this size, the interaction of the polymer with the surface of the particle becomes very important.

[**] For spherical particles, a simple calculation shows that the specific surface area will scale as $1/R$, where R is the particle radius. If the particles have a density of 2 g/cm^3 and a 10-nm radius, the specific surface area will be 150 m^2/g without any porosity at all. Similarly, the interparticle spacing of the particles d will scale with radius at fixed volume fraction. For a

surprising to find that the rheological properties are influenced, sometimes adversely. To wit, it may be difficult to justify the enhanced properties of nano-composites if their processing is very difficult or impossible.

While the hydrodynamic theories of non-interacting, spherical particles do not predict a specific dependence on particles size, it is clear that the ratio of polymer radius of gyration and particle size must have an effect. The viscosity of nanoparticle suspensions with a Newtonian matrix has been investigated experimentally, but the interpretations are clouded by the fact that the nanoparticles tend to form strings[tt] of particles.[3] The strings increase the viscosity of the suspension markedly. If the particles are stabilized to prevent chaining, they then have repulsive interactions that change the hydrodynamics.

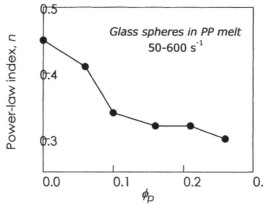

Figure 12-2. Data of Faulkner and Schmidt[4] showing a decrease of n, and the consequent increase in pseudoplasticity, with particle volume fraction ϕ_p for suspensions of glass spheres in polypropylene melt.

Repulsive interactions can be studied most easily using water suspensions, where the usual stabilization mechanism is electrostatic repulsion between like charges on the particles' surfaces. The repulsion can be changed by changing pH, which changes the surface charge on the particles, or by the addition of salt, which shields the charge. Surface charge can also be changed by the addition of ionic species that populate the surfaces of the particles.

---

volume fraction of 0.5 and a radius of 10 nm, the interparticle spacing is only ~10 nm (BCC lattice), which means that polymer molecules can easily bridge the particles. This implies that *every* polymer molecule will be influenced by the surface. In general, $d = 2\left(\sqrt[3]{\phi_{max}/\phi_p} - 1\right)R$.

[tt] Why strings instead of clusters? Once two particles form a pair, the symmetry changes from spherical to axial, and the attractive potential is modified likewise.

2. Nonlinear rheological properties of polymer-based suspensions

The nonlinear rheological properties of particle suspensions can be exceedingly complex, and cannot be covered in any detail here. Suffice it to say, the response depends upon all of the many structural variables, including size distribution, stabilization and concentration. Phenomena include shear thinning, shear thickening, time dependence, and dilatancy.[5]

Figure 12-2 shows the power-law exponent for a polymer melt filled with spherical glass beads, perhaps one of the simplest suspensions imaginable. Recall that the power-law exponent varies from 1 for a Newtonian fluid to 0 for a very shear-thinning material. Thus the data in Figure 12-2 suggest that adding particles increases the pseudoplastic nature of the suspension. Some possible reasons for this are the presence of higher-than-average shear rates between the particles, and the tendency of particles to organize in layers during steady shearing. One needs to be careful, however.

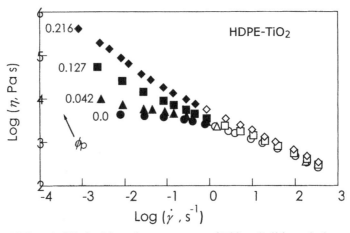

Figure 12-3. HDPE melt filled with various amounts of TiO$_2$. Solid symbols are data gathered using a cone-and-plate rheometer; open, using capillary. The upturn in viscosity at low rates is taken as an indicator of the presence of a yield stress. (Adapted with permission of John Wiley & Sons, Inc. from Minagawa and White.[6])

A common observation associated with particle-filled polymer solutions and melts is the development of yield at rather modest volume fractions of filler. Figure 12-3 gives an example; the sample is a high-density polyethylene melt filled with titanium dioxide. A number of GNF-type equations have been used to describe the yield behavior; a few are summarized in Table 12-2.

Table 12-2. A selection of GNF-type equations for systems with a yield. [a]

Name	Equation
Bingham	$\eta = \dfrac{\sigma_Y}{\dot{\gamma}} + \eta_\infty$
Generalized Bingham [b]	$\eta = \dfrac{\sigma_Y}{\dot{\gamma}} + \eta_{GNF}(\dot{\gamma})$
Casson	$\eta = \left[\eta_\infty^\alpha + \left(\dfrac{\sigma_Y}{\dot{\gamma}} \right)^\alpha \right]^{1/\alpha}$, $\alpha = 1//2$
Generalized Casson	$\eta = \eta_{GNF}(\dot{\gamma}) \left[1 + \left(\dfrac{\dot{\gamma}_Y}{\dot{\gamma}} \right)^\alpha \right]^{1/\alpha}$
Herschel-Bulkley [c]	$\eta = \dfrac{\sigma_Y}{\dot{\gamma}} + m \dot{\gamma}^{n-1}$
Papanastasiou [d]	$\eta = \eta_\infty + \dfrac{\sigma_Y(1 - e^{-\dot{\gamma}/\dot{\gamma}_c})}{\dot{\gamma}}$

[a] σ_Y is the yield stress and η_∞ is a viscosity parameter.
[b] η_{GNF} stands for any viscous GNF model. Note that the Herschel-Bulkley model is a special case using the power-law model. Usually this is adequate.
[c] Spellings are correct.
[d] The exponential "softens" the yield as the shear rate is decreased. By replacing η_∞ with η_{GNF}, the equation can be generalized to cover non-Newtonian behavior at high rates.

3. Particle anisotropy

As mentioned above, particles that form string-like clusters will give suspensions of high viscosity, for a given volume fraction of solid. This phenomenon has been known for years by the rubber industry; their "high-structure" carbon black products comprise strings of carbon nanoparticles. Similarly, silica nanoparticles are known to string.

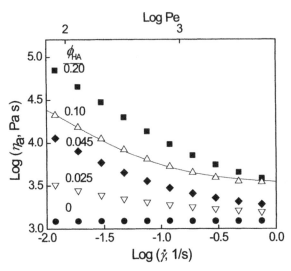

Figure 12-4. Polycaprolactone (PCL) melt filled with various amounts of hydroxy apatite nanoneedles. Tests were run at 120 °C (the melting point of PCL is about 60 °C). The line is a fit to the $\phi_p = 0.1$ data with the Papanastasiou model, Table 12-2. **Pe** in the upper scale stands for the Peclet number, which is the ratio of shear rate and Brownian rotational rate of the needles. A large value, such as on this graph, indicates that Brownian motion is too slow to affect the orientation of the nanoneedles. (Reproduced from Sun et al.[7] with permission of Springer-Verlag, © 2011..)

Why is this? First of all, the large specific values of interfacial energy for small-particle suspensions almost guarantee that the particles will agglomerate, absent a strong, long-range repulsive force. The reason for the string, rather than a clump, is connected with the particularly strong axially directed attraction of the end particle in a string for unattached particles nearby.

The rheological effects of long particles in polymers can be profound. Viscosity in the Newtonian regime for the matrix will be far higher than that predicted by equations for spherical particles and the suspension will no longer be Newtonian. The mechanisms for energy storage expand to include bending of particles, leading to large normal stresses. In fact, Cogswell[8] reported that an extrudate with chopped glass could expand enough on exiting the capillary to form a foam! The proposed mechanism was recovery of the fibers from their bent shape in the capillary.

4. Stabilizing particles in polymer suspension

Can agglomeration be prevented? Not easily, at least without modifying the properties of the matrix as well. Here's why. First of all, the particle surface area is large, and the interfacial energy is also large. To reduce the interfacial energy between the particles, the usual route is to add surfactants, creating an "inverse" emulsion of a polar particle in a continuous organic matrix—the polymer. Surfactants can help, but they must have low solubility in the organic matrix, and have hydrophilic heads that are of low charge; otherwise, unfavorable electrostatic interaction is likely. Nonionic surfactants are likely to be too soluble in the matrix. A workable compromise is a quaternary ammonium (cationic) surfactant substituted for the cations on a negatively particle surface (Figure 12-5). Most often, two of the four alkyl groups are short, and two are more bulky.

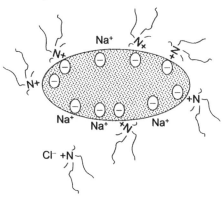

Figure 12-5. Reduction of interfacial energy by attaching a cationic surfactant to a negatively charged particle surface in place of the original inorganic cations. The particle acquires a more hydrocarbon-like character due to the alkyl groups on the quaternary ammonium cations that have displaced, e.g., sodium cations (Na^+) from the particle's surface.

An alternative to ion substitution is to graft alkyl-substituted silanes onto a neutral particle surface. With some particles, e.g., glass or silica, this works well; on others, not so well. An important benefit of silane treatment is a water-resistant interface.   Otherwise, the mechanical properties of the composite in moist applications can deteriorate significantly.

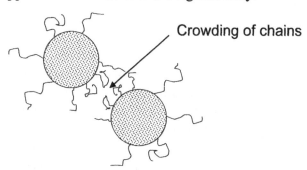

Crowding of chains

Figure 12-6. Schematic showing crowding of polymer chains on particle surfaces, leading to protection against particle agglomeration.

Particles can also be stabilized entropically by attaching polymer groups the surface (Figure 12-6). Even gold nanoparticles can be stabilized in this fashion.

The idea is to have enough chains on the surface so, when two particles approach, the degrees of freedom of the interacting chains are reduced, pushing the particles apart. This works best if the chains are expanded when the particles are isolated. In turn, this implies that the chains on the surface are miscible with the matrix, a condition that may be quite difficult to achieve. Otherwise, the suspension will act like an incompatible polymer blend; the particles will lump together.

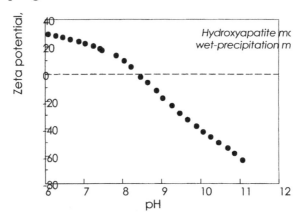

Figure 12-7. Typical determination of zeta potential for particles in a suspension. The particular results shown are for apatite nanoparticles in a water-glycerin mixture. (From Kothapalli et al. [9] Reproduced by permission of The Royal Society of Chemistry.)

Clearly, stabilization will work best if the surface modifiers are compatible with the matrix. The unfortunate consequence of this with nanoparticles is the plasticization of the polymer matrix around the particle. The effected volume fraction can become considerable. A practical consideration is the cost of treating the surface.

5. Electrostatic stabilization

Electrostatic stabilization refers to the stabilization of small particles in water or other very polar fluids due to repulsion of like charges on their surfaces. The charged surfaced may either be an inherent aspect of the chemical structure of the particles, or may introduced or enhanced by coating the particle surface with a polyelectrolyte. Neutralized poly(acrylic acid) is a favorite; if it sticks to the particle surface and is exposed to a high-pH solution, the surface can become highly anionic. Of course, the negative charges on the carboxylate groups are balanced by an equal number of cations; but, in water, these cations will be wandering around, mixing with anions in the water. The result is a net negative particle charge. The particle is, in effect, a large anion with a high charge.

Charge is characterized in terms of the zeta potential, which can be measured by examining the motion of the particle in an electric field (electrophoresis). Modern instruments do this automatically, and can be set up to change the pH of the medium. Figure 12-7 shows typical output. The pH at which the zeta potential is zero is termed the *isoelectric point*.

B. LIQUID CRYSTALLINITY AND RHEOLOGY

As nanoparticles become smaller, they begin to act more like molecules than particles, while retaining their rigidity. Particle diffusivity increases and particle-particle interactions become comparable to the entropic terms on a per mole basis. Polymers with such particles might be expected to behave like solutions, with the possibility of phases of different concentrations and structures.

The closest analogy to rigid-nanoparticle suspensions are the solutions of rigid-rod or plate-like molecules, which form liquid-crystal phases. A liquid crystal is simply a liquid that has a stable ordering of the molecules as a result, mainly, of their anisotropy. The liquid-crystal phase may be a single component or a mixture. The former are referred to as thermotropic liquid crystals; the latter as lyotropic. Interestingly, there are relatively few systems showing both behaviors.

Rod-like polymer molecules also form liquid-crystal solutions that can have very complex rheological properties. One widely studied system comprises poly(benzyl glutamate) dissolved in one of a number of suitable solvents in spite of its rigid helical structure. (The bulky benzyl group imparts high solubility in many common polymer solvents.) But perhaps the most widely exploited lyotropic liquid-crystal polymer (LCP) is Kevlar®, which is the aromatic polyamide based on 1,4-phenylene diamine and terephthalic acid (actually, the acid chloride). Kevlar® requires a highly protic solvent, such as sulfuric acid, to break up the regular intermolecular hydrogen bonding between adjacent chains.[‡‡]

Thermotropic LCP's are quite abundant. As a rule, they tend to be aromatic polyesters with melting points that are lower than the corresponding amides due to the weaker interactions of ester groups relative to amides. Introduction of

---

[‡‡] A similar aramide, Twaron®, originally developed by AKZO, is now marketed by the Teijin Group. A clever aspect of the process is the method of dissolving the aramide in sulfuric acid, wherein the sulfuric acid is frozen and powdered, then mixed with powdered polymer and warmed. (Something to keep in mind for other difficult-to-dissolve polymers.)

flexible links or "spacers" (usually methylene sequences) can be used to lower the melting point even further.

Perhaps the most widely used of the thermotropic LCPs is Vectra®, a product of Ticona. Vectra® is a polyester based on hydroquinone and a mixture of parahydroxybenzoic acid (HBA) with hydroxynapthoic acid (HNA). The HNA has its two carboxyl groups in the 2 and 6 positions. The ratio of these two can be varied to give different properties. A common variant, Vectra A950®, has a ratio of 73/27 HBA/HNA and exhibits a melting point at around 250 °C.[§§]

A thermotropic LCP softens to a workable fluid at a temperature where the solid crystalline phase converts the liquid-crystal form. Often, a nematic phase forms, wherein the stiff chains line up in the same direction. Semetic phases, in which stiff portions of the LCP chains line up to form semetic sheets, are also observed The practical impact of this from a rheological point of view is a very low melt viscosity due to the ease with which the rods slide over each other and the absence of entanglements. Compared with random-coil polymers at their respective processing temperatures, the LCP may have a viscosity that is five to ten times lower, a huge advantage in molding thin objects. If the same polymer is brought to its random state, which can be done with some structures, the viscosity usually increases. This behavior is shown in Figure 12-8.

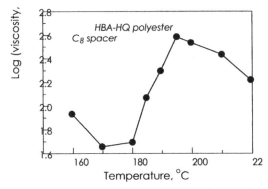

Figure 12-8. Influence of temperature on viscosity through the clearing temperature for a polyester with flexible spacers. Note the viscosity increase. (Reproduced with permission of J-H. Lee.[10], © 2004)

The thermodynamic transition from a liquid crystal to an isotropic liquid is known as the clearing point, as the milky appearance of the nematic phase disappears. This temperature can be inordinately high, often higher than the

---

[§§] The polyester $(A)_x(B)_{1-x}$ copolymers have a distribution of A and B sequences. In HNA/HBA, the HBA sequences can lead to microphase separation of hard-to-melt crystals of HBA.

polymer's decomposition temperature. Why is this? Superficially, one might expect the liquid crystal to be a very delicate structure with little holding it together. Indeed, the enthalpy change associated with the melting of the liquid crystal is very low. The thermodynamic explanation is very straight forward. At the melting temperature, the liquid crystal and isotropic fluid are in equilibrium and thus $\Delta G = 0$, where ΔG is the free energy difference between the two phases. This implies that

$$T_m = \frac{\Delta H}{\Delta S} \tag{12-3}$$

since $\Delta G = \Delta H - T\Delta S$. Because ΔH is small, ΔS must be really small to give a high clearing temperature. In fact, absent expansion and for rigid rods, ΔS will be nearly zero. Why? Because the rods gain only one rotational degree of freedom in going from the nematic state to the isotropic liquid state.

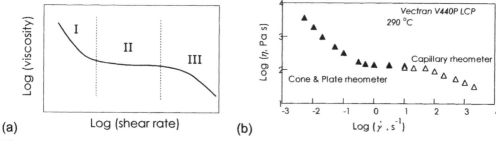

(a)   (b)

Figure 12-9. (a) Schematic showing the three zones of flow behavior. Not all three zones are present (or observable) for all LCPs, but are present in (b), which shows data for a commercial thermotropic LCP. (Adapted with permission of John Wiley & Sons, Inc. from Guo et al. [11])

When examined as a function of shear rate, the typical LCP has a more complex behavior than the typical random-coil polymer. In general, there may be up to three flow zones, which have been numbered I, II and III.[11] A schematic of this behavior is shown in Figure 12-9.

A highly recommended summary of the complexities of LCP rheology has been written by Wissbrun.[12] Needless to say, the many, many complexities of LCP flow are too involved to be covered here.

C. POLYMERS WITH MICROPHASE SEPARATION IN MELTS OR SOLUTIONS

One expects microphase separation to be associated mainly with polymer structures that have variation in constitution or configuration. It's really no surprise that flexible polyurethanes with hard aromatic segments, block copolymers and even random copolymers exhibit microphase separation, and the corresponding changes in rheology.

What is really surprising is the rheological behavior of solutions of polymers that have a uniform structure with random configuration and no evidence of crystallinity.

Discussed below are some of the phenomena resulting from physical associations between the polymer chains.

1. Hydrogen bonding

Hydrogen bonds are of intermediate strength, usually around 30 kJ/mol for bonds typical of the structures found in polymers. At 200 °C, RT is about 4 kJ/mol. Thus we see that, while weak compared to a covalent bond, hydrogen bonds can easily persist in the melt of many polymers.

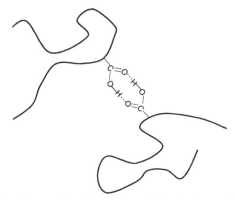

Figure 12-10. Schematic of a cyclic dimer formed by hydrogen bonding of two carboxyl groups covalently bonded to the polymer chains

A classical example is that studied a number of years ago by Blyler and Haas[13] at Bell Labs in New Jersey. They were interested in ethylene copolymers, including poly(ethylene-co-acrylic acid) (EAA) and its esters. Carboxylic acids have the ability to form ring-like structures involving two H-bonds (Figure 12-10) that persist to quite high temperatures. The fortuitous feature of this dimer is its IR spectrum, which has a distinct mode well removed from other absorption bands. Thus, the spectral changes with temperature provide an independent evaluation of the concentration of dimer. The observations in the study indicated that intermolecular hydrogen bonding is present and effective in the melt. The IR results suggested that the carboxyl groups were not decarboxylating or reacting to form anhydride linkages.

2. Physical crosslinking

Rheological phenomena involving even weaker secondary forces are quite common. The most striking is the gelation of polymer solutions of linear polymers that have low polarity. A classical example is the system comprising polystyrene in carbon disulfide.[14] Carbon disulfide is a linear molecule like carbon dioxide, and thus has no dipole moment. Polystyrene has a small segmental dipole moment and the possibility of interaction between the phenyl groups, but these weak interactions seem unlikely to lead to an abrupt rheological change.

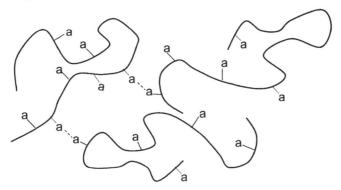

Figure 12-11. Schematic of sticker decorated polymer chains, showing some stuck but many free. The chain on the left has two stickers and thus can be mechanically active in a gel.

Some more details are in order. The gels are normally formed by slow cooling solutions of various concentrations. The dependence of the gel temperature on concentration usually is well-described by the van 't Hoff equation

$$c_{gel} = \alpha e^{-\beta / T_{gel}} \tag{12-4}$$

where c_{gel} and T_{gel} are the concentration and temperature at the gel point, and α and β are parameters. (A derivation of this equation is shown in Appendix 12-1.) It is important to point out that even if the sticker groups don't stick very well, the polymer molecules will still be fastened together if they contain a high concentration of stickers. Also, due to the dynamic equilibrium, the stickers don't stay stuck, but others form in their place. Thus, on a long polymer chain, some stickers are always fastened.

3. Microphase separation

Microphase separation refers to the separation of similarly structured parts of the polymer molecules into distinct domains, usually less than 0.1 μm in size. The separation is, of course, thermodynamically driven and must involve a significant negative enthalpy to overcome the entropically unfavorable loss of motional freedom of the polymer chains. Sources of the favorable interaction enthalpy include crystallization and reduction of the antagonistic interactions of the separating segments with the rest of the molecule. Being enthalpic, we fully expect that the segregation will depend significantly on temperature. Rheologically, the consequences are highly significant.

Block copolymers formed by living polymer techniques are a very special case in that the segregating segments are uniform in size. Likewise, in most cases, the matrix segments are also uniform in size. The result is an array of structured microphase-separated materials. Rheologically, microphase-separated AB block copolymers can flow, while ABA varieties cannot, even above the softening point of both blocks. ABA materials must, in addition, be brought above a critical temperature at which the segments dissolve in one another. This temperature is referred to as the *order-disorder* temperature, T_{OD}. $(AB)_n$ block copolymers can be very difficult to process, although some segmented polyurethanes, which belong to the $(AB)_n$ class, can flow at a sufficiently high temperature. Figure 12-12 shows the temperature dependence of a common ABA block copolymer, where the A's are polystyrene blocks, and B is a long polybutadiene block. This composition is referred to as SBS.

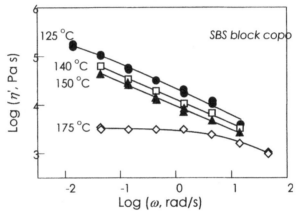

Figure 12-12. Rheological response of a SBS block copolymer as the temperature is brought through the order-disorder temperature somewhere between 150 and 175 °C. The curves are Cross-model fits. (Adapted with permission of John Wiley & Sons, Inc. from Chung and Gale.[15])

While one might expect AB block copolymers to be rheologically simple, their behavior is anything but, especially as a solution in a solvent that is very good for one block and relatively poor for the other. An example is a block copolymer made from styrene and butadiene, followed by hydrogenation of the butadiene block. The result of the hydrogenation is essentially a copolymer of ethylene and butylene, with the ratio of the two depending on the 1-2 vs. 1-4 content of the polybutadiene segment. This copolymer can be dissolved in, say, squalane,[***] which is highly specific to ethylene-butylene (EB) segment, even at high temperatures. As a result, the polystyrene blocks microphase separate, forming micelles with a highly swollen EB exterior called a corona. The important aspect of these micelles is that they all the same size. In some sense they are similar to a polymer-protected nanoparticle, but the covalent attachment to the polystyrene block guarantees that there will be no agglomeration.

At sufficiently high concentration, the AB block-copolymer micelles jam, either in a soft glassy state, or in a crystal. Both gel the solution. These gels do not depend at all on attraction, but form because of repulsive interactions, much like grid lock in a city crowded with cars. The gels are thus known as *repulsion gels*.[†††]

At the order-disorder temperature, the micelles dissolve and the solution becomes more like a normal polymer solution. However, all is not well; the polystyrene segments still tend to agglomerate. For many weak solvents, one also might expect at higher temperatures another microphase separation due to the LCST for the mixture of the solvent and the high-solubility block.

D. COVALENT CROSSLINKING OF POLYMERS

One might expect that the complications of weak interactions in polymer solutions would all disappear if the interactions were strong covalent bonds. To some extent this is true, but other complexities crop up.

There are several ways to form a crosslinked network, starting with a fluid. If the fluid comprises a polymer melt or solution, then crosslinks must be introduced in some fashion, either with a chemical curative, reaction of pre-existing groups on the chain, or by using radiation. Examples abound. If the polymer starts out as an oligomer or monomer, then polymerization and concomitant crosslinking reactions are used to form a gel. Examples include polyurethanes, silicones, epoxies, thermoset polyesters, bismaleimides and polyimides.

The important aspects rheologically are the transition from fluid to a critical gel and finally to a solid. Rheology can be used to follow the rate of cure, but this is most often done using linear viscoelastic measurements. Many attempts have been made to connect the chemistry, including reaction rates, to the

---

[***] Squalane is a hydrogenated isoprene oligomer, 2,6,10,15,19,23-hexamethyltetracosane. It is a convenient rheology solvent because of low volatility and toxicity (it's used in cosmetics).

[†††] Attraction gels with micelles and particles are, of course, possible.

rheology.   Most are terms of the reaction extent, and its relationship to the dynamic shear modulus.

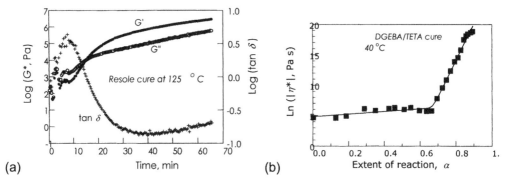

Figure 12-13. (a) Curing profiles for a phenolics (resole) resin. (Reproduced with permission of the Society of Plastics Engineers from Rose et al.[16]) (b) Relationship between the reaction $|\eta^*|$ and conversion for a curing epoxy resin.   Note the distinct transition from one scaling exponent to second much higher value around a conversion of 0.70.   (Reproduced with permission of the Society of Plastics Engineers from Shaw.[17])

Figure 12-13 (b) shows a typical result using epoxy for the relationship between the cure time and the magnitude of the complex dynamic viscosity. If conversion can be measured (DSC or IR spectroscopy), a similar scaling results that is fairly independent of cure temperature. However, if the cure temperature is too low, the polymer may not cure at all. This was studied very extensively by Gillham and coworkers,[18] who adapted what is called the time-temperature-transformation (Ttt) diagram to describe the behavior of crosslinking systems. An illustration of this is shown in Figure 12-14.

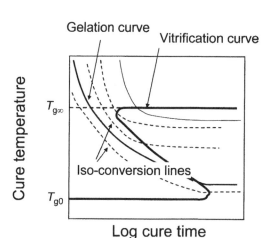

Figure 12-14. Ttt diagram for cure of epoxy resin. The way to use this drawing is to imagine an experiment at a fixed temperature. As time goes on, one passes from a fluid to a gel if the temperature is above the intersection of the gelation and vitrification curves. Below this temperature, the fluid will form a glass before curing, and may never cure. At temperatures above $T_{g\infty}$, the material cures to a very high extent because it is above the glass transition of the fully cured gel. (Adapted with permission of John Wiley & Sons, Inc. from Simon and Gillham.[19])

APPENDIX 12-1: VAN 'T HOFF EQUATION APPLIED TO GELATION

The picture of sticker-decorated polymer chains is shown in Figure 12-11. In this figure, the a's represent stickers, i.e., groups that can stick together, but not permanently. We assume that the a's are uniformly distributed, and all a–a (stuck groups) are intermolecular. Other assumptions are that the a's and the a–a's are in equilibrium with each other, i.e.,

$$a\text{–}a \leftrightarrow a + a \tag{a}$$

For this equilibrium, we define an equilibrium constant K given by

$$K \equiv \frac{[a][a]}{[a-a]} \tag{b}$$

where the square brackets designates molar concentration in, say, mol/L. A molar balance on the stickers requires that

$$[a] = [a]_0 - 2[a\text{–}a] \tag{c}$$

where $[a]_0$ is the overall concentration of the stickers. To express the sticker concentration in terms of polymer concentration, we have the relationship $[a]_0$

$= c/M_s$, where c is the mass concentration of polymer in g/L, and M_s is the molecular weight of a segment of polymer that includes one sticker. M_s has units of g/mol of stickers.

At the gel temperature T_{gel}, each polymer chain is connected to two of the a–a dimers, each of which is shared by another chain. Thus, the number of dimers in a gelled chain at the critical point is one. The molar concentration of polymer chains c/M, where M is the polymer molecular weight. Thus [a–a] $= c/M$ at the gel point. With these relationships in mind, equation (b) becomes

$$K = \frac{(c/M_s - 2c/M)^2}{c/M} = c\left[\frac{(1/M_s - 2/M)^2}{1/M}\right] \tag{d}$$

The van 't Hoff relationship describes the dependence of K on temperature T. It is

$$\frac{d\ln K}{d(1/T)} = -\frac{\Delta H}{R} \tag{e}$$

where ΔH is the enthalpy change associated with the dimerization. On integration, this becomes the familiar Boltzmann distribution

$$K = \alpha' e^{-\frac{\Delta H}{RT}} \tag{f}$$

Substituting for K using equation (d) gives the result in equation (12-4) at the gel condition, c_{gel} and T_{gel}. The multiplicative constant α is a combination of α' and the bracketed term in equation (d). (See Problem 12-9.)

PROBLEMS

12-1. Compare the equation $(1 + \phi)^{2.5}$ for the relative viscosity of a suspension to the Batchelor expression, equation (12-2). One approach is to expand the former using the binomial expansion. Do the first- and second-order terms agree?

12-2. Show that the following equations (Table 12-1) agree or don't agree with the Stokes-Einstein equation at low volume fraction:

(a) Frankel-Acrivos

(b) Maron-Pierce

(c) Mooney

(d) Roscoe

(e) Hatschek

12-3. The dimensionless parameter K_E appearing in, for example, the Krieger-Dougherty equation (Table 12-1), has been referred to as the "intrinsic viscosity." Compare this with the limiting value of intrinsic viscosity commonly used by polymer scientists.

12-4. The data below were gathered by Kunitz[20] in 1926 for a suspension of sulfur in a salt solution.

(Concentration, g/cm^3) × 100	Relative viscosity, η/η_m
1.28	1.035
3.84	1.090
7.68	1.230
15.36	1.510
24.14	1.975
30.72	2.450
48.28	5.000

For this data (as well as polymer solutions), Kunitz proposed the empirical equation

$$\eta = \eta_m \frac{1 + 0.5\phi}{(1 - \phi)^4}$$

which has only one parameter, namely, η_m.

(a) Check the agreement of this expression with the Stokes-Einstein equation at low concentrations.

(b) As Kunitz had no access to a computer (not invented yet), he could not test the fit of his equation to the data in a direct fashion. Instead, he solved for values of ϕ and consequently, the density of sulfur. The densities thus calculated were nearly constant at 1.69 g/cm^3. Check this value against those listed in standard references. If the value of Kunitz does not agree, is the discrepancy in the expected direction? Explain.

(c) Using the Maron-Pierce equation (Table 12-1) and the literature value for sulfur density, check the value of the relative viscosity at zero concentration of sulfur using the data of Kunitz. Is the extrapolated value of relative viscosity distinguishable from 1?

(d) (Computer) Using the sulfur density of 1.69 g/cm^3 found by Kunitz, examine the quality of fit of Kunitz's equation versus the Maron-Pierce equation.

12-5. Data from Kitano et al.[21] indicates that suspensions of roughly spherical particles can have maximum volume fractions ranging from 0.64 for glass spheres, to as low as 0.44 for precipitated calcium carbonate. The $\phi_{max} = 0.64$ is consistent with the expected maximum packing fraction for randomly placed spheres, but the 0.44 is not.

(a) What is a possible explanation for the observed behavior?

(b) If the volume "occupied" by an irregular particle of size L is no more than that of a sphere with volume $\pi L^3/6$ (the volume of sphere of diameter L), what would be the aspect ratio of rods of calcium carbonate needed to explain the $\phi_{max} = 0.44$? (Hint: find the ratio of the actual volume fraction to that using $\pi L^3/6$ in terms of the aspect ratio λ.)

12-6. Exfoliated clay nanoparticles are disk-like, with length L and thickness δ. Using the effective volume of $\pi L^3/6$ explained in Problem 12-5, what are the expected Einstein coefficient and ϕ_{max} for a suspension of these particles if the particles are 1 μm wide and 1 nm thick?

Log (ω, rad/s)	Log ($|\eta^*|$, Pa s) for mass fractions of	
	0	0.075
−1.000	3.625	4.089
−0.797	3.618	4.051
−0.602	3.603	3.982
−0.395	3.580	3.914
−0.200	3.549	3.838
0.003	3.511	3.770
0.198	3.466	3.701
0.405	3.413	3.633
0.599	3.352	3.565
0.802	3.276	3.481
0.997	3.208	3.405
1.204	3.116	3.329
1.399	3.025	3.238
1.602	2.919	3.139
1.797	2.805	3.041
1.996	2.684	2.919

12-7. (Challenging) The data above are for the complex viscosity magnitude of PS and 7.5 wt% exfoliated nanoclay in the same polymer.[22] The compounding was done with a miniature twin-screw extruder. The viscosities were measured at 200 °C.

(a) Using this data and a suitable GNF model, estimate the zero-shear-rate viscosities η_0 of the two melts and calculate the relative viscosity of the nanoclay-filled material.

(b) (Open end) Compare the relative viscosity from (a) with applicable theory or approximations, and attempt to explain any discrepancies. Assume the clay is similar to that described in Problem 12-6.

12-8. Moldenaers et al.[23] proposed an equation, shown below, as part of an explanation of curvature in the Bagley plot for the thermotropic LCP Vectra A950®. Their equation is

$$P = P_e e^{\beta P} - \frac{1}{\beta} \ln\left(1 - 4\beta\tau\frac{L}{D}\right) \qquad \text{(a)}$$

where P is the total pressure drop for a given L/D capillary; P_e and τ are the entrance pressure drop and shear stress at atmospheric pressure, respectively; and β is the pressure coefficient of viscosity, i.e., $d\ln\eta/dp$. A typical value of β might be 5×10^{-9} Pa^{-1}, but Moldenaers et al. found values as high as 5×10^{-7} Pa^{-1}.

(a) Explore the limit of equation (a) as β approaches zero.

(b) (Computer) Find values for β and P_e using the data shown below:

L/D	P, MPa
4.8	1.03
10	1.43
20	3.01
30	4.63
50	8.51

Data from Moldenaers et al.[23]

12-9. (Open end) Explore the literature to find if rigid aromatic polyurethanes can form liquid-crystal phases. What kinds of phases have been identified? What rheological properties do they display? What disadvantages have been reported?

12-10. Using the derivation found in Appendix 12-1, describe the influence of molecular weight on the constants α and β in equation (12-4).

12-11. (Computer) The data in the table below are gel points for a 220-kDa syndiotactic polystyrene dissolved in chloroform. To gather this data, solutions of known concentration were prepared and dissolved thoroughly at an elevated temperature. The solutions were then cooled slowly, and examined. When the gel point was reached, the temperature was gradually increased until the gel disappeared. The average temperature was then recorded as T_{gel}. Using this data, find the values of α and β in the van 't Hoff expression, equation (12-4). Comment on the fit or lack of fit.

c, g/dL	T_{gel}, °C
2.0	15
2.9	23
5.0	27
7.0	36
10.0	39.5
12.5	45
13.9	47
15.8	52.5
18.0	54
19.8	64
21.7	69
23.2	69
25.0	64

Unpublished data from Dr. Yuxian An.

12-12. (Open end) Some time ago,[24] it was proposed that the eccentric-disk geometry (a.k.a. Maxwell orthogonal rheometer) was the appropriate method for the rheological examination of gelation in polymer melts and solutions. List the advantages and disadvantages of this geometry relative to traditional dynamic analysis using the cone and plate.

12-13. An important question for many structured polymeric fluids is the behavior at very low stresses. Specifically, for materials showing evidence of a yield at low stresses, can one distinguish the observed behavior from a material with an upper Newtonian plateau? In terms of equations, we might compare a Bingham model

$$\eta = \eta_\infty \left(1 + \frac{1}{\dot{\gamma}/\dot{\gamma}_c} \right) \tag{a}$$

with a similar GNF model with an upper Newtonian behavior, e.g.,

$$\eta = \eta_\infty + (\eta_0 - \eta_\infty)/(1 + \dot{\gamma}/\dot{\gamma}_0) \tag{b}$$

(a) Plot both of these models using log-log scales assuming $\eta_0 = 30\eta_\infty$, and $\dot{\gamma}_0$ and $\dot{\gamma}_c$ are 0.1 and 1.5 respectively. Comment on the shapes.

(b) Based on these models (either of which, of course, may not describe a real material), what shear rates need be reached to arrive at a conclusion concerning the presence of a Newtonian plateau? Can such rates be reached easily on typical rheometers?

(c) Attempt to fit each of the above models to the data give below, which is for a PCL melt at 120 °C containing hydroxyapatite (HA) needles. Compare the sum of the squares of the residuals (SSE) for each fit.

Log ($\dot{\gamma}$, s^{-1})	Log (η, Pa s)
-1.93	4.05
-1.73	3.90
-1.53	3.77
-1.33	3.65
-1.13	3.55
-0.93	3.48
-0.73	3.41
-0.53	3.36
-0.33	3.32
-0.13	3.29

Data from S-P. Sun, PhD Thesis, Univ. of Connecticut, 2010.

REFERENCES

1. G. K. Batchelor, "Effect of Brownian-motion on bulk stress in a suspension of spherical particles," *J. Fluid Mech.*, **83**, 97–117 (1977).

2. For example, A. B. Metzner, "Rheology of suspensions in polymeric liquids," *J. Rheol.*, **29**, 739–775 (1985).

3. C. Y. Li, "Nanoparticle assembly: Anisotropy unnecessary," *Nature Mater.*, **8**, 249–250 (2009).

4. D. L. Faulkner and L. R. Schmidt, "Glass bead-filled polypropylene. Part I: Rheological and mechanical properties," *Polym. Eng. Sci.*, **7**, 657–665, (1977).

5. A. V. Shenoy, *Rheology of Filled Polymer Systems*, Kluwer, Dordrecht, The Netherlands, 1999.

6. N. Minagawa and J. L. White, "Coextrusion of unfilled and titanium dioxide-filled polyethylene. Influence of viscosity and die cross-section on interface shape," *Polym. Eng. Sci.*, **15**, 825–830 (1975).

7. S.-P. Sun, J. R. Olson, M. Wei and M. T. Shaw, "Rheology of Hydroxyapatite Needles Suspended in an Organic Fluid," *Rheol. Acta*, **50**(1), 65–74 (2011).

8. E. A. Cole, F. N. Cogswell, J. Huxtable and S. Turner, "Fiber-foam: A rheological phenomenon and a novel product," *Polym. Eng. Sci.*, **19**, 12–17 (1979).

9. C. Kothapalli, M. Wei and M. T. Shaw, "Solvent-specific gel-like transition via complexation of polyelectrolyte and charged nanoparticles in binary fluid mixtures: A rheological study," *Soft Matter.*, **4**, 600–605 (2008).

10. J.-H. Lee, "Development of an Electrospinning Device for Thermotropic Liquid Crystalline Polymer (TLCP) Melts," MS Thesis, University of Connecticut, 2004.

11. T. Guo, G. M. Harrison and A. A. Ogale, "Rheological behavior of thermotropic liquid crystalline copolyester Vectra A950," *SPE ANTEC Proc.* **59**, Paper #0719 (2001).

12. K. F. Wissbrun, "Rheology of rod-like polymers in the liquid crystalline state," *J. Rheol.*, **25**, 619–662 (1981).

13. L. L. Blyler, Jr. and T. W. Haas, "The influence of intermolecular hydrogen bonding on the flow behavior of polymer melts," *J. Appl. Polym. Sci.*, **13**, 2721–2733 (1969).

14. Y. S. Gan, J. Francois, J. M. Guenet, B. Gauthier-Manuel, C. Allain, "A direct demonstration of the occurrence of physical gelation in atactic polystyrene solutions," *Makromol. Chem., Rapid Comm.*, **6**, 225–230 (1985).

15. C. I. Chung and J. C. Gale, "Newtonian behavior of a styrene-butadiene-styrene block copolymer," *J. Polym. Sci., Part B: Polym. Phys.*, **14**, 1149–1156 (1976).

16. J. Rose, R. Osbaldiston, W. Smith, S. Farquharson and M. T. Shaw, "In-situ monitoring of a polymer cure using dynamic rheometry and Raman spectroscopy," *SPE ANTEC Proc.*, **44**, 939–945 (1998).

17. M. T. Shaw, "Rheology as a tool for the polymer scientist," *SPE ANTEC Proc.* , **47**, 1906-1908 (2001).

18. J. B. Enns and J. K. Gillham, "The time-temperature-transformation (TTT) cure diagram: Modeling the cure behavior of thermosets," *J. Appl. Polym. Sci.*, **28**, 2567–2591 (1983).

19. S. L. Simon and J. K. Gillham, "Thermosetting cure diagrams: Calculation and application," *J. Appl. Polym. Sci.*, **53**(6), 709–737 (1994).

20. M. Kunitz, "An empirical formula for the relation between viscosity of solution and volume of solute," *J. Gen. Physiol.*, **9**, 715–725 (1926).

21. T. Kitano, T. Kataoka and Y. Nagatsuka, "Shear flow rheological properties of vinylon- and glass-fiber reinforced polyethylene melts," *Rheol. Acta*, **23**, 20–30 (1984).

22. X. Han, C. Zeng, K. W. Koelling, D. L. Tomasko, and L. J. Lee, "Extrusion of polystyrene foams reinforced with nano-clays," *SPE ANTEC Proc.*, **61**, 1732–1736 (2003).

23. P. Moldenaers, J. Vermant, J. Mewis and I. Heynderickx, "Origin of nonlinearities in the Bagley plots of thermotropic copolyesters," *J. Rheol.*, **40**, 203–209 (1995).

24. S. J. Kurtz, personal communication. See also: L. L. Blyler and S. J. Kurtz, "Analysis of the Maxwell orthogonal rheometer," *J. Appl. Polym. Sci.*, **11**, 127–131 (1967).

Answers to Selected Problems

CHAPTER 1

1-1. Suction cups are generally used on very smooth surfaces such as glass. Thus the elastomer used to make the suction cup is in very intimate with the surface of the object to be lifted. Therefore, Van der Waals forces between the two materials will contribute to the adhesion between the two materials. Additionally, any fluid (e.g., water) will be forced to deform if the surfaces are to be separated, assuming the fluid adheres well to both materials. The situation is nearly reverse squeezing flow (see Chapter 10, Section A), with the forces varying as R^4 and inversely with H^3, where R and H are the radius and gap, respectively. The very thin gap can thus lead to very high separation forces simply because the fluid takes time to flow. This reasoning also suggests that a more modest force will eventually separate the suction cup from the substrate, given enough time.

1-2. The data set does not include the flow time for the solvent, so equation (1-5) is needed. The least-squares fit of equation (1-5) to the data is shown below. The value of $[\eta]$ from the fit is 1.6 ± 0.6 dL/g, while the Huggins coefficient is 0.43 ± 0.26, and is dimensionless.

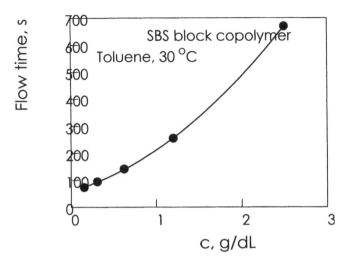

1-3. The viscometer calibration constant k gives directly the kinematic viscosity from the time, i.e.,

$$\eta/\rho = k\, t_f$$

where t_f is the flow time in seconds. Thus, the kinematic viscosity of the solvent will be derivable from the intercept of equation (1-5), which turns out to be 58.6 s. Multiplying this value by k gives the result 0.59 mm²/s (0.59×10⁻⁶ m²/s). Now, the trial-and-error selection process begins using published densities and dynamic viscosities common solvents. For example tetrachloroethylene has a density of 1622 kg/m³ and a viscosity of 0.89 cP = 0.89×10⁻³ kg/m s, giving a kinematic viscosity of 0.55×10⁻⁶. Close, but not close enough.

1-5. An ordinary electronic laboratory balance uses the local acceleration of gravity to produce a force, which is measured by the balance's load cell. Calibration with a standard mass at the same location will thus provide a direct standardization of the mass of an unknown. The local acceleration of gravity does not need to be known. If the balance is moved to a new location, presumably with a different value of g, the calibration must be repeated. This would not be the case for a beam balance, which once calibrated will be accurate even on the moon for determining mass.

1-6. An electronic laboratory balance contains a load cell, which can be used to measure force. For this application, the calibration of the load cell with weights must include the local gravitational constant. Thus, in Red River, Canada, a 25-g mass will produce (25/1000) kg × 9.824 m/s² = 0.2456 N of force. This calibration for force will be valid in Liberia, or anywhere else. However, if the balance is recalibrated with the 25-g mass in Liberia, the local gravitational constant must be used to give the correct force.

CHAPTER 2

2-1. We assume the sample mass itself does not contribute to the stress, and cylinder deforms uniformly along its length with no volume change. The stress is thus

$$\sigma_T = \frac{Mg}{\pi R^2} \qquad (a)$$

where M is the mass attached to the bottom and R is the radius of the cylinder. First, switch to SI units to end up with an answer in Pascals. Thus $M = 100$ g $= 0.1$ kg and $R = 1$ cm $= 0.01$ m. To facilitate checking, use a spread sheet to enter the data and run the calculation. The result is 3.1 kPa. This doubles when the sample stretches to twice its length.

2-2. The stress on the sample will be due solely to the mass of material below any point along the axis of the cylinder. If x is the distance from the bottom of the sample, the volume will be $\pi R^2 x$ and the force will be $\rho g \pi R^2 x$. The cross sectional area is, of course, πR^2, so the stress is simply $\rho g x$. The stress at the top will be 1000 kg/m^3 × 9.81 m/s^2 × 0.1 m $= 0.98$ kPa.

2-3. The key to a uniform stress is to have a large cross-sectional area at the top and small at the bottom. An inverted cone appears to satisfy this requirement; but, in fact, it does not work. To see this, consider that the volume of a cone below point x from the tip is $\pi R^2 x/3$, whereas the cross-sectional area is πR^2, yielding a height-dependent stress of $\rho g x/3$. Our inclination, then, is to have the area increase faster, perhaps as x^n with $n > 1$. The volume below point x will be

$$V(x) = \int_0^x (x')^n \, dx' = \frac{x^{n+1}}{n+1} \qquad (a)$$

which also doesn't work, the stress again increasing as x, the height above the tip of the cone. Clearly, an even steeper increase is needed.

Is there a solution? To some extent, yes. If one starts with a finite area, and increases this area exponentially, then the stress can be made constant. So, for example, let $A(0) = a$, and $A(x) = ae^{x/b}$, where $A(x)$ is the area at height x above the bottom. The volume is then

$$V(x) = \int_0^x A(x')dx' = \int_0^x ae^{x'/b} dx' = ab\left[e^{x/b} - 1\right] \tag{b}$$

The stress will then be

$$\sigma_T = \frac{\rho g V(x)}{A(x)} = \rho g \frac{ab\,[e^{x/b} - 1]}{ae^{x/b}} = \rho g b\left[1 - e^{-x/b}\right] \tag{c}$$

This formulation still has a height dependence, but this annoyance rapidly disappears as x increases. The stress approaches a constant value of $\rho g b$.

2-5. As the sample axle is spinning rapidly, we can neglect the downward component of gravity and assume that the acceleration is simply $r\Omega^2$, where Ω is the angular rotation rate in rad/s, and r is the radius. The force applied by the samples mass to any plane will be given by this acceleration multiplied by the mass beyond the location of the plane. (See drawing below.)

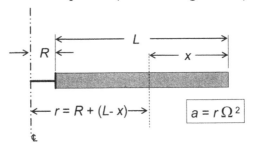

The mass is Ax, where A is the cross-sectional area of the spinning cylinder. The tensile stress σ_T then is

$$\sigma_T = \frac{F}{A} = \frac{\rho A x r \Omega^2}{A} = \rho x r \Omega^2$$

At this juncture, one can eliminate x in favor of r, R and L, or r in favor of R, L and x. As we are interested in the stress in the sample, we will use the latter. Thus, $\sigma_T = x\,[R + L - x]\rho\Omega^2$. Note that the stress is not uniform, and will have a maximum at $x = (R + L)/2$. If $R \ll L$ then the maximum stress will be near the center of the rod, which could be convenient for viewing the resulting deformation.

2-9. The density of the "Philadelphia marble" will be needed to find the load on the uniformly loaded beam with simple supports at each end. Note that the bench rotates on the supports, so the weight of the stone beyond the support will reduce the stress by applying a moment to the ends; however, we will ignore this contribution. Assuming the modern bricks in the center are standard size (length of 8 in.), we can estimate the thickness and length of the bench. To do this, scan the image and put the image into a document. Draw several lines through the thickness of the stone and measure these lines relative to the brick length.

The equation for maximum stress in a uniformly loaded beam is

$$\sigma = \frac{(d/2)sL^2}{8I} \tag{a}$$

where d is the beam thickness, s is load per unit length, L is the beam length and I is the second moment of area of the cross-sectional area of the beam. For rectangular cross sections, $I = wd^3/12$, where w is the width of the bench. The load per unit length is given by $s = \rho dw$, where ρ is the density. Thus the final equation is

$$\sigma = \frac{3}{4}\frac{d(\rho g dw)L^2}{wd^3} = \frac{3\rho g L^2}{4d} \tag{c}^*$$

Working with SI units, $L = 53'' = 1.35$ m and $d = 1.53'' = 0.0389$ m. The stress, assuming a density of 2500 kg/m^3, works out to be 0.88 MPa.

2-10. The maximum shear stress will be located at the wall of the stationary plates. As the two sides are equal, we can look at one side, which carries the load of the sample on that side plus a half of the weight of the plate. The shear stress will be given by

$$\sigma_{21} = g(\rho\delta + M_p/2) \tag{a}$$

where M_p is the area density of the plate, ρ is the fluid density, and δ is the gap. Solving for M_p gives

$$M_p = 2(\sigma_{21} - g\rho\delta)/g \tag{b}$$

Sticking to SI units and plugging in the numbers gives 193 kg/m^2, which is 19 g/cm^2.

2-11. The two are the same, as can be seen by carrying out the differentiation for both. For the first, we have

---

[*] Equation (c) in the answer to Problem 2-9 is the short explanation why the arch was invented. Making longer stone bridges and beams by finding a larger slab is futile, because the stress goes up with L^2 and down with thickness d to only the first power. Thus, to double a span without increasing the stress requires a slab four times as thick, not twice as thick.

$$\frac{1}{r^3}\frac{\partial}{\partial r}\left(r^3\tau_{r\phi}\right)=\frac{1}{r^3}\left[r^3\frac{\partial\tau_{r\phi}}{\partial r}+3r^2\tau_{r\phi}\right]=\frac{\partial\tau_{r\phi}}{\partial r}+3\frac{\tau_{r\phi}}{r} \tag{a}$$

For the second expression, differentiation gives

$$\frac{1}{r^2}\frac{\partial}{\partial r}\left(r^2\tau_{r\phi}\right)+\frac{\tau_{r\phi}}{r}=\frac{\partial\tau_{r\phi}}{\partial r}+2\frac{\tau_{r\phi}}{r}+\frac{\tau_{r\phi}}{r}=\frac{\partial\tau_{r\phi}}{\partial r}+3\frac{\tau_{r\phi}}{r} \tag{b}$$

This demonstrates the equality.

CHAPTER 3

3-1. For inward radial flow to a point sink, spherical coordinates are needed. For strictly radial flow, the $r\phi$ term of the rate-of-deformation tensor (Appendix 3-1) is

$$\dot\gamma_{r\phi}=\frac{1}{r\sin\theta}\frac{\partial v_r}{\partial\phi}+r\frac{\partial}{\partial r}\left(\frac{v_\phi}{r}\right) \tag{a}$$

In accord with Figure 3-3, we can cross out v_ϕ, because the only velocity component is v_r. Due to the symmetry of the flow, there can be no changes in the velocity field except in the r direction; thus, the derivative with respect to ϕ must be zero. The conclusion is that the $r\phi$ component of the rate-of-deformation tensor is zero. (In fact, all off-diagonal components are zero.)

3-3. The appropriate coordinate system for this problem is the cylindrical system. We assume that the inner diameter is maintained constant; thus, the outer diameter will decrease as the tube thins. The continuity equation in Appendix 3-2 shows that

$$\frac{\partial V_r}{\partial r}+\frac{V_r}{r}=-\frac{\partial V_z}{\partial z} \tag{a}$$

Because the inner tube diameter is being held constant the second term on the left will be very small as long as the tube is fairly thin. Thus, we will assume that it is zero.

Inspection of the components of the rate-of-deformation tensor in Appendix 3-1 shows that the $r\theta$, θz, and zr components will all be zero. If we define $\dot\varepsilon=\partial V_z/\partial z$, then

$$\dot\gamma_{zz}=2\dot\varepsilon$$
$$\dot\gamma_{rr}=-2\dot\varepsilon \quad \text{(using assumption from above)}$$

$\dot{\gamma}_{\theta\theta} = 0$ (again, using assumption above)
The rate-of-deformation tensor then becomes

$$\dot{\gamma} = \dot{\varepsilon} \begin{pmatrix} 2 & 0 & 0 \\ 0 & -2 & 0 \\ 0 & 0 & 0 \end{pmatrix} \qquad (b)$$

which is a state of pure shear or planar extension. Note that if the tube has a thick wall relative to its diameter, the result will be more complex.

3-5. The two tensors will have the following appearance:

$$\dot{\gamma} = \dot{\varepsilon} \begin{pmatrix} 2 & 0 & 0 \\ 0 & -1 & 0 \\ 0 & 0 & -1 \end{pmatrix} \qquad (a)$$

for simple extension and

$$\dot{\gamma} = \dot{\varepsilon} \begin{pmatrix} 2 & 0 & 0 \\ 0 & -2 & 0 \\ 0 & 0 & 0 \end{pmatrix} \qquad (b)$$

for planar extension, where $\dot{\varepsilon}$ is the velocity gradient in the principle flow direction. Note that the trace (sum of diagonals) is zero in each case, which is consistent with the assumption of incompressibility.

3-6. (e) First of all, we define $\dot{\varepsilon} = \partial V_z / \partial z$ in the usual fashion. The $\theta\theta$ component of the rate-of-deformation (Appendix 3-1) is

$$\dot{\gamma}_{\theta\theta} = 2\left(\frac{1}{r}\frac{\partial V_\theta}{\partial \theta} + \frac{V_r}{r} \right) = 2\frac{V_r}{r} \qquad (a)$$

because there are no changes in the θ direction. We need to relate this to the relative rate of change of the circumference in the following fashion

$$\frac{1}{C(r)}\frac{\partial C(r)}{\partial t} = \frac{1}{r}\frac{\partial r}{\partial t} = \frac{V_r(r)}{r} = \frac{\dot{\gamma}_{\theta\theta}}{2} \qquad (b)$$

As the flow is homogeneous, we conclude that the relative change in circumference is independent of radius.

3-7. With the cavity and roller diameters given, we conclude that the clearance is $(3.97 - 3.81)/2 = 0.08$ cm $= 0.8$ mm. The shear rate can be approximated by the relationship

$$\dot{\gamma} = \frac{\Omega(R_i + R_o)/2}{H} \tag{a}$$

where H is the gap, Ω is the angular speed, and R_i and R_o are the inner and outer radii, respectively. This equation assumes the gap is small compared to the inner radius, which is the case. The relationship between Ω and N, is $\Omega = 2\pi N/60$, where N is the rpm. Entering this information gives the working equation

$$\dot{\gamma} = \frac{\pi N(R_i + R_o)}{60H} \tag{b}$$

which shows the expected proportionality of shear rate to rpm.

Entering the numbers provided, and using consistent units gives the dimensional equation

$$\dot{\gamma} = 2.5N \tag{c}$$

or a value for K of 2.5 s^{-1}/rpm. This is somewhat higher than the calculation of Goodrich and Porter, who accounted for the larger clearance between the rotor lobes.

3-8. (a) The information provided for the velocity profile can be summarized as follows: (1) parabolic; (2) $V_z(R) = 0$; and (3) $V_z(0) = V_0$. Due to the symmetry of the geometry, we know the parabola must be centered on the axis of the tube; thus we can write the form of the equation immediately as

$$V_z(r) = V_0 - ar^2 \tag{a}$$

We need to find the parameter a. To do this, we use the last bit of information that $V_z(R) = 0$. Thus

$$0 = V_0 - aR^2 \quad \text{or} \quad a = V_0/R^2 \tag{b}$$

The final equation is the familiar

$$V_z(r) = V_0\left[1 - \left(\frac{r}{R}\right)^2\right] \tag{c}$$

(b) The shear rate is simply the derivative of the equation (c). The sign will be negative.

3-10. The usual definition for the extensional strain rate is

$$\dot{\varepsilon} = \frac{\partial V_1}{\partial x_1} \tag{a}$$

where the x_1 is along the stretching direction. On the other hand, the component of the rate-of-deformation tensor (rectangular coordinates) is twice this value, i.e.,

$$\dot{\gamma}_{11} = \frac{\partial V_1}{\partial x_1} + \frac{\partial V_1}{\partial x_1} = 2\dot{\varepsilon} \tag{b}$$

The main aspect of this problem is getting this gradient as a function of time. To do this, we write the velocity of the clamp as V_{XH}; and, assuming uniform deformation along the entire sample length L,

$$\frac{\partial V_1}{\partial x_1} = \frac{V_{XH}}{L(t)} = \frac{V_{XH}}{L_0 + V_{XH}t} \tag{c}$$

With this in hand, and using the usual assumption of incompressibility, we can write the entire rate-of-deformation tensor as

$$\dot{\gamma} = \begin{pmatrix} 2V_{XH}/(L_0 + V_{XH}t) & 0 & 0 \\ 0 & -V_{XH}/(L_0 + V_{XH}t) & 0 \\ 0 & 0 & -V_{XH}/(L_0 + V_{XH}t) \end{pmatrix} \tag{d}$$

Note that the 22 and 33 components are identical in spite of the different initial width and thickness. Obviously we have used a key, but dubious, assumption that the flow is uniform extension throughout the length of the sample, which will not obtain except in regions well away from the clamps. If this geometry is used (and it is) for gathering extensional data, it is advisable to video the sample as it is being deformed, so the true deformation vs. time can be found.

CHAPTER 4

4-1. (a) Illustrated is a case of plane stress. Mohr's circle would apply and would show that at 45° the normal stresses would be completely absent; thus, a state of pure shear.

(b) Nothing will happen.

(c) The total stresses σ_{ii} are given by the forces and areas. In the principal extension direction (shown as x on the drawing), $\sigma_{xx} = +10$ Pa. On the other edge, $\sigma_{yy} = -10$ Pa, while the faces have no force; therefore $\sigma_{zz} = 0$. The pressure $p = -(\sigma_{xx} + \sigma_{yy} + \sigma_{zz})/3$ is zero. Since $\sigma_{ii} = \tau_{ii} - \delta_{ii}\, p$, and $p = 0$, the extra stresses will be the same as the total stresses. The viscosity is not needed.

(d) For a Newtonian fluid,

$$\tau_{ij} = \eta \dot{\gamma}_{ij} \tag{a}$$

where the $\dot{\gamma}_{ij}$ are the components of the rate-of-deformation tensor. Thus, the extra-stress components simply need to be divided by the viscosity to get the rates of deformation. Thus, $\dot{\gamma}_{xx} = 5$ s^{-1} and $\dot{\gamma}_{yy} = -5$ s^{-1}. All the other components are zero.

4-2. The plots are shown below. The intercept of the log-log plot using dimensional variables has no particular meaning, as it depends upon the dimensions used.

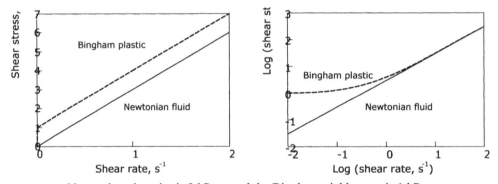

Newtonian viscosity is 3 kPa s, and the Bingham yield stress is 1 kPa.

The Bingham plastic shows up on both plots as having a limiting shear stress at low shear rates and similar behavior at high shear rates. The slopes of both on the log-log plot are 1.0 at high shear rates. On the linear plots, the slope is 3 kPa s, corresponding to the limiting viscosity η_∞. The viscosity shows up on the log-log plot as a vertical shift of log (3), which is about half a decade.

4-3. (a) The motion will be exactly like the solid cylinder. The reason for this is clear upon examination of the equation of continuity (Appendix 3-2). This is reproduced below:

$$\frac{\partial V_r}{\partial r} + \frac{V_r}{r} + \frac{1}{r}\frac{\partial V_\theta}{\partial \theta} + \frac{\partial V_z}{\partial z} = 0 \tag{a}$$

This equation states that the axial velocity gradient, which is constant across the tube wall, will depend on both the radial velocity and the radial velocity gradient. If we have a radial velocity equal to zero in one place and not in another, then the velocity gradient must change also with position. This, in turn means the extra stress will also change, because

$$\tau_{rr} = \eta \dot{\gamma}_{rr} = 2\eta \frac{\partial V_{rr}}{\partial r} \tag{b}$$

for a Newtonian fluid. As the total radial stress applied to the sample surfaces on both sides is zero, τ_{rr} must be uniform and equal to the pressure in the sample. To keep either surface fixed at a constant radius, pressure must be applied to the inside of the tube. (See Problem 3-3.)

(b) As we have decided the situation is the same as simple extension of a rod, the rates of deformation and the associated extra stresses are straightforward. Thus

$$\tau = \eta \begin{pmatrix} 2\dfrac{\partial V_z}{\partial z} & 0 & 0 \\ 0 & 2\dfrac{\partial V_r}{\partial r} & 0 \\ 0 & 0 & 2\dfrac{V_r}{r} \end{pmatrix} = \eta \begin{pmatrix} 2\dot{\varepsilon} & 0 & 0 \\ 0 & -\dot{\varepsilon} & 0 \\ 0 & 0 & -\dot{\varepsilon} \end{pmatrix} = \begin{pmatrix} 20 & 0 & 0 \\ 0 & -10 & 0 \\ 0 & 0 & -10 \end{pmatrix} \text{Pa} \tag{c}$$

The first matrix is simply a repeat of the rates of deformation for cylindrical coordinates (see Appendix 3-1), while the second matrix uses the definition of the extension rate $\dot{\varepsilon}$ being the velocity gradient in the stretch direction (here, z). The equality of V_r/r and $\partial V_r/\partial r$ is not so obvious. However, we know that both are independent of radius and independent of position along the tube. This means $\partial V_r/\partial r$ can be integrated to give a velocity that is linear with radius, i.e., $V_r(r) = ar + c$, where c is the constant of integration. At $r = 0$, V_r is zero, so c is zero. This means that both V_r/r and $\partial V_r/\partial r$ are equal to a, and by continuity $a = -\dot{\varepsilon}/2$.

4-5. (a) If the lubrication is perfect, the flow will be identical to flow into a point sink. Thus, by inspection,

$$\mathbf{V} = (-Q/A, 0, 0) \tag{a}$$

where Q is the flow rate (positive) and $A = 2\pi r^2(1 - \cos\theta_0)$ is the area through which the melt is flowing. The minus sign in equation (a) signifies inward flow.

(b) We will assume that the fluid is incompressible and inertial terms are negligible. Then the r component of the equation of motion (Appendix 2-2) becomes simply

$$0 = \frac{\partial \sigma_{rr}}{\partial r} + \frac{\sigma_{rr} - \sigma_{\theta\theta}}{r} + \frac{\sigma_{rr} - \sigma_{\phi\phi}}{r} \qquad \text{(ssc) (b)}$$

As it turns out, the last two terms are the same.

(c) The rates of deformation are given in Appendix 3-1. Using equation (a) above, and simplifying the terms, gives the array

$$\dot{\gamma} = \begin{pmatrix} 2\dfrac{\partial V_r}{\partial r} & 0 & 0 \\ 0 & 2\dfrac{V_r}{r} & 0 \\ 0 & 0 & 2\dfrac{V_r}{r} \end{pmatrix} = \begin{pmatrix} \dfrac{2Q}{\pi r^3(1-\cos\theta_0)} & 0 & 0 \\ 0 & \dfrac{-Q}{\pi r^3(1-\cos\theta_0)} & 0 \\ 0 & 0 & \dfrac{-Q}{\pi r^3(1-\cos\theta_0)} \end{pmatrix}$$

$$\text{(c)}$$

It is evident from this result that the flow is simple extension. The extra stress tensor τ is simply the above result multiplied by the viscosity η (ssc).

(d) The total stresses are given by $\sigma_{ii} = \tau_{ii} - p$ (ssc). To find any of these, we need to know (or define) one normal stress or the pressure at some point in the flow. We will assume that $\sigma_{rr} = 0$ at the exit of the die[†] at radius $r = B$.

Transforming the equation of motion to the extra stresses gives

$$0 = \frac{d\tau_{rr}}{dr} - \frac{dp}{dr} + \frac{\tau_{rr} - \tau_{\theta\theta}}{r} + \frac{\tau_{rr} - \tau_{\phi\phi}}{r} \qquad \text{(ssc) (d)}$$

where it has also been recognized that the stresses vary only with radial position. Using equation (c) along with the viscosity provides numbers for the stress terms. Putting dp/dr on the left-hand side and pulling out the constant factor $Q/[\pi(1-\cos\theta_0)]$ gives for the right-hand side

$$\frac{dp}{dr} = \frac{Q\eta}{\pi(1-\cos\theta_0)}\left(\frac{-6}{r^6} + \frac{3}{r^6} + \frac{3}{r^6}\right) \qquad \text{(ssc) (e)}$$

[†] This assumption may not be justified, especially at high flow rates.

which is zero. Thus, we have the rather amazing fact that for this flow, the pressure drop is zero. If we chose the pressure to be zero at the exit, the pressure is zero throughout and the total and extra stresses are the same everywhere.

Of course, something has to push the melt through the die. This is the total radial normal stress, which must be higher [more negative for (ssc)] at the entrance than at the exit. To plot the stresses, we will assume the radial stress is zero at the exit, which means the pressure is $2Q/[B^3 \pi(1 - \cos\theta_0)]$ at this point. This value provides a starting point from which all the stresses can be found. Moreover, it can serve a reasonable normalizing factor.

A plot of the stresses is given below, where the stresses have been divided by pressure, and the distance from the exit in terms of B, the radial position of the exit.

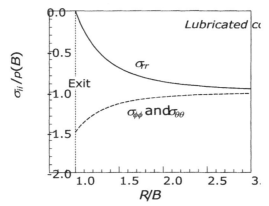

Observe that at high radius, the three stresses approach the hydrostatic state where they are all equal. The negative average of the three (i.e., the pressure) is constant throughout the die. As described, we have used the boundary condition $\sigma_{rr}(B) = 0$. If we chose $p(B) = 0$ instead, $\sigma_{rr}(R)$ will be tensile, while the other normal stresses will remain compressive. A way to picture this situation is to imagine a well-lubricated rubber rod being pulled through the conical die.

Note that to solve this problem, we have used (1) the continuity equation, (2) the equation of motion and (3) a constitutive equation. All three are needed, in spite of the relatively simplicity of the flow.

4-6. The paper cited was based on information from the literature and did not include a side-by-side investigation of the two materials. Nevertheless, the data they gathered makes a very convincing case for a difference. Narrowly distributed resins are generally made using a living polymerization; for styrene, anionic catalysts are used. One possibility is that the two resins had different degrees and sequencing of tactic diads, triads, etc., due to differences in catalyst or polymerization temperature. When the paper was published, ^{13}C NMR was known and used, but the authors of the cited papers, which were published even earlier and did not include such information.

CHAPTER 5

5-1. The results are summarized in the table below:

Model	$\sigma(\dot{\gamma};\dot{\gamma}_0,\eta_0,n)$	$\dot{\gamma}(\sigma;\dot{\gamma}_0,\eta_0,n)$	$\eta(\sigma;\dot{\gamma}_0,\eta_0,n)$
Ferry[a]	$\sigma_0\left(\dfrac{-1+\sqrt{1+4\dot{\gamma}/\dot{\gamma}_0}}{2}\right)$	$\dfrac{\sigma}{\eta_0}\left(1+\dfrac{\sigma}{\sigma_0}\right)$	$\dfrac{\eta_0}{1+\sigma/\sigma_0}$
Cross	$\dfrac{\eta_0\dot{\gamma}}{1+(\dot{\gamma}/\dot{\gamma}_0)^{1-n}}$	Implicit[b]	Implicit
Carreau	$\eta_0\dot{\gamma}/[1+(\dot{\gamma}/\dot{\gamma}_0)^2]^{(1-n)/2}$	"	"
Eyring	$\sigma_0\sinh^{-1}(\dot{\gamma}/\dot{\gamma}_0)$	$\dot{\gamma}_0\sinh(\sigma/\sigma_0)$	$\sigma/\dot{\gamma}_0\sinh(\sigma/\sigma_0)$
Ellis	Implicit	$\dfrac{\sigma/\eta_0}{1+(\sigma/\sigma_0)^{(1-n)/n}}$	$\dfrac{\eta_0}{1+(\sigma/\sigma_0)^{(1-n)/n}}$

[a] In the Ferry, Eyring and Ellis equations, the symbol σ_0 has been used in place of $\eta_0\dot{\gamma}_0$ to simplify the equation.
[b] For special values of n, the equations can be solved explicitly.

A conclusion that one might reach from this analysis is that equations with viscosity explicit in stress (e.g., Ellis, Ferry) can be easily written as $\dot{\gamma}(\sigma)$, whereas those with viscosity explicit in shear rate can be written easily as $\sigma(\dot{\gamma})$, but not the other way around.[‡] Two-parameter equations (e.g., Ferry, Eyring) are usually formulated so they can be expressed in either fashion.

5-4. (b) The Carreau model (Table 5-1) has the form

$$1/[1+\dot{\gamma}_R^2]^{(1-n)/2} \tag{a}$$

for the relative viscosity η/η_0. An equivalent form of this is

$$[1+\dot{\gamma}_R^2]^{(n-1)/2} \tag{b}$$

[‡] Often the troublesome term, e.g., reduced shear rate taken to a non-integer power, can be expanded in a series to give, after inversion of the series, an explicit result in a converging series form.

which can be expanded using the binomial expression to give

$$1 - \frac{1-n}{2}\dot{\gamma}_R^2 + \frac{(1-n)(3-n)}{8}\dot{\gamma}_R^4 \cdots \tag{c}$$

This expansion is a special case of the more general equation (5-21) in the problem statement. (Note that the two will have different values of the characteristic shear rate $\dot{\gamma}_0$.)

(c) The Cross model equation (5-1) is not quadratic at low shear rates. Instead, the expansion of this model gives the result

$$1 - \dot{\gamma}_R^{1-n} + \dot{\gamma}_R^{2(1-n)} + \cdots \tag{d}$$

(d) Comparing the binomial expansion in part (b) with that given in equation (5-21) leads to the relationship $\alpha = (3 - n)/[2(1 - n)]$. For the usual values of n $(0 < n < 1)$, α will be positive. See comment in the answer to part (b).

5-5. For uniaxial extension,

$$\dot{\gamma} = \dot{\varepsilon} \begin{pmatrix} 2 & 0 & 0 \\ 0 & -1 & 0 \\ 0 & 0 & -1 \end{pmatrix} \tag{a}$$

Using the definition of $I_3 = \text{tr}(\dot{\gamma} \cdot \dot{\gamma} \cdot \dot{\gamma})$ given in equation (5-6), we find that

$$I_3 = 6\dot{\varepsilon}^3 \tag{c}$$

which is non-zero.

5-7. At high shear rates, the value of log σ will approach either $-\infty$ or $+\infty$, neither of which is realistic. The sign of a_2 determines which value will be approached. At some shear rate, log σ will either go through a maximum or a minimum, depending on the signs of a_1 and a_2. Assuming $a_1 = 1$ and $a_2 = -0.2$, we can expect the behavior shown below, i.e., a maximum.

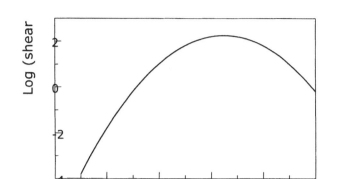

This graph points out some other issues. At low shear rates, the slope of the flow curve is greater than one, which is possible but unusual. We can see this clearly by taking the derivative of equation (5-23) in the problem statement. Thus

$$\frac{d \log \sigma}{d \log \dot{\gamma}} = a_1 + 2a_2 \log \dot{\gamma} \tag{a}$$

which describes a slope of $1 - 0.4 \log \dot{\gamma}$ for the flow curve depicted above. Thus the slope does not approach the expected 1.0 at low shear rates; instead, it approaches $+\infty$ at very low shear rates and $-\infty$ at high shear rates.

CHAPTER 6

6-1. (a) For this problem, it is important to refer to the definitions of the functions mentioned. So, $N_1 \equiv \sigma_{11} - \sigma_{22} = \tau_{11} - \tau_{22}$ and $N_2 \equiv \sigma_{22} - \sigma_{33} = \tau_{22} - \tau_{33}$ (ssc). Thus the combination reported in the table is $N_1 - N_2$, which has been referred to as N_3 (not official nomenclature).

(b) and (c) To digitize the data in the table, scan the table using optical character recognition (OCR), a program that comes with most scanners. Check the accuracy of the recognition. Convert the viscosity units to SI,[§] and calculate the shear stress, σ_{21}. (As the viscosity is "corrected," we can assume that the values are those at the rim shear rates.) Checking the regression coefficient for $\log N_3$ vs. $\log \sigma_{21}$ and its 95% confidence interval yields $1.28 \pm$

[§] Many online unit conversion sites can be found on the internet, e.g., onlineconversion.com.

0.07. Thus, we can conclude with a high degree of confidence that the observed slope is different than 2.0, based on the data at hand.[**]

(d) The graph of the data used for part (b) is shown below:

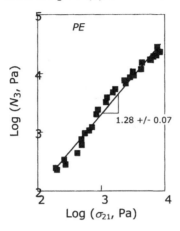

6-4. For this problem, we need to estimate η_0 and $\dot{\gamma}_0$ from the viscosity data. Using the Carreau equation, we get the curve shown in the graph below:

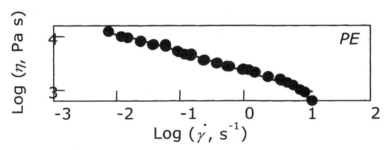

As can be seen, the data do not show much curvature, which is typical of LDPE resins. Thus both η_0 and $\dot{\gamma}_0$ will be less precise than what we might like. Nevertheless, we calculate a value of $\Psi_{1,0} = 1.5$ MPa s². Using this results and the approximation in equation (6-13) gives the result shown in the figure below. Obviously, the agreement is not very satisfactory except perhaps at low rates. Thus, approximations such as equation (6-13) should be used with extreme caution.

[**] The analysis shown depends upon the points being independently determined. As they undoubtedly are not independent, the confidence interval is likely to be higher than that shown. The best approach, then, is to replicate the experiment several times and accumulate a set of slopes to analyze.

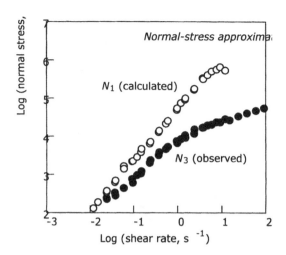

6-6. The first step toward a solution of this problem is to gather the data. This can be done by scanning Figure 6-11 and digitizing the data points. (If worse comes to worse, scan the figure as a bitmap image and open with Microsoft Paint® or similar; the pixel location of the cursor is displayed, from which the values of the points can be found.) To check the scaling exponent suggested in equation (6-5), we need a log-log plot, shown below. Fortunately, the normal- and shear-stress data were gathered at the same strains.

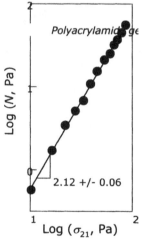

Curiously, the data presented suggest a slope slightly greater than 2.0, which is the result expected from theory [equation (6-5)]. In fact, these precise data suggest that there is only a 0.11% probability of having a slope this high by chance alone, assuming the slope is actually 2.0. However, before writing up a grant proposal, one must remember that a systematic error (e.g., crosstalk between the torque and normal force transducers) could easily produce this result. Thus, the first step is to check for artifacts produced by the equipment and/or the experimental protocol.

6-7. (a) The equation for the linear stress-growth function is

$$\eta^+(t;\dot\gamma) = \frac{\sigma^+(t;\dot\gamma)}{\dot\gamma} = \frac{1}{\dot\gamma}\int_{-\infty}^{t}G(t-t')\dot\gamma(t')dt' \qquad (a)$$

where the constant shear rate has been removed from the integral. For linear response with a steady shear rate applied at time zero, the expression simplifies to

$$\eta^+(t;\dot\gamma) = \int_{0}^{t}G(t-t')dt' \qquad (b)$$

which implies that the stressing viscosity is independent of shear rate in the linear regime. For the shear stress-relaxation function provided, we need integrals of following integrands:

1. $1/(1+x^{1/2})$
2. $1/(1+x)$
3. $1/(1+x^{3/2})$
4. $1/(1+x^2)$

In these expressions, $x = t/\tau$, where τ is the relaxation time. We can try the general expression $1/(1+x^n)$, which gives an integral function—not what we want! Turning to the specific cases give the results

1. $2[x^{1/2} - \ln(x^{1/2}+1)]$
2. $\ln(x+1)$
3. $-\dfrac{2}{3}\ln(\sqrt{x}+1) + \dfrac{1}{3}\ln\left(x-\sqrt{x}+1\right) + \dfrac{2\tan^{-1}\left(\dfrac{2\sqrt{x}-1}{\sqrt{3}}\right)}{\sqrt{3}}$
4. $\tan^{-1}(x)$

Note that since $x = t/\tau$, $dt = \tau dx$, so the factor $\tau G(0)$ comes out in front of the integral if the complete relaxation function is used. Thus the functions shown are already in their reduced or dimensionless form.

(c) Formulas 1 and 2 do not approach a steady state viscosity; thus, they cannot be used to represent fluids. Formula 4 approaches $\pi/2$ as $x \to \infty$. Formula 3 is a bit harder to see, but the first two terms cancel at high values of x, and the third term becomes simply $\pi/\sqrt{3} = 1.8138$. Thus we can fairly safely conclude that n must be greater than 1.

(d) The Gleissle mirror relationship (Table 6-1) says simply

$$\eta(\dot{\gamma}) \approx \eta^+(t)\Big|_{t=1/\dot{\gamma}} \tag{c}$$

So, the viscosity function, in reduced form, is

$$\eta_R(\dot{\gamma}) = \tan^{-1}(1/\dot{\gamma}_R)/(\pi/2) \tag{d}$$

A plot of this is shown below:

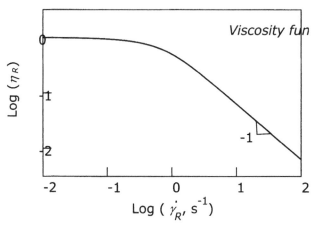

6-10. We start with equation (6-27b), which is reproduced below:

$$\eta(T) = \eta_R \exp\left\{ a_1' \left[\left(\frac{T_R}{T} \right)^{a_2} - 1 \right] \right\} \tag{6-27b}$$

This expression can be rewritten as:

$$\eta(T) = \eta_R e^{-a_1'} e^{b/T^{a_2}} \tag{a}$$

where $b = a_1' T_R^{a_2}$. Thus the final result is

$$\eta(T) = \eta_\infty e^{b/T^\alpha} \tag{b}$$

which is a three-parameter equation describing the dependence of viscosity on temperature. A better form of this equation might be

$$\eta(T) = \eta_\infty e^{(T_0/T)^\alpha} \tag{c}$$

where T_0 is a characteristic temperature where the viscosity is e times higher that the viscosity at infinite temperature η_∞.

CHAPTER 7

7-1. (a) The conversion of particle count to velocity is going to be linear in all involved variables, and can be written directly as

$$V = \frac{n}{N\delta_s}$$

(b) We will normalize the velocities, and thus the counting rate, by the center-plane velocity. Consideration of shell force balance for slit flow reveals that the stress will fall linearly from a maximum at the wall of the slit to zero at the midplane, similar to tube flow. Thus, for Newtonian fluids, the shear rate dV_x/dy will also fall linearly. When integrated to get the velocity, $V_x(y)$, the result will be a quadratic. As we don't need to express the velocity profile in terms of the total flow rate, we can simply write down the normalized quadratic velocity profile as

$$V(y) = V_{max}\left[1 - \left(\frac{y}{H/2}\right)^2\right]$$

where y is the cross-flow coordinate with origin at the midplane, V_{max} is the midplane velocity, and H is the total slit gap. As can be seen by examining this equation, the velocity is indeed predicted to be V_{max} at $y = 0$ (the midplane) and zero at $y = H/2$ (the wall). The normalized counting rates are thus expected to be as shown in the figure below:

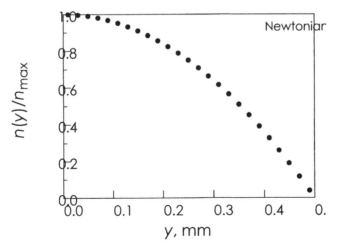

7-4. The viscosity is approximately 630 Pa s at a corrected shear rate of 115 s^{-1}.

7-5. Following on Example 7-1, we differentiate the velocity profile given in equation (7-65) to get the shear-rate magnitude. The derivative goes as follows:

$$\frac{dV_z(r)}{dr} = -V_0 k \left(\frac{r}{R}\right)^{k-1}$$

The magnitude of the shear rate is merely the right-hand side without the negative sign.

For a power-law fluid, $\tau = m\dot{\gamma}^n$; or, more properly, $\tau_{21} = m\dot{\gamma}^{n-1}\dot{\gamma}_{21}$ for (ssc) where $\dot{\gamma}$ is the shear-rate magnitude. However, the first expression will do fine here. We know that for any fluid in steady flow, the shear stress is linearly proportional to radius, i.e., $\tau = \tau_W r/R$. Equating this expression with the equation for a power-law fluid gives

$$\frac{\tau_w r}{R} = m\left[V_0 k \left(\frac{r}{R}\right)^{k-1}\right]^n \qquad \text{(ssc)}$$

where we have dropped the negative sign. We can see immediately that the two sides will be consistent only if $n(k-1) = 1$. This means that the expression for k is

$$k = 1 + \frac{1}{n}$$

Substituting this into the expression for τ_W gives

$$\tau_W = m\left(\frac{V_0 k}{R}\right)^{\frac{1}{k-1}} \qquad \text{(ssc)}$$

For these problems, a good check is to see if the Newtonian limit is achieved, that is, find the expressions for $n = 1$ or $k = 2$. The above expression gives $\tau_W = 2mV_0/R$, which agrees with that in equation (7-20) if $m = \eta$.

7-6. (a) Plot below indicates slip is small, if present at all.

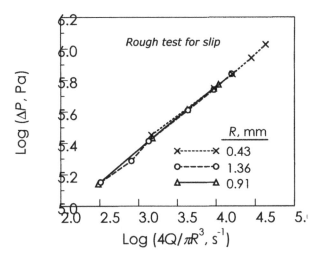

(b) The Bagley plot will require interpolation to obtain data at constant values of $4Q/\pi R^3$. As with differentiation, this can be a somewhat delicate process. In most cases, it probably best to use a simple linear interpolation, letting the Bagley plot reduce the noise in the corrected shear stress. As we have only two L/D values, it is not apparent that we will enjoy the advantages of noise reduction. However, as there are three capillaries at the two aspect ratios, the slope does indeed benefit. The Bagley plot is shown below.

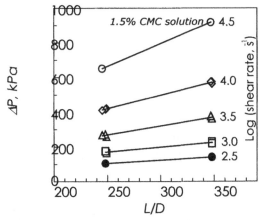

As the aspect ratios are very high, the pressure intercepts are not very precise; however, they all are positive, indicating a need for an end correction.

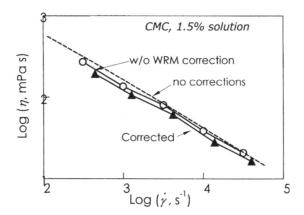

When viscosity is plotted vs. stress (from the Bagley plot), the WRM correction shows as a vertical shift; the stress stays the same. A correction according to the single-point approximation results in a −0.097 horizontal shift; the viscosity stays the same. This is illustrated below. The conclusion is that the −0.097 shift, in this example, is a reasonable approximation.

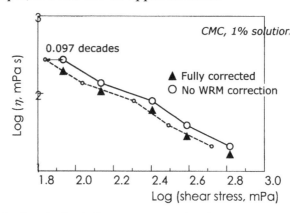

7-15. To establish the desired flow pattern, the lip of the cone should be at a radius R from the origin. The height of the lip above the plate, H_R will then be given by

$$H_R = R \sin \alpha$$

where α is the cone angle. In this case, $\alpha = 0.0401$ rad, giving an outside gap of 1.003 mm. Pythagorean's theorem is used to give for the disk radius, and doubling this gives a diameter of 49.97 mm. In view of the variances of the measurements, this result should tentatively be accepted as consistent with the measured value.

Note that gap at R is quite small, which is helpful to reducing inertial effects and for keeping the material from flowing out of the gap.

7-20. Equation (7-56) is reproduced below

$$V = \frac{d^2(\rho_B - \rho)g\sin\theta}{15\eta}(1-\alpha)^{5/2} \qquad (7\text{-}56)$$

Before plugging in numbers, we should check the dimensions to see if they are consistent; otherwise, we will need to assume SI units. Using SI units on the right-hand side gives

$$(m^2)(kg/m^3)(m/s^2)/(kg/m\ s)\ [=]\ m/s$$

Thus, the dimensions are consistent, so we can use any convenient, but consistent set.

Looking up some data, we find that tungsten carbide ball bearings have a density of 0.54 $lb_m/in.^3$, which converts to 15 g/cm^3. The density of a similar silicone (NIST SRM-260-147) at 25 °C is 0.978 g/cm3, with a coefficient of expansion of 9.2×10^{-4} K^{-1}, which reduces the density to 0.973 g/cm^3 at 30 °C. With the viscosity estimate of 2 kPa s, which is equal to 2×10^5 P, we can calculate the velocity in cm/s using equation (7-56). The result is about 4.6 h. Thus the rolling- or falling-ball geometries are better for polymer solutions.

7-22. As stated, the plane-polarized light entering the sample at 45° to the flow direction, can be regarded as two waves of lower amplitude, one with the electric field in the flow (1) direction, and the other in the transverse (3) direction. At the point of entry, the two will be exactly in phase. Both travel through the same distance L, but the component with its electric field in the flow direction will travel slower. As its frequency does not change, its phase must shift. So, the transit time of the light in general will be L/c, where c is the speed of light. From basic physics, the speed of light polarized in the flow (1) direction will be $c_1 = c/n_1$, where n_1 is the refractive index in this direction. Likewise for the transverse direction, $c_3 = c/n_3$. Thus the difference in transit times will be

$$t_1 - t_3 = L\left(\frac{1}{c_1} - \frac{1}{c_3}\right) = \frac{L}{c}(n_1 - n_2) \qquad (a)$$

The phase change in general will be $\delta = \omega t$, where ω is the angular frequency. ω is equal to $2\pi f$, where f is the cyclic frequency (Hz). The speed of light c is equal to $f\lambda$, where λ is the wavelength. In the case of a He-Ne laser, the wavelength is 632.8 nm. In terms of l, $\omega = 2\pi c/\lambda$. The phase shift is simply the difference in total phase changes for each, giving

$$\Delta = \delta_1 - \delta_3 = \omega(t_1 - t_3) = \frac{2\pi c}{\lambda}\frac{L}{c}(n_1 - n_3) \qquad (b)$$

(a) If Δ is going to be π, then the refractive index difference—the birefringence—will be equal to $\lambda/2L$, and in general

$$\Delta n = \Delta \frac{\lambda}{2\pi L} \qquad (c)$$

(b) If the stress optical coefficient is 3.6×10^{-9} Pa^{-1}, the stress difference is:

$$\sigma_{11} - \sigma_{33} = \frac{\Delta n}{C} = \frac{0.6328}{2 \times 2000} \frac{1}{3.6 \times 10^{-9}} = 44 \text{ kPa} \qquad (d)$$

7-23. The single-point described in Chapter 7 suggests that for the parallel-plate geometry, the viscosity can be estimated at a shear rate of $4/5\, \dot\gamma_R$ using the simple Newtonian formula, i.e.,

$$\eta_N = \frac{2MH}{\pi R^4 \Omega} = \frac{2M}{\pi R^3 \dot\gamma_R} \qquad (7\text{-}36)$$

where $\dot\gamma_R$ is the shear rate at the rim. As the shear rate varies linearly with radial location, the shear rate for this viscosity will be found at $4/5\ R$. Thus, this is a good location for the optical path, as the viscosity and shear rate are known here, at least within the accuracy of the single-point estimate. As the shear stress is given by the general equation $\tau = \eta \dot\gamma$ (ssc), we have the result that $\tau = 8M/5\pi R^3$ (ssc).

CHAPTER 8

8-1. (a) We are assuming linear response, so

$$\eta_E^+ = 3G_0 \int_0^t e^{-(t-t')/\tau} dt' = 3G_0 e^{-t/\tau} \int_0^t e^{t'/\tau} dt' = 3G_0 \tau (1 - e^{-t/\tau}) \qquad (a)$$

Differentiating this expression is straightforward, recognizing that

$$\frac{d \log \eta_E^+}{d \log t} = \frac{d \ln \eta_E^+}{d \ln t} = \frac{t}{\eta_E^+} \frac{d \eta_E^+}{dt} \qquad (b)$$

which gives the result

$$\frac{d \log \eta_E^+}{d \log t} = \frac{t}{\tau} \frac{e^{-t/\tau}}{1 - e^{-t/\tau}} \qquad (c)$$

The combination of t in the numerator and the $1- e^{-t/\tau}$ in the denominator creates a possible problem at $t \to 0$ because both go to zero. This can be resolved by using l'Hôpital's rule, which gives a limiting slope after one differentiation, as shown below:

$$\frac{te^{-t/\tau}}{\tau(1-e^{-t/\tau})}\bigg|_{t\to 0} = \frac{t/\tau}{e^{t/\tau}-1}\bigg|_{t\to 0} = \frac{1/\tau}{e^{t/\tau}/\tau}\bigg|_{t\to 0} \Rightarrow 1 \ @ \ t=0 \qquad (d)$$

The plot of the results using dimensionless variables is shown below:

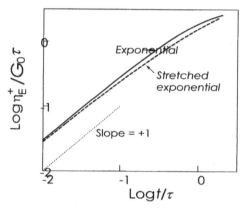

(b) On examining tables of integrals, we can locate the following integral

$$\int e^{-\sqrt{x}}dx = -2e^{-\sqrt{x}}(\sqrt{x}+1) \qquad (e)$$

This will prove to be very helpful, although the algebra is clearly going to be more complex than for Part a. Proceeding, nevertheless, gives:

$$\eta_E^+(t) = 3G_0 \int_0^t e^{-\sqrt{\frac{t-t'}{\tau}}} dt' \qquad (f)$$

It is clear that substitutions will simplify the problem considerably. We will do this in two steps. First let

$$x = (t - t')/\tau, \ \text{or} \ t' = t - x\tau, \ \text{and} \ dt' = -\tau\, dx \qquad (g)$$

Additionally, we note that at $t' = 0$, $x = t/\tau$ and at $t' = t$, $x = 0$. On making these substitutions, and switching the integration limits to take care of the signs, we have

$$\eta_E^+ = 3G_0\tau \int\limits_0^{t/\tau} e^{-\sqrt{x}}\,dx \tag{h}$$

which is just the form given in the integral tables. This is easily integrated to give

$$\begin{aligned}\eta_E^+ &= 3G_0\tau\left(-2\left[e^{-\sqrt{x}}(\sqrt{x}+1)\right]_0^{t/\tau}\right)\\ &= -6G_0\tau\left[e^{-\sqrt{t/\tau}}(\sqrt{t/\tau}+1)-1\right]\\ &= 6G_0\tau\left[1-e^{-\sqrt{t/\tau}}(\sqrt{t/\tau}+1)\right]\end{aligned} \tag{i}$$

This is the expression needed for plotting the results, and is shown in the figure above. The curve is labeled "stretched exponential" because the relaxation function is a special case of the general expression

$$G(t) = G_0 e^{-(t/\tau)^\beta} \tag{j}$$

The reason for the description "stretched" is that for $\beta < 1$, this equation describes a relaxation process covering a greater span of time. The final step is finding the slope of this expression at the beginning of the deformation. To do this, we need to take the derivative of the above expression with respect to time. Clearly, this is going to result in a very complicated expression, so a substitution is in order. A reasonable choice is $u = \sqrt{t/\tau}$, to give the expression

$$\eta_E^+ = 6G_0\tau\left[1 - e^{-u}(1+u)\right] \tag{k}$$

which is much more approachable. However, the chain rule requires we also have at hand the expression $du/dt = 1/(2\tau u)$

$$\frac{d\eta_E^+}{dt} = \frac{d\eta_E^+}{du}\frac{du}{dt} = \frac{1}{2\tau u}6G_0\tau\left[e^{-u}(1+u)-e^{-u}\right] = \frac{3G_0}{u}\left[ue^{-u}\right] = 3G_0 e^{-u} \tag{l}$$

Progressing along the lines of Part a, we have

$$\frac{d\log\eta_E^+}{d\log t} = \frac{t}{6G_0\tau[1-e^{-u}(1+u)]}3G_0 e^{-u} = \frac{\tau u^2}{6G_0\tau[1-e^{-u}(1+u)]}3G_0 e^{-u} \tag{m}$$

Canceling terms and multiplying top and bottom by e^u gives

$$\frac{d\log\eta_E^+}{d\log t} = \frac{u^2}{2[e^u - u - 1]} \tag{n}$$

Again, both numerator and denominator go toward zero as t approaches zero, requiring the use of l'Hôpital's rule. At this point we can switch back to the variable t, but since the time derivative of the numerator and denominator will both require a factor of du/dt, the result will be the same as differentiating with respect to u. The result of this operation is shown below:

$$\frac{d\log\eta_E^+}{d\log t} = \frac{2u}{2(e^u - 1)} = \frac{u}{e^u - 1} \tag{o}$$

Unfortunately, this quotient is still indeterminate; thus, another differentiation is required. This final step produces a slope of 1. Based on the results of parts a and b, we can draw a very tentative conclusion that the slope of all stress vs. time responses in the linear regime will initially be 1.

8-3. (a) Our first job is to recall the definition of ε_i. With a reference state at the initial configuration,

$$\varepsilon(0,t) = \int_0^t \dot{\varepsilon}(s)ds = \int_0^t \frac{\partial V_1(s)}{\partial x_1}ds \tag{a}$$

Assuming a uniform velocity gradient which is constant over the interval 0 to t, we have the velocity gradient equal to the velocity of sample in the moving clamp divided by the sample length. As velocity times time is equal to distance, we have the expected result that $\varepsilon_1 = (L - L_0)/L_0$. For the width-wise strain, we have $\varepsilon_2 = -v\varepsilon_1 = (W - W_0)/W_0$, where W is the width. If we hold that the extra stresses developed by these deformations are still given by the Hooke's law, equation (8-32a), we have

$$\tau_{11} = G\left(\frac{\partial u_1}{\partial x_1} + \frac{\partial u_1}{\partial x_1}\right) = 2G\frac{L - L_0}{L_0} = 2G\varepsilon_1 \tag{ssc} (b)$$

and for the transverse direction

$$\tau_{22} = G\left(\frac{\partial u_2}{\partial x_2} + \frac{\partial u_2}{\partial x_2}\right) = 2G\varepsilon_2 = -2Gv\varepsilon_1 \tag{ssc} (c)$$

where the displacements u_i are the same as the changes in length. The tensile stress σ_T is given by the usual expression

$$\sigma_T = \tau_{11} - \tau_{22} = G(2G\varepsilon_1 - (-v\varepsilon_2)) = 2(1+v)G\varepsilon_1 \qquad \text{(ssc) (d)}$$

This gives the desired result after dividing through by ε_1, per the definition of Young's modulus.

(b) The definition of bulk modulus is given below, along with the approximation at very small strains

$$K = -\frac{\partial P}{\partial \ln V}\bigg|_T = -\frac{\Delta P}{\Delta V / V_0} \qquad \text{(e)}$$

As we are starting at rest, ΔP is simply the pressure after straining, and ΔV is the final volume minus the initial volume $V - V_0$. We will assume, as above, that the sample before deforming is a square bar of length L_0, and width W_0. Thus the initial volume is $V_0 = L_0(W_0)^2$. Using the findings from part (a) we can write the final length as $L = L_0(1 + \varepsilon_1)$ and the final width as $W = W_0(1 + \varepsilon_2) = W_0(1 - v\varepsilon_1)$. Thus, for the change in volume we have

$$\Delta V = LW^2 - L_0 W_0^2 = L_0(1+\varepsilon_1)[W_0(1-v\varepsilon_1)]^2 - V_0 \qquad \text{(f)}$$

The squared term can be expanded at low strain to terms linear in ε_1 to give

$$\frac{\Delta V}{V_0} = [(1-2v\varepsilon_1)(1+\varepsilon_1)] - 1 = \varepsilon_1(1-2v) \qquad \text{(g)}$$

which gives us the denominator of equation (e). The pressure change is minus a third of the tensile stress (ssc), given in equation (d), so the end result is

$$K = \frac{\frac{1}{3} 2G(1+v)\varepsilon_1}{\varepsilon_1(1-2v)} = \frac{2}{3} G \frac{1+v}{1-2v} \qquad \text{(h)}$$

Note that if $v = \frac{1}{2}$, corresponding to no volume change, the bulk modulus is infinite.

8-4. As in the answer to Problem 8-3, we assume the sample is a square bar of length L and width W. The nomenclature for the initial and final states will be as in the answer to Problem 8-3. So, equating the initial and final volumes gives

$$W_0^2 L_0 = W^2 L = W^2 L_0 \lambda \qquad \text{(a)}$$

where $\lambda = L/L_0$. Comparing the very left-hand side and very right-hand side shows that we can cancel L_0 and solve for the ratio W/W_0. The answer is

$$\frac{W}{W_0} = \frac{1}{\sqrt{\lambda}}$$ (b)

8-5. The product $s^2 \exp(-s/\tau)$ can be written as

$$\frac{s^2}{e^{s/\tau}}$$ (a)

To apply l'Hôpital's rule, we need to take derivatives of numerator and denominator until the ratio of the derivatives shows a clear trend.

Thus, for the first round, we have

$$\frac{2s}{e^{s/\tau}/\tau}$$ (b)

which is still not clear. One more derivative gives

$$\frac{2}{e^{s/\tau}/\tau^2}$$ (c)

It is now very apparent that the denominator will approach infinity as $s \to \infty$, giving zero for the quotient, assuming a finite value of τ.

For the numerical approach, the following results are offered:

τ	s	$s^2 e^{(-s/\tau)}$
1	1	3.68E-01
	10	4.54E-03
	100	3.72E-40
	1000	0.00E+00
10	1	9.05E-01
	10	3.68E+01
	100	4.54E-01
	1000	3.72E-38

The trend toward zero with increasing values of s is abundantly clear.

8-6. The laboratory strain program for this problem is illustrate below:

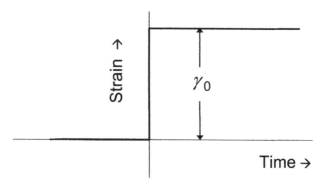

Clearly, this needs to be treated in two different pieces, that from $-\infty < t' \leq 0$ and that from $0 < t' \leq t$. The displacement functions are

$$x_1' = x_1 - \gamma_0 x_2'; \quad x_2' = x_2 \text{ and } x_3' = x_3 \text{ for } -\infty < t' \leq 0 \tag{a}$$

$$x_1' = x_1; \quad x_2' = x_2 \text{ and } x_3' = x_3 \text{ for } 0 < t' \leq t \tag{b}$$

where γ_0 is a positive number.

The shear stress-relaxation modulus requires the calculation of the shear stress as a function of time. For the LEF, we use equation (8-46), which is repeated below

$$\tau(t) = \int_{-\infty}^{t} M(t - t') \left[\mathbf{C}^{-1}(t, t') - \boldsymbol{\delta} \right] dt' \qquad \text{(ssc)} \quad (8\text{-}46)$$

As the sample is in its reference position from 0 on, all the strains will be zero over the range $0 < t' \leq t$. Thus the only contribution to the stress will (oddly enough) be from the range $-\infty < t' \leq 0$ where the lab strain is zero. Using equation (8-40), along with the displacement functions, gives

$$C_{21}^{-1} - \delta_{21} = \gamma_0 \tag{c}$$

over this time range. The stress during relaxation will thus be given by

$$\tau_{21}(t) = \int_{-\infty}^{0} \frac{G_0}{\tau} e^{-(t-t')/\tau} \gamma_0 \qquad \text{(ssc)} \quad (d)$$

Taking constants outside the integral gives the result

$$\tau_{21}(t) = \frac{G_0 \gamma_0}{\tau} e^{-t/\tau} \int_{-\infty}^{0} e^{t'/\tau} dt' \qquad \text{(ssc)} \quad (e)$$

Integration results in a single term τ, which cancels the τ in the denominator. As the shear stress-relaxation modulus $G(t)$ is defined as $\tau_{21}(t)/\gamma_0$, the result is the expected $G(t) = G_0 \exp(-t/\tau)$. This is independent of strain; thus, the LEF exhibits linear behavior in shear, which is consistent with its constant shear viscosity.

8-11. (a) The variable a <u>is</u> the breaking strain, as can be seen by setting $t_B = a/\dot\gamma_0$ and multiplying through by the shear rate.

8-13. (a) Analysis of the stress-growth experiment for an LEF has been reviewed in Example 8-2. The result is an exponential rise in stress given by equation (8-31), which is repeated below:

$$\tau(t) = \dot\gamma_0 \tau G_0 (1 - e^{-t/\tau}) \qquad \text{(ssc)} \quad (8\text{-}31)$$

This is the same as the linear viscoelastic response.

(b)

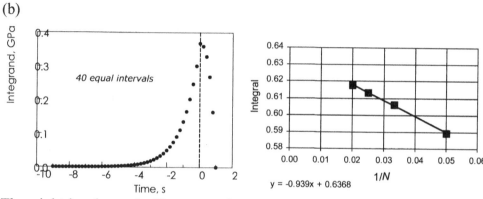

The right-hand graph illustrates the Romberg integration process with 4 iterations. The intercept is the estimate of the true value of the integral. The results exhibit a slight downward curvature, suggesting that the linear extrapolation is likely to be on the high side. On the left-hand side is a plot of the integrand, the area of which is the result at $1/N = 0.02$.

(c) The exact solution is $(1 - e^{-1}) = 0.632$ GPa (ssc), corresponding to $t = \tau = 1$ s. The above graph shows the Romberg integration with an intercept of 0.637 Pa, which as suspected is indeed a bit higher (by 1%) than the exact result. Clearly, a better strategy would be to split the integral into two parts (as is done analytically) and integrate each separately.

8-14. (a) The LEF model with a breaking time $t_B = a/\dot\gamma$ calculated from the start of the deformation ($t' = 0$) will be given by the general equation

$$\tau(t) = \int\limits_{-\infty}^{t_B} M(t-t')\left[C^{-1}(t,t') - \delta\right]dt' \tag{a}$$

where δ is the unit tensor (i.e., ones on the diagonal). For the steady-shear start-up experiment, we are interested in the 2,1 component of this equation, so

$$\tau_{21}(t) = \int\limits_{-\infty}^{0} M(t-t')\dot\gamma_0 t\, dt' + \int\limits_{0}^{t_B} M(t-t')\dot\gamma_0(t-t')dt' \tag{b}$$

where we have assumed the deformation starts at zero time, and continues to time t at a shear rate of $\dot\gamma_0$. The integration, however, stops at $t_B = a/\dot\gamma_0$, as depicted in equation (a).

Exploring on, we enter the single-exponential memory function, and remove constants from the integrals, yielding

$$\tau_{21}(t) = \frac{G_0\dot\gamma_0}{\tau}e^{-t/\tau}t\int\limits_{-\infty}^{0}e^{t'/\tau}dt' + \frac{G_0\dot\gamma_0}{\tau}\int\limits_{0}^{t_B}e^{-(t-t')/\tau}(t-t')dt' \qquad \text{(ssc) (c)}$$

Instead of substituting for $t - t'$ in the second integral, we will instead divide it into two integrals, one of which can be combined with the first. The result of this is

$$\tau_{21}(t) = \frac{G_0\dot\gamma_0}{\tau}e^{-t/\tau}\left[t\int\limits_{-\infty}^{t_B}e^{t'/\tau}dt' - \int\limits_{0}^{t_B}t'e^{t'/\tau}dt'\right] \tag{ssc) (d}$$

where we have also pulled out the common factor $e^{-t/\tau}$. The first integral is quite easy, while the second may require a peek at the tables or a search for equation (8-23). The integration leads to

$$\tau_{21}(t) = \frac{G_0\dot\gamma_0 e^{-t/\tau}}{\tau}\left\{t\tau e^{t_B/\tau} - \tau\left[e^{t_b/\tau}(t_B - \tau) + \tau\right]\right\} \tag{ssc) (e}$$

On collection of similar terms, we have the result

$$\tau_{21}(t) = G_0\dot\gamma_0 e^{-t/\tau}\left[e^{t_B/\tau}(t - t_B + \tau) - \tau\right] \tag{ssc) (f}$$

This result looks either like we have made some bad assumptions, or a mistake. How does one differentiate these two? One step is to see how the expression behaves at known times. For example, if $t = t_B$, the expression should revert to the normal LEF result

$$\tau_{21}(t_B) = G_0\dot{\gamma}_0\tau\left(1 - e^{-t_B/\tau}\right) \qquad \text{(ssc) (g)}$$

because the network has not yet broken. Are there any other key times to try? Well, at $t = 0$, one expects the stress to be zero, but instead it is

$$\tau_{21}(0) = G_0\dot{\gamma}_0\left[(\tau - t_B)e^{t_B/\tau} - \tau\right] \qquad \text{(ssc) (h)}$$

which is not zero. The reason is not an error, as we have forced the integration to go t_B even if $t < t_B$. This is certainly not a realistic answer.

(b) Starting with the Boltzmann equation for shear, but running the integration to t_B instead of t, gives the starting equation

$$\tau_{21}(t) = \int_{-\infty}^{t_B} G(t - t')\dot{\gamma}(t')dt' \qquad \text{(ssc) (i)}$$

This needs to be split into two integrals, one over the range $-\infty < t' \leq 0$, and the other for $0 < t' \leq t_B$. The first integral is zero because the shear rate is zero. The second, with the usual substitutions, looks like

$$\tau_{21}(t) = G_0\dot{\gamma}_0 e^{-t/\tau}\int_0^{t_B} e^{t'/\tau}dt' \qquad \text{(ssc) (j)}$$

which integrates easily to

$$\tau_{21}(t) = G_0\dot{\gamma}_0\tau e^{-t/\tau}\left(e^{t_B/\tau} - 1\right) \qquad \text{(ssc) (k)}$$

Again, we can see that at $t = t_B$ we agree with the linear result. However, we again notice problems at $t = 0$ (stress not equal to zero) and as $t \to \infty$ (stress is driven to zero).

So, we have concluded by these exercises that the model should be as stated in Problem 8-11.

CHAPTER 9

9-1. The viscosity of the Rouse model is given by equation (9-3). If this equation is used directly, the result will be the identity $\eta = \eta$, because this procedure was used to eliminate unknown constants. Thus, one needs to go back to the fundamental expression for the time constants,

$$\tau_p = \frac{\rho_s a^2 z^2}{6kT\pi^2 p^2} \qquad \text{(a)}$$

Combining this result with equation (9-3) and canceling like terms gives the expression

$$\eta = NkT \sum_{p=1}^{z} \tau_p = \frac{\rho \rho_s a^2 z^2}{6\pi^2 M} \sum_{p=1}^{z} \frac{1}{p^2} \qquad \text{(b)}$$

where ρ/M has been substituted for N. The sum, at least for large values of z, is independent of z, becoming instead $\pi^2/6$. When this is substituted, the result is

$$\eta = \frac{\rho \rho_s a^2 z^2}{36M} \qquad \text{(c)}$$

As $z \sim M$, the result is $h \sim M$. While not the best outcome, it is at least in accord with the absence of entanglement effects.

CHAPTER 10

10-1. (a) Setting up the inequality $R\Omega^2 \leq g/10$ and solving for Ω gives the result $\Omega = (g/10R)^{1/2}$. With $g = 9.81$ m/s^2 and $R = 2$ cm $= 0.02$ m, we find $\Omega \leq 7$ rad/s ≈ 70 rpm.

10-2. (b) The film is not accelerating, so the stress is shear and due to the weight of the film. The maximum shear stress τ_{max} at the glass surface will be

$$\tau_{max} = \frac{\rho W H g}{WH} \delta = \rho g \delta$$

where δ is the thickness of the film. The stress will fall off linearly to zero at the outside surface of the film.

10-3. (a) The bubble has a density that is negligible compared to the fluid, so the buoyancy will be

$$F = V_B \rho_F g_r = \left(\frac{4}{3} \pi R_B^3\right) \rho_F g_r = \left(\frac{4}{3} \pi R_B^3\right) \rho_F \Omega R^2$$

where R_B is the bubble radius, ρ_F is the fluid density, Ω is the spin rate and R is the cylinder radius.

10-4. (a) Beer's law states $I/I_0 = 10^{-\alpha l}$, where l is the path length and $\alpha = 200$ mm^{-1} is the absorption coefficient. Solving for l gives

$$l = -\log(I/I_0)/\alpha \tag{a}$$

If the permissible minimum in I/I_0 is 0.01, then the maximum path length is $l = 2/200$ mm $= 0.01$ mm $= 10$ μm.

(b) The objective is to squeeze a 1-mm-thick disk into a 0.01-mm-thick disk. Using equation (10-10) for squeezing flow at constant volume, we can predict the closing curve for the process, in differential form, as

$$-\frac{dH}{dt} = \frac{2\pi H^5}{3\eta V^2} \tag{b}$$

(The negative sign is due to the fact that the gap is getting smaller with time.) The above equation can be quite easily integrated to give

$$H(t) = H_0 \left[\cfrac{1}{1 + \cfrac{8\pi F H_0^4}{3\eta V^2} t} \right]^{1/4} \tag{c}$$

where H_0 is the initial gap. It is very important to check dimensions in the algebraic result before plugging in the numbers, which in this case is a matter of checking that $F H^4 t / \eta V^2$ is dimensionless. This is easily seen.

Entering the data and converting to SI units goes as follows:
$H_0 = 1$ mm $= 0.001$ m
$V = (10 \text{ mg}) / (10^6 \text{ mg/kg}) / (1000 \text{ kg/m}^3) = 10^{-8} \text{ m}^3$
$F = 5000 \text{ lb}_f \times 4.448 \text{ N/lb}_f = 22240$ N
$\eta = 10000$ Pa s
$H(t) = 0.01$ mm $= 10^{-5}$ m
$t = ?$

While the equation can be solved algebraically for t, it is often easier set the equation up as written and enter the numbers. First, the coefficient for t is calculated.[††]

[††] Calculations involving more than a couple of variables should be done on a spread sheet rather than with a calculator.

$$\frac{8\pi F H_0^4}{3\eta V^2} = \frac{8\pi}{3} \frac{22240\,\text{N} \times (0.001\,\text{m})^4}{10000\,\text{N s/m}^2 \times (10^{-8}\,\text{m}^3)^2} = 1.86 \times 10^5\,\text{s}^{-1} \qquad (d)$$

Combining with $H(t)$ and H_0 gives the equation

$$\frac{0.01\,\text{mm}}{1\,\text{mm}} = \left[\frac{1}{1+1.86\times10^5 t}\right]^{1/4} \qquad (e)$$

which can be easily solved for t, giving the result $t = 540$ s, a very reasonable value. However, this ~10 min of exposure may lead to some degradation, suggesting that more force or a smaller sample might be appropriate.

10-6. (a) and (b) The easy way to demonstrate the validity of equation (10-4) is to assume it is correct and use it to calculate the stress, which must rise linearly from zero at the center plane to a maximum at the wall of the slit.

For reference, equation (10-4) is reproduced below:

$$V_r(r,z) = \frac{3}{2}\overline{V}(r)\left[1-\left(\frac{z}{H/2}\right)^2\right] \qquad (10\text{-}4)$$

The local shear rate will be the derivative of $V_r(r,z)$ with respect to z, giving

$$\frac{\partial V_r(r,z)}{\partial z} = -\frac{3}{2}V(r)\left[2\frac{z}{(H/2)^2}\right] = -3V(r)\frac{z}{(H/2)^2} \qquad (a)$$

The important aspect of this is that indeed the velocity gradient rises linearly from zero at the center plane ($z = 0$). Thus the shear stress will also rise linearly, if the fluid is Newtonian. At the wall ($z = H/2$), the shear-rate magnitude is $6Q/WH^2$.

10-8 (b) One can argue against lockup by considering the force contribution of each circle of material as shown by equation (10-8). The shear rate $\dot\gamma$ will be proportional to radius, meaning the shear stress with go as $\dot\gamma^n$ with $n < 1$. The pressure increment will scale as the stress, thus as r^n instead of the r dependence in equation (10-6) for Newtonian fluids. Integration of this will to give a pressure profile that goes as $1 - (r/R)^{n+1}$, which will lead to finite pressure, not an infinite one. With any fluid, the differential area against which the pressure works decreases with r as the center is approached. Thus the increments of force not be infinite, even though the viscosity goes to infinity.

CHAPTER 11

11-3. The best one can do with the shear viscosity data alone is to calculate the viscosity at a shear stress equivalent to that in the melt-indexer geometry, i.e., 19.4 kPa (see Example 11-1). The shear rate at this shear stress is 220 s^{-1}, which was obtained from a Cross-model interpolation of the data as shown below:

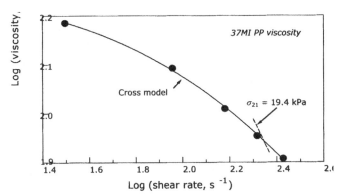

From this, one can estimate the flow rate in the melt-indexer geometry (Figure 11-1) using equation (7-30), which is reproduced below:

$$\dot{\gamma}_W = \frac{4Q}{\pi R^3}\left[\frac{3}{4} + \frac{1}{4}\frac{d\ln\left(4Q/\pi R^3\right)}{d\ln\tau_W}\right] \qquad (7\text{-}30)$$

The problem here is the determination of the second term in the square brackets. This slope will be very close to the slope on a log-log plot of the actual shear rate vs. shear stress, which can be found from the Cross-model fit to the data. It's about 1.83. Solving (7-30) for the flow rate Q gives

$$Q = 220\ \text{s}^{-1}\times3.14159\times[(2.095/2)/10]^3\ \text{cm}^3/[0.75+1.83/4]/4 = 0.164\ \text{cm}^3/\text{s} \quad (a)$$

The next step is to find the density of PP at 230 °C. As with the example in Appendix 10-1, we select the Sanchez-Lacombe equation of state, which is reproduced below:

$$\tilde{P} + \frac{1}{\tilde{V}^2} + \tilde{T}\left[\ln\left(1-\frac{1}{\tilde{V}}\right)+\frac{1}{\tilde{V}}\right] = 0 \qquad (b)$$

Checking in the reference cited in Appendix 10-1, the parameters $P^* = 356.1$ MPa, $V^* = 1.1211$ cm^3/g, and $T^* = 676.4$ K are found for isotactic PP. With an average pressure of 0.148 MPa in the capillary, the reduced pressure is $0.148/356.1 = 0.0042$. The reduced temperature is $(230 + 273.2)/676.4 =$

0.744. Solving for the reduced volume gives 1.21, corresponding to a specific volume of 1.21×1.1211 = 1.36 cm^3/g or a density of 0.737 g/cm^3. Thus the estimated MFI is 0.164 cm^3/s × 600 s/10 min × 0.737 g/cm^3 = 73 g/10 min. This is higher than that found (37.1 g/10 min), probably because the calculation from the flow curve does not include the entrance-pressure loss, which can be considerable in the melt indexer with its very short capillary.

11-5. The Mooney viscometer geometry is shown in Figure (11-3); for this problem, we assume the torque is produced solely by flat surfaces, i.e., two parallel-disk fixtures. The "Mooney" number is simply the torque on the disk in N-m divided by 0.083. Assuming Newtonian response, the viscosity from the parallel-disk geometry is

$$\eta = \frac{2MH}{\pi R^4 \Omega} \tag{a}$$

which appeared in equation (7-36). The torque M is 45 × 0.083 = 3.735 N-m, while the rotation rate Ω is equal to 2 rpm × 2π rad/rev / (60 s/min) = 0.21 rad/s. The gap H is 2.52 mm = 0.00252 m, while the radius R is (38.10/2) mm = 0.01905 m. Putting all these numbers, in consistent units, into equation (a) yields

η = {2 × 3.735 N-m × 0.00252 m / [3.1459 × (0.01905 m)^4 × 0.21 rad/s]}/2

where the final 2 accounts for the fact that there are two gaps. The viscosity is then estimated as 108 kPa s, a rather high value. If the melt is non-Newtonian (likely), we can use the ¾-th's rule to assign this viscosity to a shear rate of (0.75 × 0.01905 m × 0.21 rad/s) / 0.00252 m = 1.2 s^{-1}. The shear stress at this radius is 1.3×10^5 Pa, which is likely to cause slip on smooth surfaces.

11-9. (a) D4989

(b) The nominal shear stress in the melt indexer for the low load is 19.4 kPa. The nominal shear stress for the PPP can be estimated by multiplying the viscosity from equation (10-9) by the nominal shear rate using the slit-flow equation applied at the exit. So, starting with equation (10-9) we have

$$\eta = \frac{2H^3 F}{3\pi R^4 \dot{H}} \tag{a}$$

The equation for nominal shear rate in a slit, as applied to the outside rim is

$$\dot{\gamma} = \frac{6Q}{WH^2} = \frac{6(\dot{H}\pi R^2)}{2\pi R H^2} \tag{b}$$

11-13. We will assume the results listed are not paired, i.e., the 1.05 value in the first column has no more connection with the 1.07 than with the 1.09 in the second column. With Excel®, list the data and choose the formula "TTEST" for a blank cell. The questions asked will guide the process; they include the range of the two lists of numbers to be compared, the tails (choose 2) and the type of test (choose 2). After starting the calculation, the blank cell will contain a probability, which is $p = 0.155$ for this example. This means that there is a 15.5% chance of being in error by rejecting the hypothesis that the two resins are taken from the same population (e.g., the same bag). This probability is generally considered too high to state the Resin A is different than Resin B. If a resolution of the two is needed, then more replications will be needed.

Selecting "Tails = 2" suggests that there is no *a priori* reason to think that Resin B should have a higher MI than Resin A. The choice of "Type = 2" assumes the scatter of values in the two sets are essentially the same.

CHAPTER 12

12-1. Following the suggestion, we expand to the second-order term and compare the coefficient with the 6.2 found in the Batchelor equation. The binomial expansion for the relative viscosity η/η_m is:

$$\frac{\eta}{\eta_m} = 1 + 2.5\phi_p + \frac{2.5(1-2.5)}{2!}\phi_p^2 + \cdots = 1 + 2.5\phi_p + 1.875\phi_p^2 + \cdots \qquad (a)$$

The coefficient for the second-order term is much lower than the 6.2 of the Batchelor equation, so this expression can be expected to under predict the viscosity of suspensions at moderate particle volume fractions.

12-2. The key here is to find expansions of the equations that are valid at low ϕ_p. An example is shown below
(a) The Frankel-Acrivos equation is

$$\eta = \frac{9}{8}\eta_m \frac{(\phi_p/\phi_{max})^{1/3}}{1-(\phi_p/\phi_{max})^{1/3}} \qquad (a)$$

Rearranging this into a series gives

$$\frac{\eta}{\eta_m} \approx \frac{9}{8}\left(\frac{\phi_p}{\phi_{max}}\right)^{1/3}\left[1+\left(\frac{\phi_p}{\phi_{max}}\right)^{1/3}\right] \qquad (b)$$

It is apparent from this form that this equation will be useful only for suspensions at higher concentrations, as behavior at low concentrations is power-law in nature.

(e) The Hatschek equation is

$$\eta = \eta_m \big/ \left(1 - \phi_p^{1/3}\right) \tag{a}$$

Expanding the relative viscosity for low concentrations gives

$$\eta_{rel} = 1 + \phi_p^{1/3} \tag{b}$$

leading to a specific viscosity $\eta_{rel} - 1$ proportional to the particle volume fraction to the 1/3 power, rather than the linear dependence called for by the Stokes-Einstein relationship.

12-5. (a) There are three reasonable explanations for the higher viscosity of the CaCO$_3$ suspension: particle agglomeration, oblong particles, and porous particles.   Distinguishing these usually requires careful microscopy. Sometimes agglomerates can be broken up at higher shear rates, resulting in a decrease of viscosity, but this is easily confused with other sources of non-Newtonian behavior.

(b) The maximum packing fraction $\phi_{max,s}$ for randomly arrayed spheres of diameter D is 0.64.   In turn, this is equal to the volume of the spheres themselves divided by the total volume, i.e.,

$$\phi_{max,s} = 0.64 = (N_s \pi D^3/6)/V_T \tag{a}$$

Doing the same operation for the rods, we find that the ratio of number of rods N_r to number of spheres in the fixed volume V_T is $(D/L)^3$. The maximum volume fraction of rods $\phi_{max,r}$ is then

$$\phi_{max,r} = \frac{N_r (\pi D^2 L/4)}{V_T} = \frac{(D/L)^3 (\pi D^2 L/4) N_s}{V_T} \tag{b}$$

Substituting for N_s/V_T in equation (b) gives

$$\phi_{max,r} = (D/L)^3 (\pi D^2 L/4) \frac{0.64 \times 6}{\pi D^3} = \frac{3}{2} \left(\frac{D}{L}\right)^2 0.64 \tag{c}$$

If $\phi_{max,r} = 0.44$, then the aspect ratio might be as low as 1.48. However, as is often the case, rods can occupy less volume that $\pi L^3/6$ before interfering badly with each other.

12-8. (a) The first term of the equation simply approaches P_e. The second term is a bit more of a problem, as it looks indeterminate, i.e., 0/0. However, we can take advantage of the expansion of $\ln(1-x)$ at low values of x, which is $-x - x^2/2 + \cdots$ So, the result works out to be

$$P = P_e + \frac{1}{\beta}\left[\frac{4\beta\tau L}{D} + \frac{1}{2}\left(\frac{4\beta\tau L}{D}\right)^2 + \cdots\right] \rightarrow P_e + 4\tau\frac{L}{D} \tag{a}$$

This expression we recognize as the usual expression for the pressure drop through a capillary.

12-11. Equation (12-4) can be linearized to give the form

$$\ln c_{gel} = \ln \alpha - \beta/T_{gel} \tag{a}$$

However, we know that the variable with the principal error is T_{gel}, so we solve for $1/T_{gel}$ in terms of concentration to give

$$1/T_{gel} = (\ln \alpha)/\beta - \ln c_{gel}/\beta \tag{b}$$

Doing the linear regression of $1/T_{gel}$ vs. $\ln c_{gel}$ gives values for 4690 K for β and 12.5 for $\ln \alpha$. The variable c_{gel} has been converted to weight fraction.

Similarly, nonlinear fitting methods can be use directly on the T_{gel} vs. c_{gel} relationship:

$$T_{gel} = \beta/\ln(\alpha/c_{gel}) \tag{b}$$

This gives somewhat different values of 4550 K for β and 12.0 for $\ln \alpha$.

12-13. (a) and (b) A plot is shown below. These models indicate that shear rates two decades below the characteristic rates would be needed to gain a hint of either a Newtonian plateau or yield behavior at low shear rates. Most rotational rheometers can reach these low rates, but the polymer must be stable at the test temperature as the times required to reach steady state can be very long.

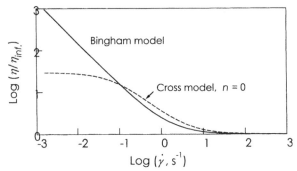

Comparison of Cross and Bingham models with limiting Newtonian behavior.

(c) The graph below are the fits of the given equations to the data provided. The SSE for the Bingham and Cross models are 0.0018 and 0.00058, respectively, giving the Cross model a distinct win. However, we need to recognize that the Cross model has one more parameter than the Bingham. A generalized form of the Bingham model with an infinite viscosity at $\dot{\gamma} = 0$ but no distinct yield point would be

$$\eta = \eta_\infty \left[1 + \left(\frac{\dot{\gamma}_c}{\dot{\gamma}} \right)^\alpha \right] \qquad \text{(a)}$$

PCL w/HA needles ($\phi_p = 0.045$)
120 °C

—— Bingham model
---- Cross model, $n = 1$

Log (η, Pa s)

Log ($\dot{\gamma}$, s^{-1})

A fit of this 3-parameter equation gives SSE = 0.00018, which is lower than that for the 3-parameter Cross model. However, our conclusions would be much more sound if the shear rates were carried to lower values and the presence of a yield was checked with a stress-growth test where the quiescent melt was subject to slowly increasing shear stress until flow was established.

Nomenclature

$\mathbf{C}$	Cauchy relative strain tensor.
$\mathbf{C}^{-1}$	Finger relative strain tensor.
f	Cyclic frequency, Hz or cycles per second (sometimes seen as cps).
F	Force, including normal force for parallel-plate and cone-and-plate fixtures.
G	Shear modulus.
$G(t)$	Shear stress relaxation modulus.
H	Total gap height for a slit die, or parallel-plate fixture.
H	Damping function (dimensionless) for the memory function.
K	Bulk elastic modulus.
k_B	Boltzmann constant.
LHS	Left-hand side of an equation.
(fsc)	Fluids sign convention: compressive stress is positive.
M	Torque.
M	Memory function, appearing usually as $M(t'-t)$.
M	Molecular weight.
M_c	Critical molecular weight for entanglement.
N	Rotation rate in revolutions per minute (rpm).
N_1	First normal-stress difference.
N_2	Second normal-stress difference.
r	Radial coordinate.
R	Radius of fixture, pipe, capillary, etc.
R	Gas constant.
RHS	Right-hand side of an equation.
(ssc)	Solids sign convention: tensile stress is positive.

t	Time.
T	Temperature.
V or v	Velocity. V is also used for volume.
x	Distance.
$x_1, x_2,$	
x_3	Coordinate distances in the flow, gradient and neutral directions.
z	Axial coordinate.
z	Number of physical segments in a Rouse chain.

GREEK SYMBOLS:

α	Placement of scale parameter in stretched exponential. Thermal expansion coefficient.
β	Compressibility. Shape parameter in stretched exponential.
ε	Tensile strain. Equal to integral of $\dot{\varepsilon}$ over time from reference state to present state.
$\dot{\varepsilon}$	Extension rate in tension. Equal to velocity gradient in stretching direction.
ϕ	Azimuthal direction in spherical coordinates.
ϕ_i	Volume fraction of component i in mixture.
ϕ_m	Volume fraction of matrix in composite mixture.
ϕ_p	Volume fraction of solids (e.g., particles) in composite mixture.
γ	Shear strain, equal to the integral of $\dot{\gamma}$ over time from reference state to present state.
$\dot{\gamma}$	Rate-of-deformation tensor.
$\dot{\gamma}$	Shear rate. Also, magnitude of rate-of-deformation tensor.
$\dot{\gamma}_{ij}$	Any component ij of rate-of-deformation tensor.
$\dot{\gamma}_c$	Characteristic shear rate for fluids with a yield point.
$\dot{\gamma}_0$	Characteristic shear rate for Generalized Newtonian Fluid models.
η	Viscosity.
η_0	Zero-shear-rate viscosity.
η_E	Extensional (tensile) viscosity.
η_E^+	Transient extensional viscosity after startup at fixed extension rate.
λ	Extensional stretch ratio.
λ	Aspect ratio.
ν	Poisson's ratio.
ν	Kinematic viscosity.
θ	Azimuthal direction in cylindrical coordinates, inclination direction (from pole toward equator) in spherical.

ρ Mass density.

σ Total or applied stress.

$\boldsymbol{\sigma}$ Total- or applied-stress tensor.

σ_{ij} Component ij of the total- or applied-stress tensor.

τ Characteristic time. Shear stress.

τ_W Wall shear stress in capillary. τ_W is taken as <u>positive</u>.

$\boldsymbol{\tau}$ Extra stress tensor.

τ_{ij} Component ij of the extra-stress tensor

ω Radial frequency in rad/s; $\omega = 2\pi f$, where f is the cyclic frequency in Hz.

Ω Rotation rate in rad/s. $\Omega = \pi N/30$, where N is rpm (revolutions per minute).

Ψ_1 First normal-stress difference coefficient.

Ψ_2 Second normal-stress difference coefficient.

Author Index

Abbott, L. E., 194
Abdel-Khalik, S. I., 123
Akinlabi, A. K., 308
Allain, C., 335
Alsamarraie, M. A. A., 290
An, Y., 193, 333
Anthony, S. M., 195
Armstrong, R. C., v, 13, 96, 193, 238, 259
Asama, H., 193
Astarita, G., 96, 123
Aubert, J. H., 309
Aymard, C., 308

Bae, S. C., 195
Baek, S.-G., 154, 194
Bagley, E. B., 130-133, 139, 160, 188,
    189, 190, 193, 195, 332, 359–360
Baird, D. G., 193, 194, 290
Barker, D. A., 10
Batchelor, G. K., 312, 330, 335, 377
Becker, G. W., 21, 44
Beiner, M., 44
Bernstein, B., 222
Bird, R. B., v, 13, 54, 95, 96, 105, 107,
    122-123, 212, 220, 238, 259, 307, 309
Blyler, L. L., 55–56, 58, 324, 335–336
Bogue, D. C., 194
Bonfils, F., 308
Brandrup, J.,

Brother, G. H., 295, 308
Brown, R. A., 193
Bueche, A. M., 248
Bueche, F. 85, 91, 121, 246–249, 251–252,
    257, 259
Burghardt, W. R., 194

Callaghan, P. T., 167, 195
Carreau, P. J., 13, 54, 73, 85, 92–94, 96,
    350, 353
Cauchy, A.-L., 219–222, 234–235, 381
Chan, T. W., 194
Chao, K. K., 194
Chapman, R. N., 309
Chapoy, L. L., 86, 93, 96, 105–106, 120,
    123, 236, 257
Chen, I.-J., 194
Chhabra, R. P., 13
Christianson, D. D., 195
Chung, C. I., 326, 335
Chung, S. C.-K., 55, 57
Clark, E. S., 122
Cogswell, F. N., 155, 318, 335
Cohen, C., 308
Cole, E. A., 335
Cole, R. H. (Cole, K. S.), 227, 238
Collias, D. I., 290
Collier, J. R., 194
Colwell, R. E., 195

Subject Index

Printed and bound by CPI Group (UK) Ltd, Croydon, CR0 4YY

16/04/2025

14658354-0003